FORENSIC

FORENSIC COLONIALISM

Genetics and the Capture of Indigenous Peoples

MARK MUNSTERHJELM

McGill-Queen's University Press
Montreal & Kingston • London • Chicago

ISBN 978-0-2280-1688-5 (cloth)
ISBN 978-0-2280-1689-2 (paper)
ISBN 978-0-2280-1815-5 (ePDF)

Legal deposit third quarter 2023
Bibliothèque nationale du Québec

Printed in Canada on acid-free paper that is 100% ancient forest free (100% post-consumer recycled), processed chlorine free

This book has been published with the help of a grant from the Canadian Federation for the Humanities and Social Sciences, through the Awards to Scholarly Publications Program, using funds provided by the Social Sciences and Humanities Research Council of Canada.

We acknowledge the support of the Canada Council for the Arts.
Nous remercions le Conseil des arts du Canada de son soutien.

Library and Archives Canada Cataloguing in Publication

Title: Forensic colonialism : genetics and the capture of Indigenous peoples / Mark Munsterhjelm.
Names: Munsterhjelm, Mark, 1966– author.
Description: Includes bibliographical references and index.
Identifiers: Canadiana (print) 20230183115 | Canadiana (ebook) 2023018345X | ISBN 9780228016885 (hardcover) | ISBN 9780228016892 (softcover) | ISBN 9780228018155 (PDF)
Subjects: LCSH: Human genetics—Moral and ethical aspects. | LCSH: Human genetics Social aspects. | LCSH: Forensic genetics—Moral and ethical aspects. | LCSH: Forensic genetics—Social aspects. | LCSH: Human gene mapping—Moral and ethical aspects. | LCSH: Human gene mapping—Social aspects. | LCSH: Indigenous peoples—Medical examinations—Moral and ethical aspects. | LCSH: Imperialism and science.
Classification: LCC QH438.7.M86 2023 | DDC 174.2/910181663—dc23

This book was typeset in 10.5/13 Sabon.

Contents

Tables and Figures

TABLES

FIGURES

Abbreviations and Glossary

1000 Genomes Project: Project and resulting database.
AIM: Ancestry informative markers.
AISNP: Ancestry inference single nucleotide polymorphism.
ALLELE: A sequence of DNA at a particular location (loci) in which there is one or more different versions, such as substitution of a guanine (G) for cytosine (C).
ALFRED: Allele Frequency Database run by Kidd Lab at Yale University.
BIG: Beijing Institute of Genomics.
Bingtuan: Mandarin Chinese colloquial term meaning "soldier corps" for the Xinjiang Production and Construction Corps (XPCC), which is a paramilitary corporate entity involved in the settler colonization of Xinjiang since the 1950s.
CAS: Chinese Academy of Sciences.
cell line: Genetic materials, typically blood samples, are infected with Epstein-Barr virus to stop cell apoptosis (cell death) allowing the resulting virally transformed cell lines to be grown easily under laboratory conditions.
CODIS: Combined DNA Index System, developed and run by the FBI.
Diversity Project: Human Genome Diversity Project.
DNA: Deoxyribonucleic acid. DNA is composed of four bases called adenine (A), cytosine (C), guanine (G), and thymine (T). These form into base pairs in which A pairs with T and C pairs with G. The human genome is composed of about 3 billion base pairs.
EUROFORGEN: European Forensic Genetic Network of Excellence.
EVC: External visible characteristic, e.g., eye colour.
GWAS: Genome-wide association study.
IISNP: Individual identification single nucleotide polymorphism.

Institute of Forensic Science (公安部物证鉴定中心): The Chinese Ministry of Public Security's Institute of Forensic Science in Beijing. It is also often directly translated from Chinese into English as the Material or Physical Evidence Identification Center. Li Caixia and Ye Jian both work there.

loci: Refers to locations on chromosomes within the genome.

NIH: US National Institutes of Health.

NIJ: US Department of Justice's National Institute of Justice.

NSF: US National Science Foundation.

PCA: Principal components analysis.

PCR: Polymerase chain reaction is a process used to make copies of ("amplify") genetic sequences for further analysis.

Partner Institute: The Chinese Academy of Sciences–Max Planck Society Partner Institute of Computational Biology operated from 2005 to 2020 and was located in Shanghai.

PISNP: Phenotype inference single nucleotide polymorphism.

PRC: People's Republic of China.

RFLP: Restriction fragment length polymorphism.

RMP: Random match probabilities.

SNP: Single nucleotide polymorphism. These are single letter changes in the DNA such as a C for a T or a G for an A.

STR: Short tandem repeat. A type of marker used in individual identification, like the FBI's CODIS database, typically in a set of thirteen to twenty STR markers.

VISAGE Consortium: Visible attributes through genomics.

VisiGen Consortium: International visible genetic traits.

VNTR: Variable number tandem repeats are sequences that repeat in a genome. Because these vary between individuals, early forensic genetic tests used them.

FORENSIC COLONIALISM

INTRODUCTION

Assemblages of Capture

Forensic genetics technologies are popularly touted as science in the service of state institutions of justice, law, and sovereignty to protect the People from dangerous criminals and terrorists. However, in this book I analyze a set of histories of forensic genetics research, development, and implementation involving leading scientists, biotechnology firms, and state security and police agencies in the United States, the European Union, and the People's Republic of China (PRC) and argue that these histories have been shaped by genocidal settler colonialism and allow for the sustained systemic violation of Indigenous peoples' self-determination, rights, and dignity. My central argument and hypothesis is that racially configured hierarchies have permeated forensic genetic technologies' research, development, and implementation. In particular, forensic genetics as technologies use various racially inflected conceptions (e.g., statistical differences in prevalence of genetic markers between "Caucasian" and "Black" populations) to help police, state security agencies, and judiciaries identify, capture, convict, and/or kill those deemed criminals and terrorists. These technologies have in turn involved scientific assemblages (networks) that rely on maintaining various Indigenous peoples in a state of capture and genetic servitude to serve scientists as resources of genetic diversity and difference. This hierarchy involves the research and development of forensic genetic technologies that seek to advance the biopolitical security of privileged populations in core regions of capitalism based on the repression of Indigenous peoples in which systemic racism imposes divisions (caesurae) between who lives and who dies (Dillon 2008, 169–70, 177–8; Foucault 2003, 255; Mbembe 2003, 17).

Death is not only physical but also social and civil, imposed through the systemic nonrecognition and often-violent suppression of the rights, dignity, and self-determination of Indigenous peoples that is foundational to settler colonialism in countries like Brazil, the US, and the PRC (Foucault 2003, 256; TallBear 2013, 150–1; Weheliye 2014, 37–8). Guided by what anthropologist Patrick Wolfe (2006, 387) termed a logic of elimination, these settler colonial states govern through dominant rationalities (what Foucault termed governmentality) that position "its security in relation to perceptions of indigeneity as insecurity. Coded as sources of threat, expressions of indigenous values, practices, knowledges and subjectivities must be eliminated through various mechanisms of security to ensure the prosperity of settler society" (Crosby and Monaghan 2012, 423). Similarly, in Xinjiang, Dibyesh Anand (2019, 129–30) argues that "The Chinese state and its (Han) majoritarian nationalism brands Uyghurs and Tibetans as sources of insecurity. Mass demonstrations in Tibet in 2008 and Xinjiang in 2009, as well as incidents of violent and non-violent protest, are understood not as the results of legitimate grievances but as products of separatism, extremism and terrorism associated with Uyghur and Tibetan identities. This representation of Xinjiang and Tibet as sites of existential threat legitimizes massive investment in security apparatus and violence against inhabitants there."

This social and civil death and denial of Indigenous peoples' self-determination has also been routine in international forensic genetic research assemblages. In these assemblages (networks), scientists effectively own many Indigenous peoples' genetic materials and data, which they frequently refer to as *resources*. This long-term usage relies on routine acceptance in forensic genetics of claims that decades old informed consent are still valid over tens of thousands of samples and other genetic materials from Indigenous peoples. The claims and usage of these genetic materials as what the involved scientists term resources violates contemporary research ethical and legal norms, which require ongoing informed consent over secondary usages, and international agreements such as the UN Declaration on the Rights of Indigenous Peoples (2007) that place control of genetic resources under Indigenous self-determination. The Declaration's Article 31 states, "Indigenous peoples have the right to maintain, control, protect and develop their cultural heritage, traditional knowledge and traditional cultural expressions, as well as the manifestations of their sciences, technologies and cultures, including human and genetic resources, seeds, medicines, knowledge of the properties of fauna and flora, oral traditions, literatures, designs, sports and traditional games and visual and performing arts."

While genetic researchers utilize the sovereignty-neutralizing term *population*, the sovereignty implications of genetics have long been a central focus of Indigenous peoples' criticisms, beginning in the early 1990s with their transnational organizing against US government agency patents and the Human Genome Diversity Project (hereafter, the Diversity Project), using the term *biocolonialism* (Barker 2004; Harry 1995, 2009, 2011). In this critique, biotechnology is but the latest wave of technological innovation to exploit Indigenous peoples in the name of Humanity (L.T. Smith 1999). This resistance has been rooted in strong Indigenous critiques and ontologies, or worldviews, that conceptualize genetics in terms of a deeply held sacred respect that defines and guides responsibilities and duties to ancestors and living and future generations, thereby challenging dominant Western-rooted market friendly ontologies (Mead 2007; L.T. Smith 1999; Tsosie 2011). In some settler states, Indigenous peoples' political organizing and lobbying in the wake of major ethics scandals has led to various settler colonial states and Indigenous jurisdictions passing legislation and implementing regulations that recognize Indigenous self-determination and control over research involving their communities (Arbour and Cook 2006, 153–4; Harry 2009; Munsterhjelm 2014, 218–20).[1]

However, such Indigenous criticisms and international legal and ethical norms, legislation, and regulations have been effectively ignored by involved scientists who have used their extensive collections of Indigenous peoples' genetic materials as resources in the research and development of forensic genetic technologies. Indifferent to Indigenous peoples' ontologies, sovereignty, and rights, these forensic genetic research assemblages rationalize their quest through a range of rhetorical claims about improving public security and so serving the People and Humanity while also having the informed consent of their Indigenous research subjects, a seemingly win-win situation. Involved scientists routinely refer to their collections or public biobanks of blood samples, derived cell lines, and associated data as human genetic diversity resources, but these are also effectively forms of property owned by participating scientists, which are exchanged along with the involved scientists' expertise and labour for research funding and other institutional and regulatory support (Ong 2013, 78). This property relation takes various forms, such as private property of collections owned by various laboratories like the Diversity Project in a Paris suburb and Kidd Lab at Yale University, and communal commons type properties over which public access rights are claimed, such as DNA sequences of Indigenous peoples in publicly accessible databases run by both of these institutions, the high resolution DNA genome sequencing of the Karitiana and Surui by the Diversity Project, and the thousands of sequence entries from Indigenous peoples in the Allele Frequency Database (ALFRED) developed and run by Kidd Lab. In both of these forms, there are claims being made by outsiders on Indigenous peoples' genetic materials and data; claims that are made in the name of Humanity and serving the Greater Good. The fact that these Indigenous peoples' genetic materials were taken many years ago, such as the blood samples taken from the Karitiana and Surui by American and Brazilian scientists in 1987 and from the Ticuna of what is now western Brazil in 1976 has been routinely ignored.

In its main organizing narratives, forensic genetics seeks to assist police forces to identify, track, and capture/kill suspects and so is distinct from medical genetics, which seeks to manage or improve individual and population health in a politics of life, often termed biopolitics (Foucault 2003, 243–4; Toom et al. 2016, e2). In mass media, scientific discourses, and security discourses, forensic genetics is propagated as a means of protection against dangerous Others; as technologies of justice that guide the sovereign violence of law and sovereign violence in the name of the People and/or Humanity, thereby securing privileged populations against those deemed dangerous Others, something that tends to work along racial and class lines. These technologies have been developed within assemblages where scientists in major research centres in the US, Europe, and China have routinely violated Indigenous peoples' rights, sovereignty, and dignity; for example, asserting that informed consent given for other research projects decades ago is still valid. These violations can be analyzed through the concept of racializing assemblages, in which race functions "not as a biological or cultural classification but as a set of sociopolitical processes that discipline humanity into full humans, not-quite-humans, and nonhumans" (Weheliye 2014, 4). These sociopolitical processes are evident in forensic genetic research where many Indigenous peoples have been in a state of biotechnological servitude as a lesser species of human who has not been endowed with meaningful rights and carry extremely onerous obligations to serve Humanity by serving scientists as cryogenically preserved blood samples and/or virally transformed cell lines (derived from blood samples) and data in perpetuity (Munsterhjelm 2015, 41–2; Kowal and Radin 2015, 63).

As part of the expansion of the US security state under the George W. Bush administration after the 9/11 attacks, the 2004 US$1 billion President's DNA Initiative program run by the US Department of Justice stated, "Alternative genetic markers and assays can potentially provide further information about biological samples under investigation, such as an estimation of ethnic origin, physical characteristics, and skin, hair, or eye color" ("Alternative genetic markers" n.d.). A number of scientists responded to this call for the expansion of forensic genetic manhunting technologies to include race and ethnicity estimates. For example, in a 2008 US Department of Justice research funding report, Kenneth Kidd showed how human genetic variation (diversity) research was co-opted: "*Our collection* of population samples also provides *a unique resource* for validating SNPs that can be used in investigations to identify the ethnic ancestry of the individual leaving a DNA sample at a crime scene ... *Our populations* provide an excellent global overview of human variation as shown in various publications" (Kidd 2008, 17–18, my emphasis). According to *Webster's II New College Dictionary* (Berube 2005, 966), the word *resource* has various meanings including: "Something that can be looked to for support or aid"; "An accessible supply that can be withdrawn from when necessary"; "An ability to handle a situation in an effective manner"; "Means that can be used profitably"; and "Available capital: assets"; all of which are evident in Kidd's Department of Justice funding reports. Kidd's statement shows how human genetic variation was co-opted into manhunting research and development. Kidd Lab's collection includes a large number of Indigenous peoples, such as the Karitiana and Surui of western Brazil who were controversially sampled in 1987 by the Canadian epidemiologist Francis Black of Yale University (Borofsky 2005, 228–9; Cultural Survival 2004; Kidd 2008, 17–18; Rohter 2007; Santos 2002, 98n3; Vander Velden 2004). In the entrepreneurial networks of contemporary forensic genetic research and development, scientists exchange their expertise, labour, and collections of Indigenous peoples' cell lines and data (and that of non-Indigenous peoples) for funding and resources in research cooperation with government security agencies and the corporate sector, interactions mediated through the sharing of the manhunt organizing narrative of improving security by identifying criminals and terrorists. These manhunt narratives therefore have a crucial central role in organizing the production assemblages of forensic genetics.

WHY ARE THESE TECHNOLOGIES SIGNIFICANT?

From its beginnings in the mid-1980s, forensic genetics has incorporated concepts of race and ethnicity as legitimate categories in improving the human hunting capacities of its technologies. Use of forensic genetics as evidence for matching criminal suspects to crime scenes began with the development of so-called DNA fingerprinting, which was invented in 1984 and patented by the British geneticist Alec Jeffreys.[2] The controversies over the introduction of forensic genetics as evidence into the US and Canadian courts during the late 1980s and early 1990s dealt with whether the limited number of genetic markers being tested might differ in their prevalence between different racialized and regional populations. However, the 1990s involved a period of consultation, standardization, and consolidation of forensic genetics as a standard investigative practice. Today, the FBI's Combined DNA Index System (CODIS) and similar databases in other countries contain the searchable information of millions of convicted criminals and in many jurisdictions also detainees or suspects. CODIS was initially based on thirteen and later twenty genetic markers called short tandem repeats (STRs) that can identify individuals with a high degree of certainty.[3]

However, forensic genetic proponents ask: What if the sample does not match any existing individual profile in a database? The use of racial or ethnic categories was proposed shortly after the invention of DNA fingerprinting. For example, during a roundtable discussion at a 1988 forensic genetic conference in the US, Jack Ballantyne of the Suffolk County Crime Laboratory in New York State suggested the future potential of forensic genetic testing in determining "precise racial data"; for Kenneth Kidd it was "racial origin," something reiterated by other scientists, including DNA fingerprinting inventor Alec Jeffreys, in the early 1990s (Bartram, Plümecke, and Schultz 2021, 5; Track et al. 1989, 344). After 9/11, scientists began to expressly focus on racial taxonomies using conventional categories like Caucasian or Native American and euphemisms of phenotype for visible appearance and ancestry in the expansion of research and development. Early efforts to provide estimates of an unknown suspect's race and appearance date back to the early 2000s. As manhunting technologies, these forensic genetic technologies are resurrecting what had been largely discredited categories of race as somehow biologically inscribed and so scientifically valid (Bliss 2012, 202; Duster 2015; Fullwiley 2014; Obasogie 2012; Dorothy Roberts 2011a, x–xi, and 261–4).[4] A number of authors have argued that because these technologies reify racial categories, they will likely disproportionately target racialized minorities (Ahuriri-Driscoll, Tauri, and Veth 2021, 12; Duster 2015, 3–4; Fullwiley 2014, 808–11; ; M'charek 2008 520–1; M'charek, Toom, and Prainsack 2012, e16–17; Dorothy Roberts 2011a, 263–4; Toom et al. 2016, e2). In the US context, legal scholar and bioethicist Osagie Obasogie (2012) proposes that new biotechnologies using race-related categories must complete "race impact assessments" of the benefits of using racial categories while also acknowledging the costs and potential harms of reifying race as legitimate.

ASSEMBLAGES OF CAPTURE

In this book, I use the concept of *assemblages of capture* to understand how Indigenous peoples have been integrated into forensic genetics, an area of research that has been largely overlooked.[5] An assemblage, a term often used in actor network theory and security studies, is a network of organizations oriented towards a particular goal, it is generally hierarchical in structure, and it is governed by various political economies. Assemblages are a useful analytical concept because they avoid the overemphasis on the state that has been typical of international relations and sociology approaches (see, for example, Callon 1986; Dillon 2008).[6] The assemblages in this book are international heterogeneous networks that coordinate a range of human and nonhuman, individual and collective actants (actors), including state police and security agencies and research institutions, all guided by discourses of public security. In these assemblages, scientists use material equipment like DNA analyzers in university or police agency labs to process stored blood samples and derived cell lines from Indigenous peoples and non-Indigenous peoples and in so doing organize the international production of forensic genetics.

These international research and development assemblages exceed the boundaries of involved states, but state institutions retain a decisive role, contrary to some globalization theses. Dean and Villadsen (2016) argue that Foucault and his interpreters have frequently engaged in "state phobia," denying the relevance of the state or otherwise marginalizing it from theory and analysis. They argue against this analytical tendency and instead for the continued importance of the state, which they borrow from Max Weber's *Economy and Society* to define as:

> The primary formal characteristics of the modern state are as follows: It possesses an administrative and legal order subject to change by legislation, to which the organized activities of the administrative staff, which are also controlled by regulations, are oriented. This system of order claims binding authority, not only over the members of the state, the citizens, most of whom have obtained membership by birth, but also to a very large extent over all action taking place in the area of its jurisdiction. It is thus a compulsory organization with a territorial basis. Furthermore, today, the use of force is regarded as legitimate only so far as it is either permitted by the state or prescribed by it. (1978, 56; as quoted in Dean and Villadsen 2016, 20)

Crucially, the state remains ideologically and civil religiously important through nationalism since forensic genetics justifies itself by claiming to serve law and sovereignty. In this view, police manhunts are a legitimate use of force in which police risk and sometimes sacrifice their lives in the protection of the People of the nation and by extension Humanity against internal and external threats, with police killed in the line of duty honoured in various public memorials as sacred dead heroes (Marvin and Ingle 1999, 65 and 72; Anthony Smith 2000).[7]

VIOLENCE WORK

In this book, I draw on recent critical scholarship that conceptualizes the police as armed petty sovereigns acting in the service of the state through violence-backed discretionary and prerogative powers purportedly in pursuit of public security but in reality imposing and maintaining capitalist social relations (Arnold 2007, 3; Neocleous 2021, 21–2; Seigel 2018, 9–10). Based on anthropological studies, including in-depth interviews with Brazilian police officers involved in torture and extrajudicial killings, Huggins, Haritos-Fatouros, and Zimbardo (2002) developed the concept of *violence work* as a form of labour. US historian and social theorist Micol Seigel (2018, 9–10) calls police *violence workers*; they are an immediate everyday manifestation of state sovereignty and they can enact, "the potential violence that is the essence of their power. Yes, the violence of the police is often latent or withheld, but it is functional precisely because it is suspended." I argue that forensic genetic researchers are also violence workers through both their research and development technologies and their participation in investigations and court cases. Their efforts are guided by the express aim of improving the manhunting capacities of police and security agencies. By extension, Indigenous peoples' genetic materials and data (as well as those of non-Indigenous people, such as those in the well-known Twins UK project) become *violence work resources* that scientists use in research and development of manhunting technologies.

This book looks at the scientific cooperation between US, Chinese, and European Union security apparatuses in forensic genetic research and development of ancestry and phenotype genetic marker systems during the 2010s and how they were eventually disrupted from 2019 onwards. The concept of national security apparatuses is useful for long-term geopolitical networks of state institutions, including their personnel and the material infrastructure that is organized through their respective political economies, discourses, values, ideologies, myths, and histories (Seigel 2018, 21). However, to deal with the larger scale and often short-term networks that make up forensic genetics, I use the concept of assemblages as a more flexible mid-level concept that covers the often global production networks in forensic genetics and genetic research. These assemblages can be short-term, producing, for example, a coauthored study and research paper involving dozens of scientists from many different research institutions and security agencies. They can also be long-term, such as the international biobank the Diversity Project, which has 1,063 cell lines representing about fifty-two populations, including many Indigenous peoples, in a cryogenic storage facility in a Paris suburb and in publicly available genetic databases.

While there is extensive literature on the use of Indigenous peoples in genetic research dating back to the 1990s, including about the Diversity Project, there has been little academic attention given to the use of Indigenous peoples in forensic genetic research and development in the US, the European Union, and the PRC. Indeed, it is only since 2017, with growing international pressure over the increasingly genocidal forced assimilation of the Uyghurs and other Turkic peoples in Xinjiang by the Chinese government, that assemblages of elite scientists at leading research universities and institutes, biotechnology firms, and police and security agencies across the US, the European Union, and China have been subjected to strong public scrutiny; something the research for this book has contributed to (Chang and Fountain 2019; Human Rights Watch 2017; Moreau 2019; Munsterhjelm 2018, 2019; Wee 2019; Wee and Mozur 2019a, 2019b). However, I will also show that such attention has been politically selective; samples from Indigenous peoples taken as long ago as the 1970s and 1980s continue to be used routinely as resources in the research and development of forensic genetics.

THE MANHUNT NARRATIVE

Forensic genetic assemblages are organized around a singular organizing narrative schema about the manhunt. In this regard, the manhunt narrative is performative of how forensic genetics participate in organizing and guiding state violence, including the violence of law and the protection of national sovereignty and public security (Butler and Athanasiou 2013, 20, 89–90). For those who accept the manhunt narrative, the research and development of manhunting technologies as a means of improving public security becomes a sign of moral and ethical merit, a key claim of the foundational epideictic rhetoric (that is, ceremonial rhetoric) that is pervasive in forensic genetics and readily translated by police and state security agencies as well as the corporate sector. For example, these claims are evident in Verogen's LinkedIn webpage. A spinoff company started by US genetic analysis equipment maker Illumina and biotechnology venture capital firm Telegraph Hill Partners, "Verogen serves those who pursue the truth. It's about advancing next-generation sequencing to help unlock the true potential of forensic genomics. Supporting labs with solutions and expert service purpose-built for the challenges of human identification. And, ultimately, committing to effectively and efficiently ensuring justice, security, and public wellness – for all" (Cage Report 2023; Verogen 2022).

This narrative is also evident in the Human Identification Solutions conferences organized since 2015 by another US scientific equipment maker Thermo Fisher Scientific with titles such as "Increasing Security Solving Crime" (Lackey 2017); "Seeking Answers Solving Crime" (Thermo Fisher 2018b); and "Unite. Together, we find the truth" (Thermo Fisher 2021).[8] Such conferences bring together a range of actors from different institutions, including keynote speakers such as well-known forensic genetic researcher Bruce Budowle, a former FBI Lab senior scientist who is now at the University of North Texas; police forensic expert Wang Le of the Chinese Ministry of Public Security's Institute of Forensic Science; and victims' rights advocate Jayann Sepich, whose daughter was brutally raped and murdered, to extoll the virtues and prospects of these technologies in improving public security (Thermo Fisher 2016; Sepich 2017).[9] In March 2015, Bruce Budowle's (2015a) presentation at the Human Identification Solutions conference in Madrid began with an expression of the foundational mythic quest of forensic genetics: "We all have reasons we do forensic genetics. And of course, the motivations are victims, finding perpetrators, exonerating the innocent, mass disasters, terrorist attacks. All of these things present challenges to society and we have to find ways to help investigate and solve those kinds of problems." Budowle's (2015b) accompanying slide, entitled "The Motivation," featured images of newspaper clippings about crimes as well as exoneration, a thief trying to open a window, school photos of victims, a decaying corpse in a body bag, and one of the planes crashing into the World Trade Center on 9/11, offering a powerful call for justice. This call to action is a persuasive form of epideictic rhetoric that seeks to invoke a sense of common purpose of a shared quest for security, a process that rhetoricians Perelman and Olbrechts-Tyteca (1969, 54–5) termed *communion*.

As Foucault (2003, 89–90) notes, in Thomas Hobbes's *Leviathan*, sovereignty is conceived of as always immanent, tenuous, and contingent, continuously threatened from within and without by three major enemy or adversarial subjectivities that are manifestations of the state of nature: the criminal, the foreign invader (to which we can add today's terrorist), and Indigenous people. In this immanent Hobbesian view, each of these manifestations constitutes an ongoing threat to sovereignty, highlighting that sovereignty is never final but is rather contingent and tenuous such that security becomes a fundamental guiding principle in governance – and the presence of these figures readily invokes a logic of emergency (Opitz 2010, 93–4; Young 2003 9–10). Just as the "fabricated dialectic between fear of crime and sovereign power saturates U.S. popular culture" (Linnemann and Medley 2021, 67), these enemy subjectivities of the criminal and the terrorist are central to the theories of sovereignty prevalent in the discourses and organizing narratives of forensic genetic technologies. These are not technologies of biopolitical optimization; rather, they are sovereignty enforcement technologies that guide decisions about friends and enemies in the exercise of sovereignty and operate within the overall biopolitics of fear (Debrix and Barder 2009, 400–1).

This biopolitics of fear includes state institutions, but this use of fear extends to entities beyond the state, which encourages "all sorts of public agents/agencies to mobilize the specter of danger, threat, insecurity, and enmity." Reiterating a well-established principle of police, the forensic genetic solution to this fear is better identification technologies (Neocleous 2021). In his 1796 book *The Science of Rights*, German philosopher Johann Gottlieb Fichte advocated the use of personal identification papers, stating that "The chief principle of a well-regulated police is this, that each citizen shall be at all times and places, when it may be necessary, recognised as this or that person. No one must remain unknown to the police" (as quoted in Neocleous 2021, 40). Today, police use a wide range of information and technology to identify, surveil, track, and capture (Neocleous 2021, 40). The political scientist Elspeth Van Veeren (2021, 27) contends that "hunting is a social imaginary … that organises, in profound ways, the social, political and economic through subjects, practices and knowledges, including in relation to security discourses," which have become pervasive since the US government's declaration of the Global War on Terror.

The discourses of post-9/11 security can be analyzed using recent retheorizations of epideictic rhetoric beyond its typical use in civic rituals like funeral orations to show its central role in public discourse. Epideictic rhetoric is one of three forms of rhetoric originally proposed by Aristotle; the other two are forensic rhetoric (accounts of the past) and deliberative rhetoric (proposed future action). Cynthia Miecznikowski Sheard (1996, 776) argues that epideictic rhetoric (sometimes termed ceremonial rhetoric) is central to such calls to action: "Much contemporary epideictic rhetoric associated with civic rituals, for instance, or even academic ones, takes as its exigency a problem to be solved, a condition to be changed, a cause to be taken up. Such exigencies would seem to make epideictic discourse preliminary to forensic and deliberative discourses and therefore indispensable (rather than inferior) to them." Through the mobilization of fear, epideictic rhetoric is able to call for greater security of the population. The titles and themes of Thermo Fisher's Human Identification Solutions conferences and Budowle's slides and narration are examples of the epideictic rhetoric of forensic genetics' central organizing narrative, a mythic quest in which scientists research and develop technologies that can help reveal the identity of murderers and terrorists and protect the living while exonerating the innocent (Budowle 2015b, 2). Scientists enact a productive subjectivity in which they research and develop technologies to measure and estimate genetic traits at the individual and collective levels to help identify, track, and capture those deemed as threats, criminals, and/or terrorists by interrogating bodies as objects, aggregates of various genetic markers, rather than as persons or citizens (Epstein 2007, 157). These sorting technologies are part of a larger push towards a comprehensive set of biometric measures that interrogate subjects in various ways to identify, distinguish, and control flows of productive and destructive bodies in contemporary capitalism (Dillon 2008, 174–5; Epstein 2007, 154).

Forensic genetic research and development assemblages are built on the underlying rationality that scientific advances in genetic research can improve the security of populations and ensure they remain safe and productive (Munsterhjelm 2015, 303–4). Ideologically, psychologically, and morally, the manhunt narrative schema in forensic genetics invokes a quest to improve security by protecting society from an evil embodied in the figures of the criminal and the terrorist who threaten to disrupt the social order and unleash a state of nature (McQuade 2020, 66–7; Munsterhjelm 2015, 302–4; Neocleous 2015). In this way then forensic genetics supports the idea that the police are the "thin blue line" that defends order against the forces of chaos (Linnemann and Medley 2021, 69; McQuade 2020, 67–8; Neocleous 2000, 110).

The epideictic rhetoric of forensic genetics promises to help police and security agencies reveal the identity of the psychopathic killer, the criminal, the terrorist, the insurgent, and in general the dangerous Other in the service of sovereign and legal order aimed at the biopolitical security of populations by catching these dangerous Others who hunt and prey on the People.[10] These technologies are organized through a narrative schema of the manhunt, which has them identifying, finding, and capturing, and/or killing those deemed as threats or enemies: "The police is a hunting institution, the state's arm for pursuit, entrusted by it with tracking, arresting, and imprisoning" (Chamayou 2012). While police perform other social regulation functions (e.g., accident response and rescue), their manhunt role is a powerful form of governance enacting the sovereign violence of the state in ordering society and populations along capitalist lines, which in colonial contexts was termed *pacification*; for example, enforcing the property rights of mining or oil and gas corporations on Indigenous territories against Indigenous protests (Crosby and Monaghan 2016, 39–42; Neocleous 2013, 16–18).[11] Hence, state security and police agencies continue to retain crucial power and significance in their exercise of capital-imposing and capital-sustaining violence (Ince 2018).

According to the French philosopher Grégoire Chamayou (2012, 87–98), police manhunts are not a state of exception like martial law (contra Agamben 1998), but rather a central sovereign exercise of violence and governance practice based on police as petty sovereigns exercising discretion by identifying particular individuals and groups as prey to be hunted (du Plessis 2015). Scientists anticipate this police exercise of sovereign power and so enable it when they conduct forensic genetics research, development, and implementation and provide expert testimony at court trials (Bourne, Johnson, and Lisle 2015, 314; Ong 2006, 101). Importantly, scientists' anticipation of how police, security forces, and judiciaries will use these technologies in manhunts to enforce law and sovereignty also makes them agents in such manhunts and legal and sovereign decisions. To understand how manhunt organizing narratives function in forensic genetics, we have to include the laboratory and the research assemblage (Bourne, Johnson, and Lisle 2015, 310–12). Expanding the scope of the assemblage is similar to the ways security studies analyze how the surveillance and immigration assemblages of private and public institutions function in managing flows of people deemed productive and those deemed dangerous (see, for example, Aradau and Van Munster 2007; Bigo 2008; Dillon 2008; Epstein 2007).

In forensic genetics, heterogeneous assemblages of state security agencies cooperate with private corporations (e.g., genetic analysis equipment makers such as Illumina or Thermo Fisher), academic publishers (e.g., Springer Nature), and researchers in scientific research institutions in the name of the People and/or Humanity and based on a shared epideictic rhetoric. The goal of these international assemblages is to develop technologies that help police and security forces to identify, capture, and/or kill those deemed threats. During the 2010s, these assemblages included state agencies like the FBI, the US Department of Justice, the Chinese Ministry of Public Security, and various European national police agencies while exceeding the state in scope (cf. Bigo 2008, 10–1). The state is understood here as a collection of heterogeneous apparatuses that include personnel, institutions, and agencies, rather than some monolithic black boxed entity, but one that is also still united through shared national myths, regulation, budgeting, legislation, and sovereignty that are crucial to its organizing (Müller 2015, 32).

In response to this organizing quest for justice and protection from unseen terrorists and criminals, forensic genetic researchers impose the productive subjectivity of Indigenous peoples as resources in the production of knowledge from collections of samples gathered in Indigenous communities and cryogenically frozen and carried by air transport to labs in the United States and elsewhere. One of the major centres has been Yale University's Kidd Lab, where scientists infected the accumulated samples with Epstein-Barr virus to create virally transformed cell lines as resources, cryogenically storing and sharing the resulting cell lines and samples, growing cell lines, and extracting DNA from these cultured cell lines for testing, as well as collecting the cell lines of Indigenous peoples contributed by other scientists. In a 2011 grant funding report to the US Department of Justice's National Institute of Justice (NIJ), noted population geneticist Kenneth Kidd explains this property relationship, and his and his colleagues' expertise and experience, through their development of panels of genetic markers called single nucleotide polymorphisms (SNP or "Snips" in genetic parlance): "We justified *our* goals of continuing to develop both IISNP [individual identification SNP] and AISNP [ancestry inference SNP] panels based on *our unique collection* of population samples (Table 4–1), *our* well-equipped molecular laboratory, *our* extensive experience in population genetics, and considerable experience testifying during the early use of DNA in forensics" (Kidd 2011, 23, my emphasis). The repetition of *our* engages the rhetorical technique of anaphora to emphasize the strengths and capacities of Kidd Lab to achieve the goals of their NIJ grant and so serve the state. In pointing to Kidd's "unique collection" in the service of the US security apparatus, Kidd expresses a form of ownership over genetic materials from a number of Indigenous peoples, including the Karitiana and Surui (Kidd et al. 1991; Rohter 2007). Originally sampled for biomedical research on disease susceptibility and stored, grown, and shared through biobanks like the Diversity Project or Kidd Lab, Kidd and other scientists have for over thirty years made extensive use of the Karitiana's, Surui's, and other Indigenous peoples' genetic materials in the research and development of forensic genetic technologies. Hence, these forensic genetics as manhunting technologies that are intended to help police identify, track, and capture, and/or kill involve their own hierarchies in which Indigenous peoples captured through genetic sampling decades ago continue to serve scientists in perpetuity.

By serving scientists, who themselves claim to be serving the People and/or Humanity, Indigenous peoples can then be understood as also serving the People and/or Humanity. In this way, genetic researchers justify their use of samples from Indigenous peoples as helping society. Analyzing the relationship between whiteness and Indigenous peoples as property in genetic research, Jenny Reardon and Kim TallBear (2012, S242) argue that this research perpetuates long established colonial relations of racial dominance: "In Moore v. Regents of California (1990), the California Supreme Court similarly argued that once tissues leave an individual, the individual does not retain property rights. Although biological anthropologists' and population geneticists' uses and claims about DNA have been allowed, when non-scientists such as tribes or individual research subjects assert claims and the right to control DNA extracted from their bodies, these claims are disallowed." Scientists claim Indigenous peoples are willing donors who provided their samples after informed consent, yet routinely speak of these Indigenous peoples' genetic materials as forms of property.

THE UNIVERSAL NARRATIVE SCHEMA

Kim TallBear (2013, 17) defines so-called Native American DNA as a "material-semiotic object with power to influence indigenous livelihoods and sovereignties, and genetic scientists and entrepreneurs as frontline agents in the constitution of that object." This set of power relations involves the coproduction of scientists as scientists and Indigenous peoples as their objects of research, as "scientists observe the movement of particular nucleotides via human bodies across time and space (between what is today Siberia/Asia and the Americas). The presence of such markers is then used to animate particular 'populations' and individuals and their tissues (both dead and living) as belonging to that identity category" (12). In this way, through the metonymy of genetics in which genetics are fundamental forms of individual and collective ancestral identity, scientists are able to represent Indigenous peoples in a scope that stretches from the present back to ancient prehistoric migrations (Munsterhjelm 2014, 5–7; Reardon and TallBear 2012, 237; TallBear 2013, 13). Scientists coproduce themselves as scientists by imposing and using this subjectivity of Indigenous peoples as property and resources. The genetic research and forensic genetic research papers analyzed in this book involve the performance of a hierarchy between privileged Western and Chinese scientists and Indigenous peoples, including the Karitiana, Surui, and Uyghurs, among others, who function as resources and objects of exchange in the production of genetic knowledge.

To analyze how scientists hierarchically coproduce themselves along with Indigenous peoples as objects and resources of their research, I will make use of organizational communications theorist Francois Cooren and his colleagues reinterpretation of the universal narrative schema initially developed by A.J. Greimas of the Paris School of Semiotics (see figure I.1).[12] Greimas posited that narratives can be broken down into a particular sequence of phases, each with their respective set of actant positions: *sender* and *receiver*, *subject* and *object*, or *helper* and *opponent* (Cooren 2000, 71–4). A single actor can have a range of these actant positions within a given narrative (72). For example, an opponent in the early stages of a film can be transformed into a loyal ally in the later stages while a helper in the early stages may later commit treason and become an opponent (Callon 1986, 219–21; Cooren 2000, 187, 194).

A major distinction in Francois Cooren's (2000) reinterpretation and synthesis of Greimas's narrative schema, actor network theory, and speech act theory is what Cooren terms the *organizing property of communication*; that is, the emphasis placed upon the hierarchical organizing aspects of these narratives. This hierarchical organizing contrasts with the more flatland networks that Michel Callon and Bruno Latour often engage with (Cooren 2000, 188–9). In particular, Cooren argues that each of the involved actors can translate the organizing narratives as they see fit. For example, an employee might view a job as a source of income, whereas the employer will view the employee as fulfilling a particular role in the organization; however, regardless of their interpretations, by doing their job the employee submits to the employer's dominant organizing narrative (192–3).

Cooren's reinterpretation of Greimas's narrative schema includes a series of five phases that can be used to analyze narratives. A crucial aspect of the narrative schema is that it helps us see the narrative as structured from beginning to end as a hierarchically organized series of steps by involved subjects (e.g., scientists) that lead to the accomplishment or failure of the overall quest for scientific knowledge (Cooren

2000, 189–90). The following framework is based on Cooren and Fairhurst (2004, 798–800) and Cooren (2000, 71–3) but I have adapted their model to forensic genetic articles, which is illustrated in figure I.1.

Manipulation phase: In this initial phase, a *sender* defines the problem or imbalance that must be dealt with. The *sender* may be human, but can also be a nonhuman or macroactor such as humanity or scientific knowledge. In this initial phase, the *sender* gives a quest to the *receiver*, an exchange that creates an overall modality, or condition, of *having-to-do*. The main quest in forensic genetic articles is for technologies that help reveal the identity of the criminal, terrorist, and/or insurgent because their anonymity allows them to be threats to social and legal order. The manipulation phase is similar to Foucault's problematization in that it defines a problem to be solved ethically and ontologically through the use of epideictic rhetoric (Osborne 2003). This process also defines both the subject and the object of the quest, because one cannot exist without the other (Cooren and Taylor 2006, 120–1; Munsterhjelm 2014, 37).[13] In forensic genetics, researchers use Indigenous peoples' genetic materials as their objects of research.

Commitment phase: If genetic researchers accept and so submit to the quest sent to them by Humanity, Science, and/or the People, they are transformed into *receiver-subjects* who have a fiduciary duty to see the quest through (though they could instead reject the quest). This submission to science involves significant moral and ethical claims and so creates a modality of *wanting-to-do*. In forensic genetics, scientists have expertise and knowledge and so they accept the quest and the moral obligation (fiduciary duty) to serve in the quest for knowledge that helps in investigations. This duty is evident in Bruce Budowle's (2015b) commentary and slide entitled "The Motivation" discussed above.

Competence phase: This phase makes up most of the narrative and provides an account of the steps through which the research was completed. It can be analyzed as a series of subnarratives embedded within the overall narrative, like scenes in a movie, in which the *receiver-subject* gains helpers and overcomes obstacles and opponents (or not). In these subnarratives, the *receiver-subject* demonstrates the modalities of *knowing-how-to-do* and *being-able-to-do*. In forensic genetic articles, the researchers engage in a series of subnarratives that explain their methodology and how the samples were obtained (generally stating informed consent was obtained), outline how they processed the samples and conducted their statistical analysis, and finally discuss their findings' significance to the field.

Performance phase: In this phase, the *receiver-subjects* succeed or fail in their quest. This phase involves the modality of *to-do*.

Kidd's translation of the US security organizing narrative schema: embedded hierarchy of organizing narratives and actant time-spaces

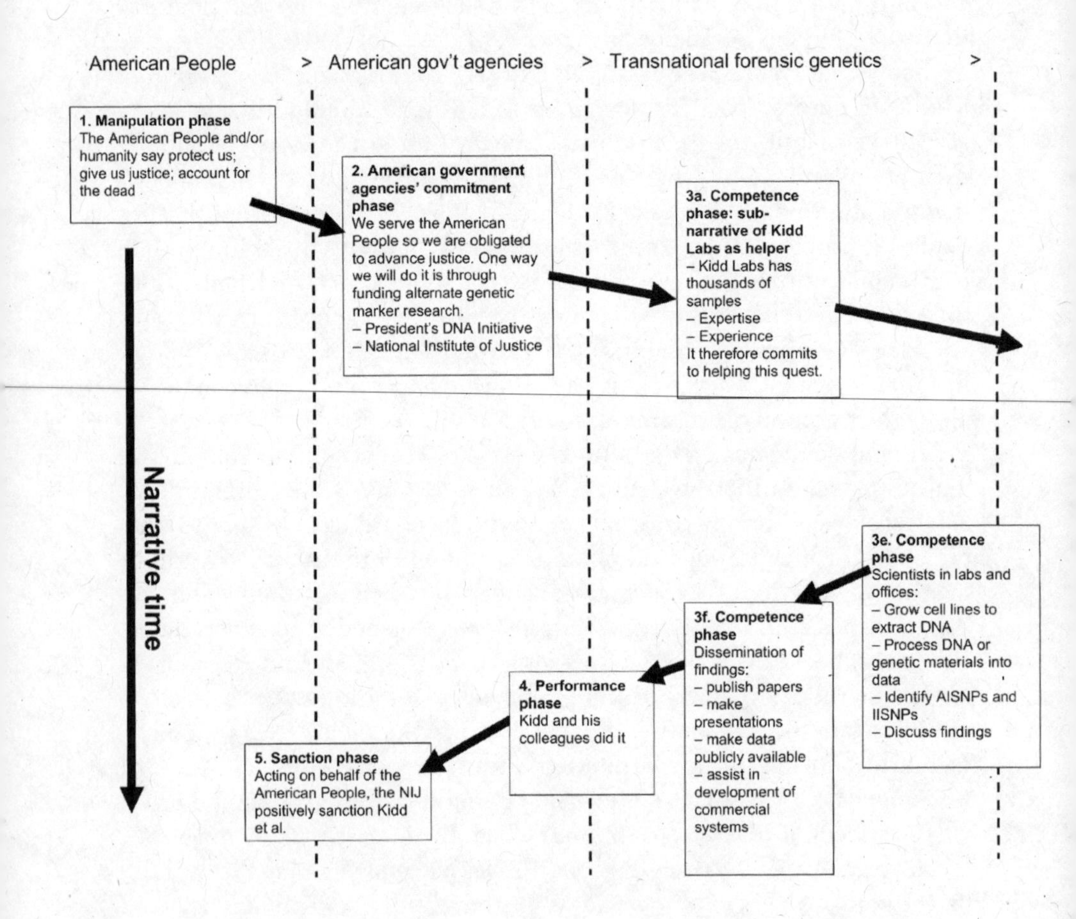

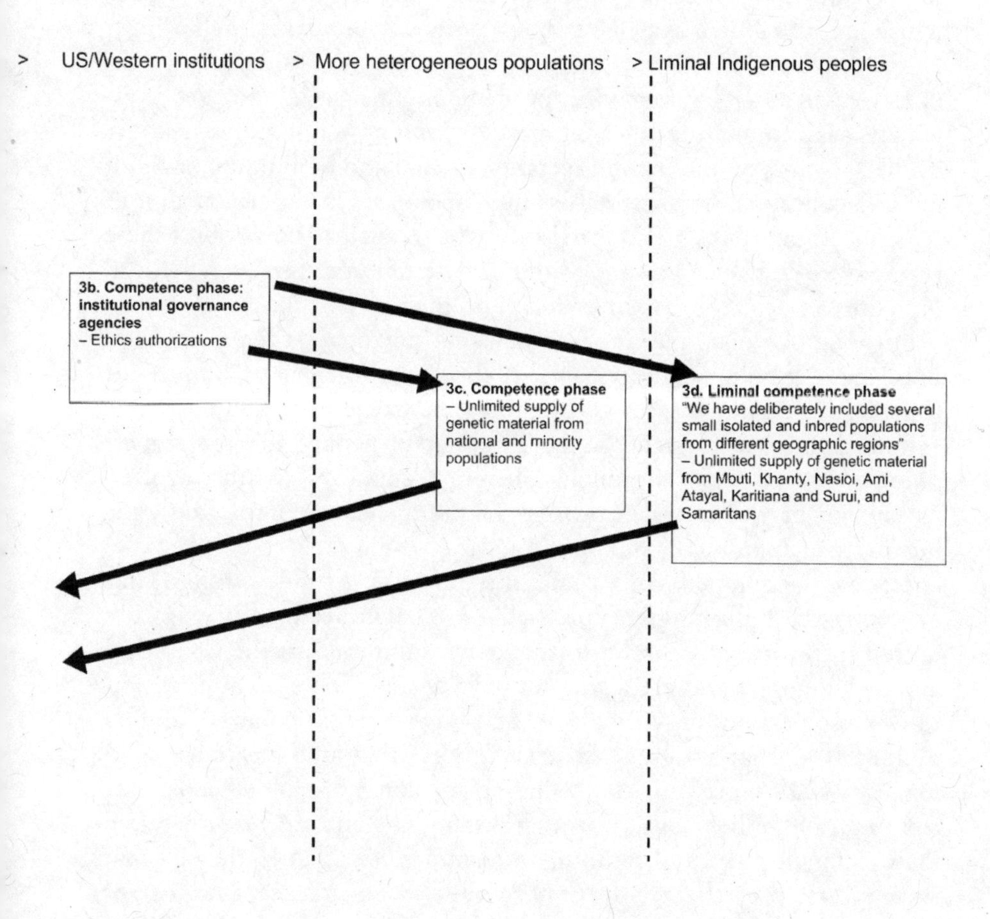

Figure I.1 | Exceeding the State: This diagram shows how Kidd Lab submits to the National Institute of Justice's quest for justice narrative schema by organizing a hierarchy of time-spaces, including state agencies, into a transnational assemblage. This assemblage uses Indigenous peoples to produce forensic genetic research to develop panels of AISNP and IISNP markers.

Sanction phase: In this final phase, the *senders* of the quest or their representatives (e.g., scientific journal editors) positively or negatively sanction the *receiver-subjects* based on the *receiver-subjects'* success or failure. This involves an *evaluative modality*. For forensic genetic articles, publication is a sign of positive sanction; the article has passed the peer review process and been published by international journals, which are distributed through corporate owned databases like Springer Nature.

As we will see in the remainder of the book, this model provides a systematic way to analyze research articles, funding reports, patents, and media accounts of how various actants, human and nonhuman, individual and collective, organize across time and space in the production of scientific knowledge. It also provides a way to analyze how robust these research assemblages are and whether external criticism and political pressure can destabilize them (which I do in chapter 11).

In the above mapping of the organizing narrative schema for Kidd Lab, the objective of the research is the quest for genetic knowledge to help investigations. This genetic research article links and hierarchically orders a set of events involving different actants in their respective time-spaces into a seemingly coherent account of how the involved researchers carried out their project. This sequence and importantly the hierarchy of submission among the actants within their respective time-spaces can be mapped and visualized (figure I.1). At the bottom of this hierarchy are Indigenous peoples, who are constructed as having submitted to the scientists, who in turn have submitted to the US People and/or Humanity and are acting on their behalf. These organizing narratives use persuasion to create an attractive passage point (cf. Callon 1986, 205–6) that involves "An *articulation* – that amounts to associating any two or more projects by finding any common point – can create a *translation* – that consists in establishing the equivalencies between these different projects – resulting in *identification*; that is, the fact that these diverse projects are now united, while each actor keeps his, her, or its own agenda" (Cooren and Taylor 2000, 185, emphasis in original). Through this shared attractive passage point, ethics review panels gave permission to the scientists, security and science agencies gave funding, Indigenous communities gave samples, and scientists processed cell lines. This narrative sequence of events form a coherent whole, a story about how the scientists conducted their research project and produced new findings that will help improve security (Cooren 2000, 189). In the organizing narratives of genetic research assemblages, scientists constitute themselves as heroic receiver-subjects of the quest by constituting

Indigenous peoples genetics as material-semiotic objects that are transformed as scientists process Indigenous peoples across disparate forms of time-space in an instance of coproduction (Munsterhjelm 2014, 51; TallBear 2013, 11–12, 17, 70–1). This circuit of production involves a series of translations that hierarchically organizes and articulates the sequence of events in different time-spaces involving different actants into a narrative form and so into a coherent whole.

There have been extensive debates over the validity of racial categories in genetic research, but it has been argued this leads to an analytical impasse that fails to address the pervasiveness and utility of racial concepts in forensic genetics (Ossorio and Duster 2005, 115–16; St Louis 2021, 209). Rather, building on Chun (2009), sociologist Brett St Louis (2021) contends, "the logical incoherence and negligible bio-materiality of race has not stymied its conventional meaning or functionality. Therefore, a central question arising regarding race as a technology is not what race is, but what does race do?" (St Louis 2021, 209). Anthropologist Duana Fullwiley (2015, 36) critiques the use of race in genetic technologies: "Race is a sorting schema that is rarely neutral. It permits people to be classed, judged, included, excluded, normalized, pathologized and, at the extreme, killed." Fullwiley analyzes biomedical and forensic genetic research and development and contends that "when technologies are born of race sorting logics, then the resultant race problems and their proposed solutions contain the same disturbing seed elements" (37).

In this book, I conceptualize this race sorting logic. This selectively draws on the extensive retheorization of synecdoche and metonymy since the 1950s (Bierwiaczonek 2020, 225–9; Nerlich and Clarke 1999, 198–201; Wachowski 2019, 54–73). These provide a useful set of conceptual tools for understanding the ways racial/ethnic categories are fundamental to forensic genetic research. Linguists Brigitte Nerlich and David Clarke (1999, 201) define the differences between metonymy and synecdoche: "To summarise, one can therefore say that metonymy is based on qualitative, synecdoche on quantitative relations, that is on set-inclusion. Metonymy is based on our world-knowledge about space and time, cause and effect, part and whole, whereas synecdoche is based on our taxonomic or categorical knowledge. Metonymy exploits our knowledge of how the world is, synecdoche of how it is ordered in our mind." Linguist Ken-ichi Seto (1999, 92) contends that synecdoche involves a conceptual transfer of meaning "between a more comprehensive category and a less comprehensive one." Synecdoche involves the use of taxonomies, in which "genus–species relations, based on set-inclusion, are quantitative, i.e., they involve classes of entities which differ in the number of their members, whereby the smaller class is included in the bigger class" (Bierwiaczonek 2020, 226). In this taxonomy:

Racial Category (biogeographic continental like "Native American")
Population (e.g., seventy-four donors represent the Karitiana)
Individual Donors (e.g., Karitiana members sampled in 1987)

The way this taxonomic hierarchy works in genetic research papers is consistent with linguist Wojciech Wachowski's (2019, 63–7) view that moving back and forth between higher level (superordinate or genus) categories, like Native American, to subcategories or species like Karitiana, which involves shifts from more inclusive to less inclusive categories, all involve synecdochical transfers of meaning, not metonymy. Therefore, in forensic genetic research articles, synecdoche as genus-species (in a taxonomic, not biological, sense) relations defines the categories of racial/ethnic taxonomies like Native American as genus that include smaller classes such as Karitiana, Surui, and Ticuna peoples as species.

Metonymy is distinct from synecdoche because, according to linguist Bogusław Bierwiaczonek (2020, 226), metonymy involves "part–whole relations, as other relations based on contiguity, [that] are essentially qualitative, i.e., they involve associated entities or concepts which are of different kinds" in which meaning is transferred from one kind to another. In forensic genetic articles, these metonymic transfers between different types of entities are evident in the seemingly contiguous or continuous relationships between donors and their blood samples, which are virally transformed into cell lines that are grown in labs then processed into DNA extract and then run through genetic analyzers or sequencers to create genetic data that is considered representative of those donors in a type of part-of (partonomy) relation. This series of transformations from donor to genetic data involves a series of metonymic transfers that form a *metonymic chain* in that the genetic data represents a donor.

Within these narratively organized assemblages, concepts of genes, genetics, DNA, and the human genome all function as objects of exchange (what Star and Griesemer [1989] call *boundary objects*) that are flexible conceptually and can be translated readily to mean and signify different things to different institutional and individual actors across time and space (Munsterhjelm 2014, 33–5; Shea 2008, 514–15; Star and Griesemer 1989). As material semiotic objects of exchange, Indigenous peoples' genetics are plastic and able to metonymically retain traces of agency as they are processed from cell lines to DNA extract to data and circulated across and so connect the disparate time-spaces (Cooren and Taylor 1997, 247; Latour 1987, 161; Munsterhjelm 2014, 33–4, 2015; Star and Griesemer 1989, 393; TallBear 2013, 17). However, they retain sufficient rigidity to still be considered real representatives of whatever they signify; in this case, genetic samples, cell lines, and data are accepted as representative of Indigenous peoples (Munsterhjelm 2014, 74–5).

Drawing on Bierwiaczonek's concept of synecdochic metonymy (2020, 228–34), this representation relies on synecdoche in its categorization and then metonymy of genetics in which genetic materials and data stand in for and provide access to Indigenous peoples, including the dead, living, and unborn, something popularized in many TV documentaries, such as has been done by genetic anthropologist Spencer Wells in the PBS Nova series *The Journey of Man* (Littlemore 2015, 9; Munsterhjelm 2014, 5, 41; Reardon and TallBear 2012). This linking together and articulation or organizing of the various stages of the research article narrative relies upon a chain of metonymy that begins with scientists taking blood samples from Indigenous peoples in their villages (Brdar 2015, 86–9). This chain of synecdochic defined metonymy will be explained in detail in the next chapter but can be summarized as:

1 Donors FOR Indigenous people
2 Blood samples FOR donors
3 Cell lines grown from blood samples FOR blood samples
4 DNA extract FOR cell lines
5 Data FOR DNA extract
6 Findings FOR Data (Brdar 2015; Denroche 2015, 120–4; Radden and Kövecses 1999, 31, 41–2).[14]

Synecdoche is involved in the racial categorization of Indigenous peoples, and the genetic materials and data representing them as metonymic chains across time and space have been fundamental to the production of considerable forensic genetic knowledge. However, as we will see, these chains are vulnerable when challenged as immoral and unjust by sufficiently strong external assemblages (see table I.1 for a summary of the forensic genetic narrative schema).

According to Bierwiaczonek (2020, 229) in analyzing synecdochic metonymy, which he terms syntonymy, "we should distinguish two kinds of synecdoche, namely, the specializing genus for species synecdoche proper (more traditionally known as specialization) and the generalizing species for genus syntonymy (traditionally known as generalization)." In the organizing narrative schema of forensic genetics,

Table I.1 | Summary of forensic genetic narrative schema

Schematic phase	*Modality*	*Form of rhetoric*	*Synecdochic and synecdochic metonymic transfers*
Manipulation Science is the sender that gives quest	Having-to-do	Epideictic rhetoric	Synecdoche of racial categories (genus) e.g., European, Native American, African East Asian, Eurasian, Oceania TO
Commitment Scientists as receiver-subjects accept their quest Receiver-subjects define how they complete quest creating fiduciary contract	Wanting-to-do	Epideictic rhetoric Deliberative rhetoric (future plan)	Synecdoche of specific populations (species) e.g., Karitiana, Surui, Uyghur, Nasioi TO
Competence	Knowing-how-to-do Being-able-to-do	Forensic rhetoric (account of past events)	Synecdochic defined metonymic chains e.g., Karitiana data FOR DNA extract FOR cell lines FOR blood samples FOR donors TO
Performance	To do	Epideictic rhetoric	Synecdoche of racial categories (genus)
Sanction	Evaluation and acknowledgment of what has been done	Epideictic rhetoric	Synecdoche of racial categories (genus)

Source: Based in part on Cooren and Fairhurst (2004, 802), Fairhurst (2007, 34), and Munsterhjelm (2014, 70–1).

Syllogism	*Predominant form of time-space (others forms have implied presence) .*
(Deductive: universal to local or specific project) Universal premise: "Science must study genetics to learn about genetic diversity and/or racial categories to create AISNP or PISNP panels to help police a nd security forces to hunt down criminals and/or terrorists"	– Transnational science
Local particular premise: "We are scientists and we want to study genetic diversity to create AISNP or PISNP panels to help hunt criminals and/or terrorists." Local premise: "Indigenous peoples and/or non-Indigenous peoples can be studied for genetic diversity and/or racial categories to create AISNP or PISNP panels" Conclusion: "We scientists will study them"	– e.g., Brazil, USA, European Union, China, etc.
(Inductive: local or specific project to the universal) Local premise: "We received authorizations." Local premise: "We enroled and sampled Indigenous and/or non-Indigenous peoples" Local premise: "We variously tested Indigenous and/or non-Indigenous peoples' samples, cell lines, DNA extract, and/or data." Local premise: "We found the following results about AISNPs or PISNPs" Local premise: "We discuss our results in relation to, and thereby associate our results with, established knowledge about genetic diversity and/or racial categories to improve AISNPs or PISNPs."	– Settler state institutions – Indigenous and non-Indigenous territories – Laboratories – Transnational science
Universal premise: "We have contributed to scientific knowledge about genetic diversity and/or racial categories to improve AISNPs or PISNPs manhunting technologies"	– Transnational science
Universal conclusion: Peer reviewers say, "You have contributed to scientific knowledge about genetic diversity and/or racial categories to improve AISNPs or PISNPs"	– Transnational science

synecdochic transfers between different taxonomic levels are crucial in forensic genetics in the initial deductive transfers from the manipulation phase TO the commitment phase, which involves moving from the universal space of forensic genetic discourses and knowledge to the specific projects of the authors. Then a series of inductive synecdochic metonymic transfers articulate the competence phase, performance phase, and sanction phase. These transfers move from the specific project contexts of the materials and methods section in which Indigenous and non-Indigenous donors are enroled in their local communities through to the authors claiming that they have contributed to universal scientific knowledge about ancestry and phenotype related SNPs and other types of markers, which is positively sanctioned by the journal publishing their article.

SCOPE OF THE BOOK AND OUTLINE

This book is not a survey of the entire field of forensic genetics. It utilizes a case study and multimodal approach to examine the first research assemblages to emerge around Yale University's Kidd Lab, as well as related assemblages, including the Human Genome Diversity Project, that engaged in early diversity collection efforts by leading scientists, including Kenneth Kidd of Yale University and Lucas L. Cavalli-Sforza of Stanford University, targeting various Indigenous peoples. It then considers how defence counsels co-opted some of these Indigenous peoples' genetic data into the US and Canadian courts during the early 1990s, and their subsequent integration as diversity resources in the post 9/11 research and development of individual identification and ancestry inference panels. It then considers influential scientists' research on phenotype (external appearance like eye colour and hair shape) inference SNP marker technologies in the US, the European Union, and China and the destabilization and reconfiguration of these assemblages using Uyghur and other Turkic peoples. Readers may find the semiotic narrative analysis of forensic genetic articles technical at times, but my intention is to both introduce these technologies and empirically show how these assemblages of research use genetic materials captured from Indigenous peoples as violence research and development resources. Genetic research involving Indigenous peoples has long been shaped by Indigenous resistance and activism (Barker 2004; Harry 2009; Munsterhjelm 2014, 49–51; Reardon and TallBear 2012). The derailment of the Diversity Project in the late 1990s finds its contemporary in the discrediting of Kidd's cooperation with the Chinese Ministry of Public Security and other such projects in the present.

The book is organized as follows. Chapter 1 traces out the conquest and colonization of the Western Amazon region by the genocidal Brazilian settler state from the 1960s through to the 1980s as a context under which US and Brazilian scientists sampled the Ticuna in 1976 and the Karitiana and Surui in 1987. I then analyze a 1991 paper by Kenneth Kidd, Judith Kidd, and other prominent researchers as an account of how assemblages of capture function as the means through which scientists enrolled the Karitiana, Surui, and Maya Indigenous peoples into a state of servitude within global genetic research and development assemblages. Chapter 2 then deals with how, during the "DNA Wars" of the early 1990s over the introduction of forensic genetic test evidence, defence counsels appropriated Kidd et al.'s Karitiana, Surui, and Maya data. The data then circulated among defence counsels, who used it for over a decade in US and Canadian courts as evidence of significant group differences in genetic marker frequencies to try to undermine prosecution's use of forensic genetic tests that matched suspects to crime scenes.

There are significant distinctions in forensic genetics before and after the 11 September 2001 attacks. The 9/11 bombings' identification problems and expansion of security state–related biometric research led to calls for the expansion of forensic genetic testing to include ancestry and phenotype. Chapter 3 shows that after 9/11, Kenneth Kidd and other major researchers integrated these Indigenous peoples into various long-term forensic genetic research projects on individual identification SNP, ancestry informative SNPs, and phenotype informative SNPs. The individual identification SNP and ancestry marker panel were integrated into the Illumina FGx forensic genetic sequencer system. This system was tested by Bruce Budowle and his colleagues in a paper on samples taken from the Yavapai Indigenous people of central Arizona before the 1990s, likely by the US National Health Service. This testing was part of the validation process of the new system, done in part with US NIJ funding. Chapter 4 provides a critical introduction to Chinese settler colonialism in Xinjiang in Northwest China, then covers how Kenneth Kidd and Bruce Budowle's cooperation and sharing of DNA extracts with scientists from the Chinese Ministry of Public Security's Institute of Forensic Science (also referred to as the Material or Physical Evidence Identification Centre) in Beijing helped develop a twenty-seven SNP ancestry panel intended to differentiate Han Chinese from Uyghurs and other Indigenous groups. Chapter 5 then analyzes how involved Institute of Forensic Science researchers utilized this research cooperation as part of a series of ancestry-related genetics Chinese patent filings that seek to differentiate Han Chinese from Uyghurs and Tibetans and other Indigenous peoples in China.

The second segment of the book shifts to an analysis of European-Chinese cooperation in the development of phenotype inference SNP genetic markers. Chapter 6 considers European Union–centred assemblages of the Visible Genetic Traits (VisiGen) Consortium and an interlocking series of projects including the Twins UK, the Rotterdam Study, and the Queensland Institute of Medical Research in Australia. Chapter 7 offers the case study of the early development of forensic genetic phenotyping involving Uyghurs and Han Chinese by researchers at the Chinese Academy of Sciences-Max Planck Society Partner Institute of Computational Biology in Shanghai. Chapter 8 shows how the VisiGen Consortium cooperated with the Beijing Institute of Genomics (BIG) and Partner Institute of Computational Biology in Shanghai in a series of four papers on appearance-related genetic phenotyping involving tens of thousands of research subjects from Latin America, Europe, Australia, the US, and China, including over seven hundred Uyghurs. These articles were published between December 2017 and November 2019, during the PRC's escalating repression of the Uyghurs and other Indigenous peoples in Xinjiang.

Chapter 9 analyzes PRC government planning documents and journal articles to show how phenotyping has been integrated into a biometric suite of security technologies under the government's 13th Five-Year Plan (2016–2020). It shows how the researchers at BIG and the Partner Institute of Computational Biology in turn began research cooperation with the Institute of Forensic Science in 2014–15. In Chapter 10, we see how this cooperation culminated in a series of papers published between 2017 and 2019 involving a new project that sampled over seven hundred Uyghurs from Tumxuk in the south of Xinjiang. Chapter 11 is about how the Uyghur diaspora, activist academics, human rights organizations, media outlets, and eventually the US government destabilized the Institute of Forensic Science research assemblages studying ancestry inference with Kenneth Kidd and those cooperating with the European Union's VisiGen phenotype inference SNP markers. It shows how international forensic genetic networks have been disrupted by growing international criticism of PRC repression in Xinjiang and by the ongoing and escalating political economic conflicts between the US, the European Union, and China, three of the world's largest economies and military/police security apparatuses. However, despite the disruptions with regard to using Uyghurs and other Turkic Indigenous peoples from Xinjiang, all three apparatuses continue to use Indigenous peoples as productive resources in forensic genetics.

1

"Kidd Lab Genehunter Center"

In 1984, Kenneth K. Kidd and Lucas L. Cavalli-Sforza of Stanford University, both important figures in the development of population genetics, were on a transatlantic flight when they had the inspiration for what became the Human Genome Diversity Project, a collection of genetic materials sampled from "disappearing" Indigenous peoples and held for the benefit of Humanity. They began asking colleagues from the fields of epidemiology, anthropology, and population genetics to gather samples from Indigenous peoples in the field to contribute to the collection (Leslie Roberts 1991a, 1615–16). Kidd later appeared in a 1995 film on the Human Genome Diversity Project controversy entitled *The Gene Hunters*. The title is based on a sign on a door at Kidd Lab in the Yale University School of Medicine that reads "Kidd Lab Genehunter Center" (the sign appears at 10:33 in the film). In the film, Kenneth Kidd is interviewed at Kidd Lab where he explains the culturing of Indigenous peoples' cell lines in a growing room and then provides commentary as he extracts DNA from cell lines. He then explains the global expansion of ancient human migrations over millennia, in part by showing the relationships between a number of Indigenous peoples. In the film, Kidd illustrates his findings in part with a computer generated tree diagram of the relationships between the Karitiana, Surui, and Ticuna along with a number of other Indigenous peoples. These initial collections provided the foundation for both the Kidd Lab collection and the controversial Diversity Project.

The Karitiana, Surui, and Ticuna cell lines continue to serve to this day. On the Kidd Lab website is a list entitled, "47 population samples routinely studied at Kidd Lab" (Kidd Lab 2017). On this list are three Indigenous peoples from western Brazil: "Karitiana (57)†," "Rondonian Surui (47)†," and "Ticuna (65)†," with a note at the bottom of the page:

"† indicates samples that may contain related individuals" (Kidd Lab 2017). The Kidd Lab's ALFRED (Allelle Frequency Database) webpage contains a description of the Karitiana that states, "The sample was collected in the Karitiana village (Rondonia Province, Brazil) by F. Black. HLA haplotypes [groups of immune system blood genes that are inherited from each parent] indicate that the Karitiana have no non-Amerindian admixture and are genetically distinct from other sampled populations in relative geographical proximity, such as the Surui" (ALFRED 2019b). The ALFRED description also mentions: "Endogamy [intermarriage with group] prevails among the Karitiana, and their documented family structure demonstrates that the entire group is essentially one family" (ALFRED 2019b). The ALFRED description of the Surui states, "The sample contains individuals from the Rondonia Province in Brazil and was collected by F. L. Black. Statistical analyses indicate many pairs of first degree relatives in the sample" (ALFRED 2019d). The ALFRED entry for the Ticuna states, "Ticuna sample collected by D. Lawrence and D. Wallace from individuals in several villages in the western part of Amazonas, Brazil (close to the border of Peru and Columbia near Rio Salimas). The total sample contains several family groups" (ALFRED 2019e). In each of these seemingly innocuous descriptions are traces of the complex history of settler colonial conquest and genocide in the western Amazon region. Researchers have shared and Coriell Cell Repositories has sold these peoples' cell lines, making these important "resources" in the international development of forensic genetics. As we will see later, the Karitiana and Surui both appeared in Chinese patent applications and journal research articles by Chinese Ministry of Public Security researchers.

The inclusion of these Indigenous peoples as resources by Kidd Lab involves science research intertwining with settler colonization of the western Amazon region from the 1960s through to the 1980s. This chapter shows how American and Brazilian scientists took advantage of Brazilian settler state's violent conquest of the Amazon region to study and sample colonized Indigenous peoples. Scientists such as Kidd and Cavalli-Sforza were guided by a salvage paradigm in which they were acting on behalf of Humanity by incorporating these Indigenous peoples into the scientific commons of genetic research before they disappeared as distinct entities due to massacre, displacement, disease, and/or assimilation (L.T. Smith 1999, 25, 100).

The story of the sampling of the Karitiana and Surui in 1987 by American and Brazilian researchers and the commodification of their cell lines and DNA in the US within the larger context of Brazilian settler colonialism has been the subject of extensive academic, journalistic, and political criticism as exemplars of biocolonialism and biopiracy (see, for example, Borofsky 2005, 228–9; Rohter 2007; Santos 2002, 98n3; Vander Velden 2004). The sampling of the Ticuna has also been the subject of some critical inquiry. These Indigenous peoples are not unique in this regard. Indeed, various university laboratories in the US contain many thousands of blood samples taken from Indigenous peoples of the Amazon region from the 1960s to the 1980s (Radin 2017). Historian of science Joanna Radin (2017, 160) points out that many scientists do not conduct research on such collections due to ethical concerns. However, such concerns have not affected the use of the Karitiana, Surui, and Ticuna Indigenous peoples' genetic materials, who have since the early 1990s played disproportionately important roles in the development of forensic genetics. Therefore, drawing upon these accounts, I will show the relation between the original settler colonial expropriation in the conquest in the Amazon region and the assemblages of large-scale technoscience that emerged during the period of post–World War II American hegemony in the region.

SETTLER COLONIALISM AS ORIGINAL EXPROPRIATION

The sampling of these three Indigenous peoples occurred during the intensified settler colonization of the Amazonian region that followed the 1964 US-backed military coup in Brazil. The Amazon regions had been the subject of various resource economic booms, such as gold and rubber, but these faded away and much of the resulting colonization of the regions was largely restricted to areas along the major rivers. An earlier effort by the populist Brazilian president Getúlio Vargas in the 1940s for a "March to the West" had little long-term effect on the ground. However, the post–World War II building of the physical infrastructure of settler colonial rule, particularly roads, became a focus of subsequent efforts. In the late 1950s, the Brazilian government under President Juscelino Kubitschek started to intensify efforts to colonize the interior regions by building the new capital city of Brasilia in the highlands of the south central savannah region and constructing highways (including BR-364 begun in 1961) into the Amazon region. Following the US-backed military coup in 1964, the new military regime intensified these efforts as part of a policy of national security, national development, and economic modernization (Tavares 2013, 213–16). In its efforts to secure and assert Brazilian sovereignty in the Amazon region against competing claims from Colombia, Peru, and Venezuela and integrate it into the national economy, in 1966 the regime began Operation Amazonia, with its system of new highways, mining exploration, commercial cattle ranching, and settlement (Early and Peters 2000, 247–8). These efforts were followed in 1970 by the National Integration Program under Brazilian president Emílio Garrastazu Médici, who stated the program was "the solution to two problems: men without land in the Northeast and land without men in Amazonia" (Early and Peters 2000, 247–8; Skidmore 1988,145–6). This solution involves a refusal to recognize Indigenous peoples as political entities and instead reduces them to merely a part of nature. This assertion that Indigenous peoples' territories were *terra nullius* is a classic expression of settler colonialism's logic of elimination (Wolfe 2006). This designation involved what Achille Mbembe calls necropolitics and involves the "creation of *death-worlds*, new and unique forms of social existence in which vast populations are subjected to conditions of life conferring upon them the status of *living dead*" (2003, 40, emphasis in original). If we apply Foucault's (2003, 241, 248) concepts of biopolitical optimization as make-live and sovereignty as take-life, then necropolitics involves not-make-live as active denial of what is necessary to live individually and collectively, which complements and reinforces the sovereign power to take-life (Munsterhjelm 2013, 4–5).

In Marx's formulations of original expropriation (which is also translated from German as primitive accumulation) in chapter 31 of volume 1 of *Capital*, he insightfully argued that violence is an important force of production in the conquest, imposition, and maintenance of colonial capitalist commodity production relations: "The discovery of gold and silver in America, the extirpation, enslavement and entombment in mines of the aboriginal population, the beginning of the conquest and looting of the East Indies, the turning of Africa into a warren for the commercial hunting of black-skins, signalised the rosy dawn of the era of capitalist production. These idyllic proceedings are the chief moments of primitive accumulation" (Marx [1867] 1996, 739).

Under settler colonialism, violence and the threat of violence have been central to the acquisition of Indigenous peoples' lands and their physical resources (Coulthard 2014, 15–16). However, contrary to Marx's emphasis on proletarianization, in which peasants are displaced from the land through enclosure and are forced to sell their labour in the market, Indigenous labour is frequently largely superfluous (though at times locally significant like peonage and slave labour in the rubber trade in Brazil) (Coulthard 2014, 9–12). Similarly, Carl Schmitt ([1950] 2003, 94) in *The Nomos of the Earth* asserted that this founding violence, "the law of the stronger," was the ultimate source of authority in European overseas imperialism and the basis of original expropriation. Europeans viewed Indigenous peoples' territories as *free land*, open to expropriation and expansion, thereby anchoring the European-centred system in part by shifting the focus of European conflicts to regions outside of Europe (Ince 2018, 889–90; Schmitt [1950] 2003, 81, 94, 161).

As a settler colonial state, Brazil (like the US or Canada) is based on this law of might because it derives its sovereignty claims as a successor to earlier imperial formations who claimed and/or conquered these regions through force. The military junta saw the colonization of the Amazon as a way to extend the sovereignty of the Brazilian state into these territories and so deal with Brazil's political, economic, and social class inequities, not through redistribution but by creating more wealth (Early and Peters 2000, 248). This colonization would settle Brazilian citizens and so reinforce Brazilian national security, particularly its sovereignty claims, at a time when other settler states such as Colombia and Peru were also beginning to colonize and economically develop their Amazon territories that bordered western Brazil.[1] The intersection of these biopolitical goals and the military dictatorship's national security–related sovereignty concerns in the development in the Amazon region was summarized in a well-known slogan "integrate not to surrender," "in other words, connecting 'Amazônia Legal' to the Brazilian economy and life to prevent a foreign invasion" (Rodrigues and Kalil 2021, 8). The racial caesurae that divides who must live from who must die involves the biopolitical imperatives to advance the well-being of select portions of the Brazilian population based upon the sovereignty founding violence of settler colonial conquest after which Indigenous peoples are subjected to necropolitical exposure to death through the expropriation and exploitation of their territories, commercial agriculture like cattle and soybeans, overhunting, mining, pollution, overfishing, and deforestation – all of which undermine Indigenous peoples' ways of life (Dillon 2008, 177–8; Mbembe 2003, 25–30).

Operation Amazonia began with the dictatorship's declaration of direct federal control over "Legal Amazonia," the Brazilian portion of the Amazon basin encompassing over 5,000,000 km^2 including Rondonia and all or part of seven other states (Tavares 2013, 215). Among the dictatorship generals, in addition to its national security implications, the colonization of Amazon was both an important ideological symbol of Brazilian national development and of massive economic importance. It was also intended to ensure economic growth without redistribution of existing wealth or land (Early and Peters

2000, 247–8). Brazil's landowning elites were among the military junta's most significant supporters and the necessity of their continued support made any redistribution of existing lands to peasants out of the question (Skidmore 1988, 145–6). Brazil's left, including labour, peasant, and communist movements, were ruthlessly suppressed by the regime during the first few years of its rule (Skidmore 1988, 131–5). Therefore, though not threatened by any viable resistance movements, the military rulers recognized how Operation Amazonia could use mass colonization to deal with the often-restive urban poor of the frequently drought-stricken northeast and farm labourers from southern Brazil who had been displaced by the mechanization of large-scale agriculture (145–6). Both problems could be spatially remedied by the settler colonization of Indigenous people's territories in what David Harvey (2003, 145–52) calls accumulation by dispossession, a biopolitical spatial fix through which these surplus populations were shifted to the newly colonized Indigenous lands, thereby allowing for further capital accumulation based on this original expropriation of Indigenous peoples' territories (Tavares 2013, 213–16). This settler state biopolitical management of surplus populations was premised on the logic of elimination, the forced assimilation and genocide of Indigenous peoples. In this way, Operation Amazonia and related settler colonization efforts increasingly integrated these regions into Brazilian political economic development and larger global circuits of capital.[2] This program of colonization in the Amazon region was the foundation upon which thousands of Indigenous people were sampled from the 1960s onwards.

HUNTING DIVERSITY IN SPACES OF SETTLER COLONIAL CAPTURE

Chamayou (2012, 16) critically reinterprets Foucault's biopolitical concept of the biblical pastoral flock who is guided, protected, and known by name by counterposing it to the biblical figure of King Nimrod as a hunter whose "Cynegetic [hunting] power gathers together what is scattered, centralizes and accumulates it in a limitless logic of annexation." If we apply both these concepts to the above brief history of the Amazon region, the biopolitics of the Brazilian settler colonial formations (in conjunction with a handful of other settler states) relied on hunting power to conquer and subjugate the numerous Indigenous peoples of the Amazon region. The sovereign violence committed during the settler colonization of the Amazon during the post–World War II period was full of such law of the stronger and original expropriation (Marx [1867]

1996, 739; Schmitt 2003, 93–4) imposed through countless massacres, skirmishes, and ambushes, which, combined with the onslaught of disease, decimated many Indigenous populations (Garfield 2001, 143–4; Lewis and McCullin 1969).[3] This hunting power was tempered at times by the pastoral activities of FUNAI, the Brazilian government ministry in charge of Indigenous peoples, with its medical care and legal demarcation of some Indigenous peoples' territories as reservations. However, FUNAI's purpose was to manage the integration of Indigenous peoples into settler colonization processes, not to stop such processes. It was under FUNAI's oversight that the 1987 sampling of the Karitiana and Surui occurred.

SCIENTISTS CALL FOR THE URGENT SAMPLING OF INDIGENOUS PEOPLES

If we apply this analysis to the history of genetic research on Indigenous peoples there emerges a particular complementary logic that guides the interaction between the global assemblages of genetic research and those of settler colonialism's elimination of Indigenous peoples. Genetic researchers have frequently naturalized, or, in Marxist terms, reified, settler colonialism as an inevitable social process in their calls for action to identify and genetically sample those Indigenous peoples whom they deem are disappearing due to these reified forces. This reasoning is evident in the 1991 Human Genome Diversity Project (the Diversity Project) call to action by a team led by influential population geneticist Lucas Cavalli-Sforza, which begins, "The Human Genome Project can now grasp a vanishing opportunity to preserve the record of our genetic heritage" (Cavalli-Sforza et al. 1991, 490).[4] Here, *grasp* means to seize or capture. *Vanishing* derives from the Latin *evanescere*, which means to "disappear, pass away, die out," while *preserve* derives from the late Latin *praeservare*, to "guard beforehand."[5] The etymologies and use of these terms suggests that scientists must actively seek out and capture Indigenous peoples (populations) before they disappear to preserve them for Humanity:

> The populations that can *tell us* the most about *our* evolutionary past are those that have been isolated for some time, are likely to be linguistically and culturally distinct, and are often surrounded by geographic barriers. Such isolated human populations are being rapidly merged with their neighbors, however, *destroying irrevocably the information* needed to reconstruct *our* evolutionary history.

> Population growth, famine, war, and improvements in transportation and communication are encroaching on once stable populations. It would be tragically ironic if, during the same decade that biological tools for understanding our species were created, major opportunities for applying them were *squandered*. (Cavalli-Sforza et al. 1991, 490, my emphasis)

Cavalli-Sforza and his colleagues in their call to action on behalf of Humanity argue that this opportunity for the capture of disappearing Indigenous peoples cannot be *squandered*, which here is an antonym of preserved.[6] In this call, the coauthors argue Indigenous peoples that had once been isolated from the forces of modernity are now disappearing as distinct entities and so, as a result, is the unique information their genetics can tell us about human evolution.[7] The metaphor of DNA as information is based on the metonymy of genetic samples FOR population. Capture is the appropriate word because Indigenous peoples' genetic materials once immortalized at centers of accumulation like Stanford and Yale will serve Humanity forever, enacting a logic of the cynegetic hunt of accumulation (Chamayou 2012, 12). The Diversity Project quest sought to enact a nomos (law) of appropriation of Indigenous peoples genetics, their viral transformation into cell lines for distribution, and their subsequent use as resources in the production of genetic research for present and future generations (Schmitt 2003, 351).

This uncritical acceptance naturalizes the aforementioned colonization of Indigenous peoples and their territories. Groups still living in these regions must be incorporated before they are assimilated or exterminated: "We must act now to preserve our common heritage. Preserving this historic record will entail a systematic, international effort to select populations of special interest throughout the world, to obtain samples, to analyze DNA with current technologies, and to preserve samples for analysis in the future" (Cavalli-Sforza et al. 1991, 490). The authors set out a research agenda that involved extended assemblages of capture deployed across the world: identify, locate, sample, and preserve. So though these peoples may afterwards be displaced, assimilated, or exterminated, their genetics will nonetheless live on as cell lines in the service of science acting on behalf of Humanity. Semiotically and rhetorically, this call to action is what sociologist Jukka Torronen (2000) calls an interrupted narrative that seeks to motivate others to join in: "We urge national funding agencies to gaze favorably on proposals to collect and preserve DNA samples from human populations worldwide" (Cavalli-Sforza et al. 1991,

491). Elaborating on Foucault's concept of biopolitics, Emma Kowal and Joanna Radin (2015, 63) analyze this and similar situations in terms of cryogenics or cryopolitics: "If biopolitical assemblages make live and let die, cryopolitical ones reveal the dramatic consequences of mundane efforts to make live and not let die"; that is, Indigenous peoples samples are kept in a state of incomplete death ready to serve Humanity. In the case of the Diversity Project, involved scientists were not going to make Indigenous peoples live but rather they were going to use cynegetic hunts of accumulation to sample five hundred to seven hundred populations and create cell lines from them that could be stored in cryogenic freezers at biobanks in an effort to ensure these Indigenous peoples' genetics will continue to serve scientists who claim to serve Humanity (L.T. Smith 1999, 101–2).

The Diversity Project call to action is not unique in its view of Indigenous peoples. Rather, as Reardon (2005, 72) points out, scientists had long used an "analytic concept of race to organize this research" that represented their object of research, Indigenous peoples, as disappearing and premodern.[8] This concept was relatively consistent from early meetings in the 1950s on Indigenous population sampling for what became known as the International Biological Program (1964–1974) through to the Diversity Project of the 1990s (Radin 2017, 161; Reardon 2005, 72).

THE 1976 *ALPHA HELIX* EXPEDITION

While American researchers obtained samples from many thousands of Indigenous peoples in Brazil during this period, only a small subset was eventually integrated into forensic genetics research in the US and globally: the Ticuna, the Karitiana, and the Surui.[9] The first of these to be sampled was the Ticuna. In July and August 1976, after extensive diplomatic negotiations with agencies of the governments of Brazil and Colombia, a research team led by the influential American geneticist James Neel participated in an expedition into the Amazon region by the US government scientific research vessel the *Alpha Helix* (Radin 2017, 146).[10] Neel's original goal of visiting and obtaining samples from a number of Indigenous peoples in the more isolated border regions of Brazil was blocked by FUNAI because some Indigenous peoples had been angered by road-building on their territories (Radin 2017, 147). Neel criticized this reasoning and, in his correspondence with one of Brazil's leading geneticists, Francisco M. Salzano, he argued that the expedition's goals had been undermined by his inability to access these isolated peoples (Radin 2017, 146).

These disputes, however, were nowhere to be seen in the upbeat 3 March 1976 news release by the University of California at San Diego that describes Neel's goals to advance the knowledge of the evolution of "primitive man" during the phase of the expedition dedicated to his research: "Phase II – July 13–August 27 (Leticia, Colombia-Manaus, Brazil); James V. Neel, M. D., University of Michigan, Ann Arbor, chief scientist. This study continues ongoing research seeking to define the genetic structure of primitive man and to develop a clear picture of the interaction of man and his environment. An investigation of health, physical characteristics, and cultural factors affecting selection and genetic drift as processes of evolution is planned during this phase" (University of California San Diego 1976, 2).

The press release then continues by describing how the expedition assemblage would utilize missionary transport infrastructure to gain access to a number of Indigenous peoples: "Utilizing *Alpha Helix* at Leticia as a base, small planes maintained by the Missionary Aviation Fellowship will aid in collecting blood and urine specimens from as many as eight Brazilian Indian groups. Such tribes include the Ticuna, Tukano, Mairona, Maruvo, and Kashiwana. The biological specimens will be flown to *Alpha Helix* for processing in the ship's laboratories" (3).

Funded by the US Department of Energy, the National Science Foundation (NSF), and the US Public Health Service, Neel and his team eventually collected some 1,760 samples from the Ticuna in eight villages on the Solimões River (Neel et al. 1980, 38, 52). During the expedition, the researchers took a total of 2,736 samples from various Indigenous peoples, including those from the Ticuna (Radin 2017, 149; Neel et al. 1980, 38). Neel et al.'s description in a 1980 paper highlights the key role of air transport and cryogenic refrigeration systems in accumulation of samples (Radin 2017, 148–9): "Venous blood samples were collected in 13 ml Becton Dickinson vacutainers containing 2.25 ml of ACD solution [an anticoagulant]. They were chilled immediately upon collection and air-lifted back to the Ann Arbor lab [in Michigan, US] as soon as possible, usually arriving within 3–5 days after collection. In Ann Arbor cells and plasma were separated and stored at either -72°C or in liquid nitrogen until the determinations were performed. In each village, samples were obtained from as many persons as possible, regardless of relationship" (Neel et al. 1980, 39).[11]

Storage in nitrogen required the addition of a chemical called glycerol that acts as a cryoprotectant and prevents the destruction of the cell tissue during freezing (Radin 2017, 35–45). Neel et al. describe the use of such refrigeration techniques in their section on typing of blood groups: "Single typings were performed on red cells that had been preserved in glycerol-sorbitol freezing solution in liquid N [nitrogen], for 18–33 months" (Neel et al. 1980, 39). Radin (2017) shows how Neel was guided by a logic of accumulation. When Neel heard talk that by finishing early he had not maximized his use of the *Alpha Helix* during his portion of the expedition, he responded in a letter to a program officer at the US NSF:

> The backbone of our work involves the collection of blood samples from these Indian populations, which samples undergo a wide variety of very sophisticated tests once they reach our base laboratory here in Ann Arbor. We collect these samples in tubes called vacutainers. On the basis of all our previous experience, I estimated that our team could count itself quite fortunate to collect as many as 2,000 such samples, which would constitute some kind of a record. However, because there is nothing more frustrating than to find oneself knee-deep in Indians and out of vacutainers, I put in an extra 800 tubes as a contingency measure. (As quoted in Radin 2017, 149–50)

Neel argued he was prepared and so was successful. Using a metaphor of quantity about having enough containers in case he was "knee-deep in Indians," he rejected accusations that he had missed an opportunity to maximize his collection efforts (Radin 2017, 149–50).

In the 1980 article by Neel et al., the concept of the Ticuna as mysterious Other was clear in its title, "Genetic studies on the Ticuna, an enigmatic tribe of Central Amazonas." In this article, Neel et al. argue that the Ticuna had continued to be isolated first because prior to colonization the areas along the major rivers were controlled by Arawak-speaking peoples and then later due to the local particularities of the rubber trade:

> With the passage of time these latter tribes were decimated-assimilated through contacts with neo-Brazilians, but the free movement of the Ticuna to the river banks was again limited, this time by the local dealers in rubber, who, establishing trading stations at the mouths of the many small tributaries to these rivers, controlled movement from the interior. Unwittingly, first these Arawaks, and then the rubber traders, provided the Ticuna a remarkable degree of protection against intermixture for a tribe so close to great waterways, so that even today, as we shall show, the genetic heritage is still approximately 98% Amerindian. (37)

This idea that the "genetic heritage" was protected "against intermixture" involves the reification of the processes of colonialism. Perhaps Neel et al. were only referring to the Ticuna located away from the rivers, because many Ticuna were effectively enslaved or pressured through debt peonage into rubber production (Nimuendaju 1952, 9; Soares 2018).

Cell lines were created from some of the Ticuna samples by Douglas Wallace of Emory University, funded by a 1987 US NSF grant (NSF 8718775) for US$193,700 (National Science Foundation 1990). Some of these cell lines eventually ended up in Kidd Lab and are on their list of populations routinely studied there, "Ticuna (65)†" (Kidd Lab 2017). In turn, some of these Ticuna cell lines were then transferred by Kidd Lab to the Diversity Project, where they continue to serve, including in the development of forensic genetic technologies. In 1976, like other Indigenous peoples in the Amazon region, the Ticuna were under increasing pressure from intensified settler state colonization. As well, at that time, Brazil was still governed by a repressive US-backed military dictatorship, so scientists' continued claims that the samples are still covered by a blanket informed consent from the Ticuna no longer accords with contemporary ethical and legal standards of ongoing informed consent and mutual reciprocity. Under such questionable circumstances, the Ticuna Indigenous people were genetically captured and integrated into the circuits of genetic research-based production and accumulation, where they continue to serve as resources.[12]

THE KARITIANA AND THE SURUI

The two other Indigenous peoples on the Kidd Lab list from what is now Brazil, the Karitiana and Surui of Rondonia state, have played a disproportionately important role in the development of forensic genetic technologies in part due to their respective historical experiences of near genocide. In 1839, Charles Goodyear developed the vulcanization process, which involved heating rubber processed with sulphur. This process transformed rubber into a durable industrial material, one that became crucial to the Industrial Revolution. The Karitiana and the Surui first came into the extended market driven violence of colonialism during the Brazilian rubber boom from the mid-1800s to early 1900s.[13] The first direct reference to the Karitiana is in a 1910 report that speaks of regular armed conflict occurring between the Karitiana and rubber tappers (Vander Velden 2010, 48). Historical references to the Karitiana territories involve a variety of locations in the region, possibly indicating they were moving around to avoid the encroaching rubber tappers and settlers (Vander Velden, 2010, 49). Based on his analysis of Karitiana oral histories, Brazilian anthropologist Felipe Ferreira Vander Velden (2010) hypothesizes that due to the incursions and violence of Brazilian and Bolivian rubber tappers in the early 1900s, the Karitiana broke into two bands, the Karitiana and the Juari.[14]

Years later, under the leadership of Antônio Morais (Moraes in some sources), the Karitiana, whose population numbers had dropped to low levels (forty-five in one report), sought out the Juari and the two bands rejoined (Vander Veldan 2010, 57).[15] Sustained contact with the Karitiana began in the late 1950s through the Indian Protection Service (SPI), which was the Brazilian government ministry in charge of Indigenous peoples, and Christian missionaries. As well, in 1959, casserite (tin ore) was discovered on the Karitiana's traditional territories causing further displacement. Storto and Vander Velden (2018) write that with the dramatic population decline:

> This situation led the group to adopt extreme measures to avoid their complete extinction. Firstly, an elderly leader, Antônio Morais, married various Karitiana women (7 or 10, depending on the different versions), including some women in principle interdicted by matrimonial rules. This event ended up generating a densely related population from the genealogical and genetic viewpoint: a study by the Federal University of Pará, in 1991, showed that the coefficient of average consanguinity – which measures the degree of genetic kinship of a population – of the Karitiana was 0.142 (between first-degree cousins, this figure is 0.125). Also according to the research, all the Karitiana below 16 years old descend from the chief Morais, very often by various genealogical lines.

The Karitiana's numbers stood at around sixty-five people according to Rachel Landin (1989, 1) when she did fieldwork on naming and kinship relations among them in the early 1970s. In 1987, when they were sampled by Francis Black there were 130 and today there are around 300 (Kidd et al. 1991; Storto and Vander Velden 2018).

Similarly, the Surui had also been engaged in conflicts with rubber tappers during the 1900s. However, in the six years following their first official contact with FUNAI officials in 1969, their numbers are estimated to have dropped to around four hundred due to influenza and other introduced diseases (Chiappino 1975, 9). With the completion of the BR-364 Highway in 1968 from the state of São Paulo to the city of Porto Vehlo large numbers of settlers began to colonize the Surui territories.

During the 1970s, the Surui engaged in a series of armed conflicts with these invaders leading to a number of deaths (Junqueira and Mindlin 1987, 35). Finally, with the help of FUNAI, in 1981 they expelled the last of the settlers and took over the coffee plantation that had been left behind (Junqueira and Mindlin 1987, 35). Furthermore, they eventually were able to acquire a reservation of just under 250,000 ha that was officially recognized in 1983. Their numbers have recovered to over 1300 (Kanindé Associação de Defesa Etnoambiental and Mindlin 2018).

The sampling of the Ticuna and later the Karitiana and Surui were conducted in well-established assemblages involving US and Brazilian researchers (often US-trained) in their universities and the Brazilian settler colonial governance apparatus in the Amazon including the Brazilian Air Force, missionaries, FUNAI, and other institutions. The sampling of the Ticuna by James Neel and his colleagues during the 1976 *Alpha Helix* expedition is well documented. In contrast, the sampling of the Karitiana and Surui is not discussed in US or Brazilian genetic researchers' accounts. I have not been able to find any detailed published accounts about it by Francis Black or any of his colleagues. Francis Black began work on Amazonian Indigenous peoples during the late 1960s. A virologist and epidemiologist, he researched the response of isolated populations to introduced diseases. His sampling of the Karitiana and Surui was part of a larger project that was studying infectious diseases among Indigenous peoples involving some 2500 individuals from over twenty different peoples (Atwood 1988; Black 1991, 767). In his 2004 dissertation based on fieldwork with the Karitiana, Vander Velden writes that Karitiana oral histories are vague about the 1986–87 sampling by Francis Black. As they tell it, "many years ago," two "skinny, deep-bellied, frog-bellied Americans" (americanos magros, de barriga funda, barriga de sapo) arrived in two planes on the village's airstrip and then took blood at the village infirmary (as quoted and recounted in Vander Velden 2004, 161–2). Claims of ethical consent are contradicted by the Karitiana, whose oral histories say they were not able to speak Portuguese well at the time and so did not understand what was going on and thought the blood sampling was medical treatment (Vander Velden 2005, 18). In a 2007 *New York Times* article on the Karitiana case, in response to claims by a US National Institutes of Health (NIH) official that appropriate protocols were followed in the collection of the Karitiana samples, Rohter (2007) writes, "The Indians themselves, however, respond that at the time the first blood samples were drawn, they had little or no understanding of the outside world, let alone the workings of Western medicine and modern capitalist economics." Brazilian FUNAI officials have denied that Black had their ministry's authorization. Hence, researchers' claims that they obtained informed consent before taking the samples from the Karitiana and Surui is contested by such accounts. Furthermore, Karitiana leaders have since publicly withdrawn their collective consent to be researched and demanded compensation, something that has been ignored by Kenneth Kidd and his colleagues and the Diversity Project collection (Storto and Vander Velden 2018). In the Karitiana worldview,

> The unauthorized collection of their blood was, therefore, an affront to the Karitiana symbolic conceptions concerning the body and its proper functioning. However, more than this, it amounted to serious moral offence: the Karitiana speak of the *tasoty*, literally "great men," not only in terms of physical size but above all in terms of wisdom, thought and work: a "great man" is one who does not "think along one path only," but "spreads out in all directions," a man who possesses wisdom and responsibility. In sum, the model of an appropriate and respected social persona: the man who "speaks well to people," welcomes them readily into his home, does not "tell lies or think and speak badly" of others, and respects the rules of reciprocity, so important to the group.

THE KARITIANA AND SURUI AS GENETIC RESOURCES

The above summary of the larger historical context within which these Indigenous peoples were sampled does not provide a detailed understanding of how they function as human objects of exchange and value within genetic research. Typical commodities from colonized Indigenous territories in the Amazon region such as timber, beef, soybeans, rubber, and metals are valued for their own inherent physical properties, so their place of origin and the conditions under which they were grown or extracted are largely irrelevant to subsequent usages. Ethical sourcing criteria can be something of a counterbalance, yet such criteria do not affect the inherent physical properties of these resources, like the protein content of soybeans for animal feed or the rust resistant properties of nickel in steel making, that make the commodity valuable in the subsequent production processes. In contrast, genetic materials from Indigenous peoples retain their value in subsequent usage as these circulate within and between genetic research assemblages (Munsterhjelm 2014, 74–6).

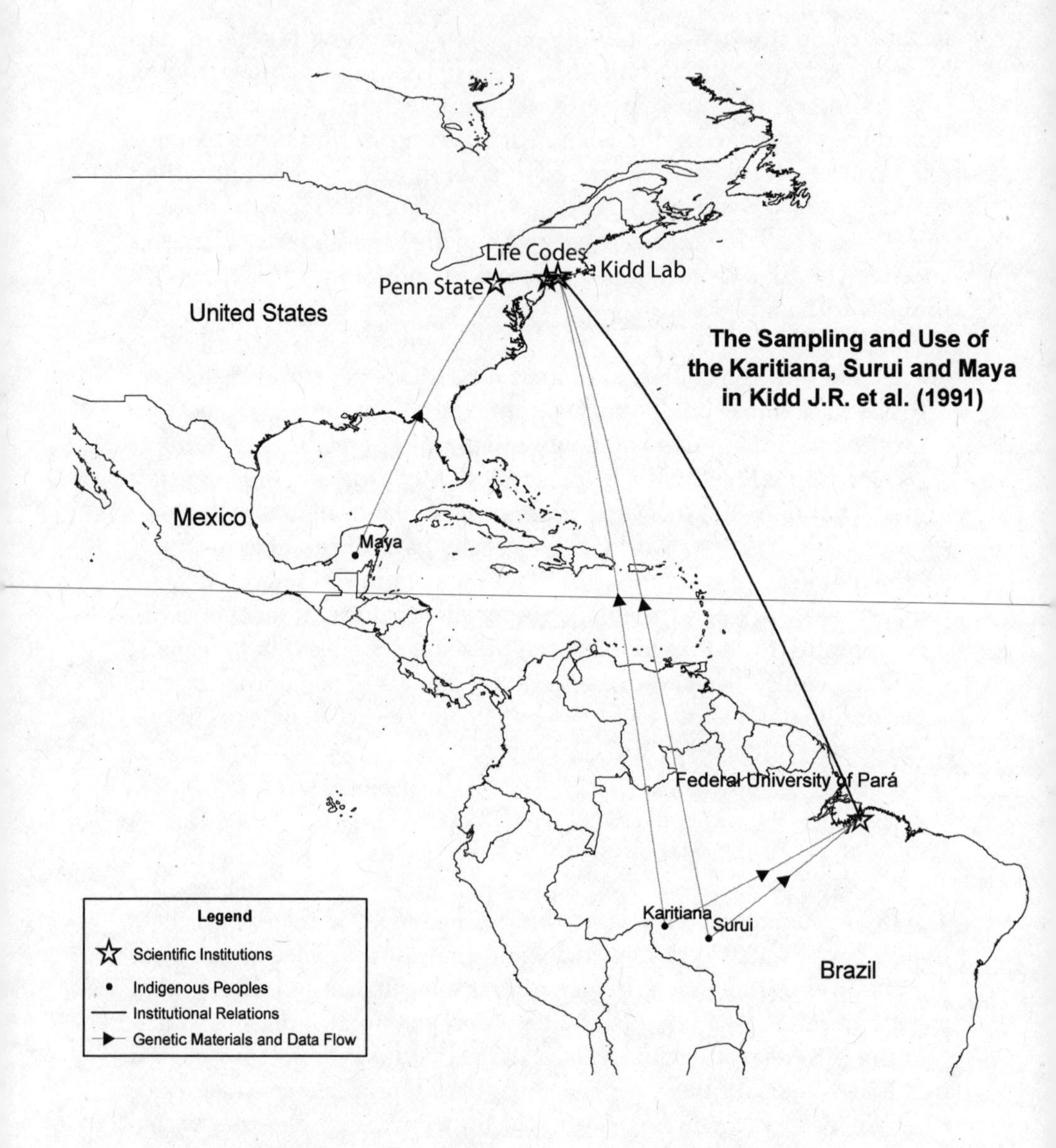

Figure 1.1 | The geographic spatial relations of the article's research assemblage. In addition to researchers at Kidd Lab, Pennsylvania State University, and Lifecodes, J.F. Guerreiro of the Federal University of Para provided assistance in obtaining the Karitiana and Surui samples.

Utilizing the narrative schema model, the following section closely analyzes this production process in a controversial 1991 scientific research article that appeared in the journal *Human Biology*, published by Wayne State University in Detroit, US. Written by Kenneth Kidd, Judith Kidd, and Francis Black of Yale University, Kenneth Weiss of Pennsylvania State University, and Ivan Balazs of the Lifecodes Corporation (an early forensic genetic testing company discussed further in chapter 2), this article is entitled "Studies of Three Amerindian Populations Using Nuclear DNA Polymorphisms." As chapter 3 will detail, the Karitiana's and to a lesser extent the Surui's respective recoveries from near genocide became important in the conflicts over the introduction of forensic genetic technologies into the US and Canadian judiciaries in 1989. The Karitiana, Surui, and Campeche Maya data used in this article was appropriated by the defence lawyers during the *US v. Yee* murder case in discovery hearings over the admissibility of genetic evidence.[16] The influence of these judicial system conflicts is evident in the participation of Balazs in this 1991 article. The credibility of the private American genetic testing company, Lifecodes, where he worked was severely damaged by defence lawyers and the judge's ruling in the 1989 *US v. Castro* case (discussed in chapter 2). In addition, this 1991 article itself has since been included in many activist and academic critiques and was cited during late 1990s in Brazilian parliamentary inquiries into biocolonialism and biopiracy over the sale of Karitiana and Surui cell lines and DNA extracts by the NIH funded Coriell Cell Repositories in New Jersey, US (Borofsky 2005, 228; Santos 2002, 98n3; Vander Velden 2005). The article's legal and political significance derives from its status as a unit of capture, production, and accumulation, a narratively constructed account written by American scientists that helped incorporate these three Indigenous peoples into the global assemblages of scientific knowledge, including the Diversity Project (see figure 1.1).

INFRASTRUCTURE OF CAPTURE AND ACCUMULATION

The research for this 1991 paper was funded in part by US NSF grants. In 1987, Kenneth Kidd obtained a US$80,000 grant from the NSF with the title "DNA Markers and Genetic Variation in the Human Species," which was followed in 1988 with a US$120,000 grant under the same title (National Science Foundation 1989). The 1987 grant application does not have an abstract, but the 1988 application's abstract explains the goal of developing a biobank of cell lines aimed at genetic variation through cooperation with anthropologists: "Dr. Kidd will continue his work to collect, preserve, and analyze DNA from a large series of anthropologically significant populations. Through collaboration with field anthropologists who will provide blood samples, Dr. Kidd will preserve material as Epstein-Barr virus transformed lymphoblastoid cell lines. The DNA of these samples will then be studied for several DNA polymorphisms (RFLPS)" (National Science Foundation 1989). Kidd in cooperation with Cavalli-Sforza had already begun the accumulation of Indigenous peoples' samples through cooperation with anthropologists and other researchers conducting fieldwork to capture their genetic materials (Leslie Roberts 1991a). Biological anthropologist Jonathan Friedlaender of Temple University provides an account of his 1985 sampling of the Nasioi of Bougainville Island (currently part of Papua New Guinea, but in 2019 they voted to become independent), who are also included on the Kidd Lab list of population resources and as part of the Diversity Project:[17]

> Ken [Kidd] and Judy [Kidd] knew that I was going back to the field in the mid-1980s, when cell transformation had matured as a technique, and Ken said, "Jonathan, it would be great if you got some blood samples that we could transform into cell lines. We think this is going to be a great resource." I said, "What do I have to do?" He said, "You've got to get the samples back here unfrozen, kept at 4°C, within 50 hours of their drawing." I said, it takes about that long to fly from Bougainville via Australia to the USA with connections – I'll see what I can do. How many?" And he said, "As many as you can get, but probably 20 to 25 per group would be enough." (Friedlaender and Radin 2009, 148–9)

The combination of rapid air transport, cold storage technologies, and immortalization of samples into cell lines using Epstein-Barr virus would allow Kidd Lab to become a centre of genetic research accumulation, distribution, and production.[18]

Another important element for Kidd Lab becoming a centre of accumulation was acquiring US government funding. Kidd's 1988 NSF funding abstract identifies three major goals:

- Creating a "data bank from as large a number of populations as possible" that could be analyzed using new genetic restriction fragment length polymorphism (RFLP) technologies to look for "affinities between different populations."
- Investigating "variability within and between groups for the purpose of inferring population demographic history."

- The "investigation of the possibility of observing the operation of natural selection in human populations."

The abstract concludes with a type of epideictic rhetoric about what Kidd's accumulation of genetic samples from Indigenous peoples in a data bank that can then be analyzed utilizing new genetic testing technologies will accomplish: "Prehistorians and historians are interested in tracing the movements of human populations. Unfortunately, written records appear only late in the human record, and archaeological evidence is often ambiguous. Many provocative hypotheses have been proposed about the relationships of different groups but with traditional data these are difficult to prove or disprove. Through analysis of DNA RFLPS, Dr. Kidd's research should help to provide answers" (National Science Foundation 1989). This final paragraph utilizes epideictic rhetoric in which Kidd's research will answer questions of human origins through analyzing Indigenous peoples' genetics, something with significant theological and philosophical implications.

TRANSLATING THE KARITIANA AND THE SURUI

The following analysis uses a synthesis of rhetorical, organization, and metonymic theory to show how Kidd et al. (1991) enrolled the Karitiana and Surui into genetic research. This 1991 scientific research article follows a heroes' quest cycle of departure, adventure, and return with knowledge that transforms the world (Campbell 2004, 227–28; Munsterhjelm 2014, 228, note 5).

Manipulation Phase

The quest for human diversity is an epideictic rhetorical call to action (Munsterhjelm 2014, 43–4).[19] The authors begin, "The past decade has seen the development of new technology for studying human variation," which sets the article in the present and defines scientific discourses of human variation and diversity as the sender of this quest for knowledge (Kidd et al. 1991, 775). The call to action creates a having-to-do modality and the authors now have a fiduciary duty to complete the quest.

The remainder of the manipulation phase involves a brief discussion of current challenges and limitations of genetic research on mitochondrial DNA (mtDNA) compared to nuclear DNA polymorphisms. The significant advantages of these new technologies are emphasized through the epideictic technique of invoking quantity. In particular, nuclear DNA polymorphisms involve a "1000-fold more genes," whereas mtDNA is limited to "a single DNA sequence and a matrilineal gene tree" that omits "gene flow through males" (Kidd et al. 1991, 775).[20]

In addition to noting the limitations of mtDNA, Kidd et al. also use epideictic rhetoric of greater quality as they argue for the improved resolution of nuclear DNA polymorphisms in studying genetic heterozygosity (variation) over classical genetic markers including "the blood groups, serum proteins, red cell enzymes, etc." (776). The technological capacity to measure heterozygosity – when a gene has two different alleles (variants) at a specific locus (location), for example, one dominant and one recessive – represents a way of learning about the more universal concept of human diversity. In terms of research methodology, measuring heterozygosity becomes a way of operationalizing the concept of human diversity.

Commitment Phase

From a general discussion of the potential advantages of analyzing nuclear DNA polymorphisms over mtDNA and classical markers like blood type in the manipulation phase, the narrative now shifts to how the authors will apply the analysis and put it into practice in their project. They begin this commitment phase by recounting how they have already begun studies on genetic variation: "Six years ago K.K. Kidd and L.L. Cavalli-Sforza began a large collaborative study to survey genetic variation in population samples from around the world," referring to the Diversity Project, and go on to describe how they have completed studies on "5 populations from 4 continents using 100 different DNA markers." They continue, "Here we present a progress report on extending this study into the Americas. We now have initial data on three separate native American populations: two Amazon basin tribes from Rondônia, Brazil, and a Mayan population from the Yucatan, Campeche, Mexico" (Kidd et al. 1991, 777). In the above, the authors, having accepted the general quest for knowledge from Humanity about human diversity, "extending this study," use deliberative rhetoric to briefly explain how they will carry out this quest through researching these three Indigenous peoples, thereby providing knowledge about human diversity in the Americas.

Competence Phase

The competence phase begins with a subnarrative in which the scientists define their quest to explain their findings: "We have found high levels of heterozygosity for most of the 37 nuclear DNA polymorphisms studied at 31 distinct loci [locations on chromosomes], thereby demonstrating the utility of these markers for studies of native American populations. The large number of independent nuclear DNA loci that can now be studied also promises new insights into the Asian origins of the Amerindians" (Kidd et al. 1991, 777–8). In effect, they make a claim of already possessing the object of value, which is the ability to invoke the presence of, and thereby represent, these Indigenous peoples of the Americas, including the living and their ancient ancestors. The authors state in advance how their findings of high heterozygosity (as a measure of genetic diversity) render it a suitable technology to study Indigenous peoples in the Americas and that these genetic markers (loci) can be used to study these peoples' prehistoric origins in Asia. Kidd et al. have made the assertion of possessing the means to obtain knowledge about human diversity; however, they have not yet fulfilled all of the able-to-do and knowing-how-to-do modalities of the competence phase.

ENROLING INDIGENOUS PEOPLE AS DIVERSE POPULATIONS

The authors begin their demonstration of the knowing-how-to-do and able-to-do modalities in materials and methods section, which begins with a description of how two of the authors sampled the three Indigenous peoples in Mexico and Brazil. Citing the results of genetic tests for HLA haplotypes (a type of immune system protein), the scientists emphasize these are culturally and genetically distinct populations, with the Mayans described as having less than 10 per cent European admixture and the Karitiana and Surui as having no European admixture.[21] In effect, diversity means difference from Europe, such that Europe defines the norm:

> Populations Studied. The Mayan sample was collected (by K. Weiss) in the Mexican state of Campeche in the Yucatan peninsula. Blood was drawn from individuals living in an ethnically and geographically isolated village of approximately 400 people. These people speak Yucatec rather than Spanish, and on the basis of blood and serum markers, European admixture is estimated to be no more than 10% (K. Weiss, unpublished data, 1989).
>
> The Amazon basin samples were collected (by F. Black) in the Rondonia province of western Brazil. The Karitiana and Rondonia Surui are Tupi speakers. Each group was sampled from a single village; the two villages are separated by approximately 420 km. HLA

> haplotypes indicate no non-Amerindian admixture, and both groups have a low probability of large-scale recent admixture with other Amazonians; HLA data show genetic differences between the two populations (F. Black, unpublished data, 1989).[22] (Kidd et al. 1991, 778)

The authors emphasize the geographic and cultural difference and isolation of these Indigenous peoples in their territories far away from Europeans. Importantly, this differentiation is both spatial and temporal and so fits with the overall dominant universal object of human diversity, since the authors seek to emphasize these "isolated" peoples are representative of ancient human diversity in the Americas because their genetics indicate low to no contact with the outside European-centred modern world (Reardon and TallBear 2012). In particular, the temporality of extending back to the pre-Columbian period or pre-colonial period is implicit, with the authors' emphasizing the villages still speak their own Indigenous languages and have low or no genetic "admixture" with Europeans. This repeats genetic researchers long established view of Indigenous peoples as a premodern Other (Munsterhjelm 2014, 5–7; Reardon and TallBear 2012, S237–8; Santos 2002, 82). These time-spaces are constructed as valuable sources of genetic diversity due to being temporally, geographically, and culturally isolated from the recent historical circuits of capital.

Significantly, there is no mention of informed consent in the article. This omission is typical of papers published before the conflicts over the Diversity Project and other research on Indigenous peoples during the mid-1990s.

IMMORTALIZATION

The next section discusses how the researchers used a virus to transform these Indigenous peoples' blood samples into stable research objects that can be grown and distributed indefinitely as productive resources within the assemblages of genetic research in a state of capture. Spatiotemporally, this next subnarrative involves a move to Yale University and to the time-space setting of the scientific lab. There, Judith Kidd infected the Karitiana, Surui, and Maya blood samples with Epstein-Barr virus, giving the white blood cells called B-lymphocytes in the samples the important research production ability "to proliferate indefinitely in a test tube" (Miller 1982, 305):

> Cell Transformation. For all three populations blood was drawn in the field and transported to the lab at Yale where the B-lymphocytes

> were isolated and transformed with Epstein-Barr virus following essentially the protocol of Anderson and Gusella (1984). We established cell lines from each of these populations on 48–54 individuals. For each population five cell lines from unrelated individuals have been deposited in the NIGMS Human Genetic Mutant Cell Repository at the Coriell Institute for Medical Research (Camden, New Jersey) and are publicly available. (Kidd et al. 1991, 778)

The virally transformed cells that can reproduce indefinitely are still considered equivalent to the original samples provided by the Karitiana and Surui donors. Judith Kidd used the Epstein-Barr virus as a transformant to immortalize the samples, creating virally transformed cell lines that could be grown under lab conditions and stored cryogenically; they were no longer only available from Indigenous donors (J.R. Kidd, Pakstis, and K.K. Kidd 1993, 26). Through the process of transformation, the immortalized cell lines became renewable productive resources within research assemblages that could be grown in unlimited quantities and stored in lab collections and biobanks, shared with other researchers, and sold by Coriell Cell Repositories for US$85 a cell line until 2015.

If we apply social philosopher Martha Nussbaum's concept of objectification to the immortalization of the Karitiana samples, immortalization involves the violability of the samples so they are "lacking in boundary-integrity," thereby rendering them *fungible*, interchangeable with other samples (Munsterhjelm 2015, 296; Nussbaum 1995, 257). That is, prior to immortalization the only source of samples were individuals from these three Indigenous peoples. However, after immortalization, based on a metonymy of the virally transformed cell lines FOR the Indigenous peoples' blood samples, endless quantities of cell lines representing these Indigenous peoples could be grown, stored, and distributed. While the Epstein-Barr virus genetically altered the cell lines' apoptosis (cell death) capacities, these transformations are not considered significant enough to change the fundamental identification of the cell lines with the Indigenous peoples.[23] These virally transformed cell lines became what Latour (1987) terms mobile, stable, and combinable and so ready for biobanking, growing in labs, and international distribution. Immortalization rendered them into objects of exchange within contemporary biotechnology accumulation processes, effectively commodities, as typified by Coriell Cell Repositories' sale of cell lines and DNA extracted from cell lines from the Karitiana and Surui from 1991 until 2015. The violation of the integrity of the Indigenous peoples' blood cells effectively captures and transforms them into cell lines, allowing their accumulation in centres like Kidd Labs, Stanford University, and the Diversity Project in Paris. The immortalization process as a violation of the cellular integrity of the blood samples combined with cryogenic storage meant that these cell lines have the status of not-let-die, a state of biological servitude (Kowal and Radin 2015, 63).

No longer constrained by apoptosis, the remainder of the materials and methods section discusses the growing and processing of immortalized cell lines into data. The section entitled "DNA Purification and Polymorphism Typing" outlines how the scientists first grew cell lines in some culture medium at 37° Celsius (usually bovine fetal serum made from cow fetuses). Once a large enough volume of cells was grown, DNA was extracted from the cells and the DNA was processed through several steps to finally produce X-ray film–like images called audioradiographs with their distinctive bands that researchers then visually interpreted to identify the polymorphisms and produce data.

Next, there is an extended discussion of how to process the data. This processing of cell lines and data takes place in the time-space of the laboratory, a sort of generic transnational scientific time-space since the narrative does not identify the particular labs these various steps were done in.

The results and discussion section involves a time-space shift from an account of the specific enactment of transnational science in the project's generic labs to transnational science as a field of knowledge. The statistics from processes in the preceding section are analyzed and discussed. There are several statistical tables of expected and observed frequencies of heterozygosity at the various loci. There is also a discussion of the technical limits of the genetic processing procedures. Kidd et al. frame their contributions as either complementing or contradicting the canon of existing knowledge. That is, they highlight how their findings fit within the field while also noting the limitations of the genetic testing techniques and results. While noting the limitations of Europeans defining the norm in genetics research, they nonetheless describe the Maya as being 7.0 per cent less genetically heterogeneous than Europeans, with the Surui 18.7 per cent less and the Karitiana 27.1 per cent less (Kidd et al. 1991, 788). These findings based on thirty non-VNTR loci (locations) on the chromosomes of Karitiana, Surui, and Maya Indigenous donors now stand FOR the heterozygosity of these peoples.[24]

With these findings on heterozygosity, the authors next move to a discussion of the ancestral history of the Karitiana, Surui, and Maya. Central is a discussion of whether or not these Indigenous peoples' ancestors encountered some type of "bottleneck," which occurs when a population is reduced to a small number by one or more events (like a natural disaster) that then reduces the overall amount of genetic variation (which heterozygosity measures) in the population thereafter. The authors also discuss founder effects – low levels of genetic diversity that result from a population descending from a small initial founding population (Kidd et al. 1991, 788). Instead, they refer to the potential that the loss in heterozygosity resulted from the interaction of the size of the initial population with the number of generations. This return of the narrative to the time-space of transnational scientific knowledge is also pivotal to the performance phase that follows since it marks the return of the receiver-subjects from their quest for knowledge about human diversity.

Performance Phase

The conclusion then reiterates the major findings and the performance of the quest: "This study demonstrates the utility of DNA polymorphisms in general for genetic studies of the Amerindian populations" (Kidd et al. 1991, 791). In particular, the authors hypothesize that the data does not indicate a major bottleneck in the peopling of the Americas, but rather that the lower heterogeneity compared to Europeans could be explained in terms of smaller initial founding populations inhabiting the region over a longer period of time (791). Through the metonymy of genetics, the scientists now speak not only for these Indigenous peoples and their ancestors, but also for those ancient peoples that crossed from Asia to North America thousands of years ago.

Sanction Phase

Acting on behalf of science, the journal's editorial board and peer reviewers positively sanctioned Kidd et al. by publishing the article (see figure 1.2). They briefly describe the process as: "Received 20 September 1990; revision received 12 April 1991" (Kidd et al. 1991, 793).

IMPORTANCE OF METONYMY IN LINKING THE DIFFERENT TIME-SPACES

Scientific research articles are narratively organized by linking together the different actions of involved actors in their respective time-spaces into a coherent whole. The scientists in their labs at Yale University, Pennsylvania State University, and the Lifecodes forensic testing

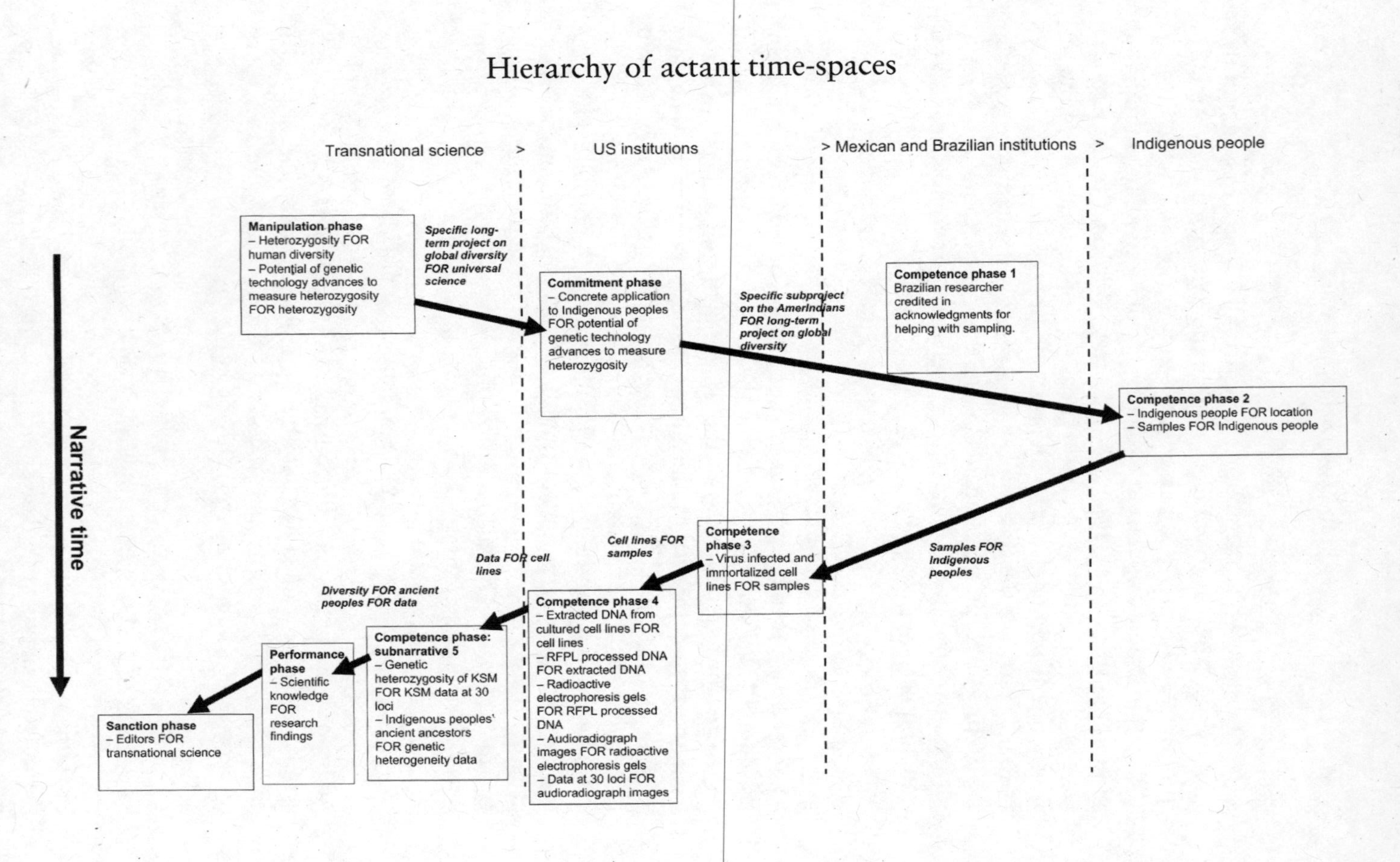

Figure 1.2. | Narrative diagram of metonymic transfers in the sampling and use of the Karitiana, Surui, and Maya in J.R. Kidd et al. (1991).

company in New York City interacted through the sharing of cell lines and data as objects of exchange. The organizing properties of these objects shared between these different scientists in their respective labs allowed the research assemblage as an "organization to transcend the bounds of local interaction, which is inherently ephemeral, transitory" (Cooren and Taylor 1997, 247). This ability to synecdochically represent Indigenous peoples as populations through the metonymy of genetics spanned the sampling of Indigenous peoples, the transformation of samples into cell lines, and research findings and data to refer to the involved Indigenous peoples, dead, living, and unborn. This equivalency between the cell lines and data and involved Indigenous peoples is made possible through a chain of metonymies that link together the different stages of the article (Brdar 2015, 86–9; Denroche 2015, 120–4). The chain of synecdochic metonymic linkages within the article is long and involves multiple incremental and seemingly contiguous shifts, which can be summarized as:[25]

1 Karitiana population FOR dead, living, and unborn Karitiana people including ancient ancestors
2 Karitiana donors FOR Karitiana population
3 Karitiana samples FOR Karitiana donors
4 Karitiana cell lines FOR Karitiana samples
5 Karitiana extracted DNA FOR Karitiana cell lines
6 Karitiana audioradiographs FOR Karitiana extracted DNA
7 Karitiana data FOR Karitiana audioradiographs
8 Karitiana statistics (allele data, haplotypes, etc.) FOR Karitiana data
9 Karitiana findings FOR Karitiana statistics

Through this synecdochic metonymic chain, findings become equivalent to ancient ancestors, living descendants, and unborn generations. Genetic research articles involving Indigenous peoples rely upon this sort of metonymic chain that links the different events of the article, from the research question through to conclusions, into a coherent whole.

CONCLUSION: ASSEMBLAGES OF CAPTURE AND ENCLOSURE

There are two complementary processes at work in the sampling of these Indigenous peoples in 1976 and 1987 that transformed them into resources used in the scientific commons. Within the settler colonial state of Brazil, the Indigenous peoples were undergoing a period of original expropriation in which their territories and sometimes labour were colonized and incorporated into the circuits of the settler state national political economy.

Layered onto this settler colonialism was a series of international scientific assemblages of Brazilian and US researchers. Under the pretense that these Indigenous peoples were disappearing, the involved research assemblages engaged in the capture and extraction of blood samples and data from these Indigenous peoples to study how the experience of undergoing colonization processes affected their immunological system responses to foreign diseases such as measles. The sampling was also for human evolution research–related sampling in which these peoples were viewed as living proxies for "primitive man." I suggest that these assemblages engaged in hunting and capture. Capture can be defined as taking into possession or control by force. These Indigenous peoples were identified in advance, and involved scientists used transport infrastructure and obtained a sufficient number of samples to represent the population. The scientists did not directly use physical force against these Indigenous peoples, but instead they used the existing hierarchical relations of settler colonial governance to do the sampling. There were substantial asymmetries in the relations between the researchers who organized the sampling process and the Indigenous peoples as the objects of research.

Social philosopher Martha Nussbaum (1995) defines seven overlapping and reinforcing processes in objectification that can all be applied in genetic research involving Indigenous peoples:[26]

- Instrumentality: Scientists use Indigenous peoples' genetics to conduct research and achieve various professional, organizational, and scientific goals.
- Denial of Autonomy: Indigenous peoples are excluded from decision-making in research, which infringes on their sovereignty and self-determination.
- Inertness: Researchers cast Indigenous peoples as ahistorical static representatives of life in the premodern past or in a primal state of nature beyond the line.
- Fungibility: Indigenous peoples lose their unique histories and contexts and become interchangeable "isolated" populations in centres of accumulation like the Diversity Project and Kidd Lab.
- Violability: Indigenous peoples' cell lines and data become alienated and can be used for all manner of research and development.

- Ownership: Scientists take on de facto ownership of Indigenous peoples' cell lines, part of "our unique collection," using them as they see fit, giving them away, and Coriell Cell Repositories even selling them.
- Denial of Subjectivity: Scientists omit consideration of Indigenous peoples' feelings and concerns (Munsterhjelm 2015, 296).

The Karitiana and Surui Indigenous peoples have repeatedly expressed their anger over what they see as a betrayal by the researchers who sampled them and those who have continued to use their genetic materials. They are also concerned with how the blood samples (and resultant cell lines) taken from them as donors in 1987 by Francis Black of Yale University affects the afterlife of donors who had since died (Rohter 2007). In a 2007 *New York Times* article about the sale of Karitiana and Surui cell lines (which in 2015 cost US$85) and DNA samples (which in 2015 cost US$55) by the NIH-funded Coriell Cell Repositories:

> But Francisco M. Salzano, one of Brazil's leading geneticists, with more than 40 years of experience in the Amazon and dealing with indigenous peoples, argues that it is acceptable to brush aside such concerns.
>
> "If it depended on religion and belief, we would still be in the Stone Age," he said in a telephone interview from his office at the Federal University of Rio Grande do Sul. "None of these samples have been used in an unethical manner," Dr. Salzano added. As for the question of informed consent, he added, "That is always relative." (Rohter 2007)

Salzano haughtily rejects the Karitiana's and Surui's concerns about the afterlives of their ancestors as something from the "Stone Age." He also dismisses the criticisms of informed consent and the controversial conditions under which they were sampled in 1987. Central to this dismissal of the Karitiana's and Surui's concerns is a strongly secular claim that religion and belief should not interfere with scientific research. Yet the post 9/11 security apparatus greatly expanded biometric research and development and resurrected racial categories under the euphemisms of biogeographic ancestry and visible phenotype research in forensic genetics. This expansion was justified by references to the problems in identifying the sacred dead victims of the 9/11 terrorist attacks, particularly in the 2000s, along with its lesser discussed reference to identifying the remains of 9/11 hijackers and keeping them separate from the victims and the more general terms of national security (which I discuss in chapter 3).

Objectification involves an array of flexible processes. Frantz Fanon contended that the rendering of colonized peoples as exotic in the sense of foreign, mysterious, and strange is central to colonialism's hierarchies: "Exoticism is one of the forms of this simplification. It allows no cultural confrontation. There is on the one hand a culture in which qualities of dynamism, of growth, of depth can be recognized. As against this, we find characteristics, curiosities, things, never a structure. Thus in an initial phase the occupant establishes his domination, massively affirms his superiority. The social group, militarily and economically subjugated, is dehumanized in accordance with a polydimensional method" (Fanon 1967, 35). Such exoticism is central to how researchers have viewed Indigenous peoples as both genetic conduits to ancient times and as resources to explore human evolutionary history (Reardon and TallBear 2012, 237). For Francis Black and other Western epidemiologists and anthropologists, Brazilian settler colonization was a type of "natural experiment" and an important opportunity for them to study how isolated populations' immune systems reacted to pathogens they had never been exposed to (Walker, Sattenspiel, and Hill 2015, 2–3). Western scientists captured many Indigenous people through international techno-science assemblages that intersected with and utilized the infrastructure of Brazilian settler colonial original expropriation (Radin 2017). There were some conflicts, such as FUNAI restricting researchers' access to some Indigenous peoples, as occurred in the *Alpha Helix* expedition, as part of its general biopolitical pastoral mandate to care for Indigenous peoples against the onslaught of colonial settler cynegetic sovereign violence.[27]

American anthropologist Jonathan Marks (2003, 3), in his critique of the Diversity Project, observes, "Blood from indigenous peoples has been a valuable scientific commodity, traded between laboratories and researchers, for different projects establishing a network of relationships, obligations, and coauthorships." In this particular political economy of scientific knowledge production, government-funded research transform these Indigenous peoples into resources that exist and circulate through the scientific commons in various forms including cell lines, samples, and genetic sequence data used in the production of hundreds of published papers. These resources can then be subject to commercialization, as we will see later. Central to this capture and subsequent resource status within scientific research assemblages are the chain of metonymic transfers that create an equivalency between the virally transformed cell lines and each of these Indigenous peoples. Kidd and other Kidd Lab associated researchers in a 1995 article advocate the research potential of the Karitiana, Surui, and other Indigenous peoples' cell lines being sold by

Coriell Cell Repositories as productive resources: "All these population samples *exist* as Epstein-Barr virus-transformed, lymphoblastoid cell lines, most of which were established by us under approved human subjects protocols. The Coriell Institute for Medical Research (NIGMS Human Genetic Mutant Cell Line Repository) in Camden, New Jersey, has available for distribution 5–10 cell lines from nine of the populations in this study: Ami, Atayal, Biaka, Mbuti, Druze, Han(s), Maya, Karitiana, and R. Surui" (Castiglione et al. 1995, 1448, my emphasis). In the first line, the scientists make a metonymic equivalency between the cell lines and populations. In the second, seven of the nine populations available for distribution are Indigenous peoples, including the Ami and Atayal sampled by Lu Ru-band in Eastern Taiwan in 1993–94 and the Mbuti sampled by Cavalli-Sforza in 1986 (ALFRED 2019c; Lu et al. 1996; Munsterhjelm 2014, 167–9). An equivalency based upon the metonymic chain allows for the distribution of cell lines representative of Indigenous peoples.[28] The assemblage's capture of these Indigenous peoples and subsequent transformation through Epstein-Barr virus involves a type of original expropriation. This capture involves a hierarchical sociopolitical relation that constitutes the Indigenous peoples as living research objects that *exist* in a state of biological servitude within the scientific commons. These research assemblages that advance scientific knowledge in the service of the biopolitical well-being of favoured Western populations accumulate Indigenous peoples' genetics from necropolitical zones.

This process is not enclosure of the commons in the sense of privatization of Indigenous peoples' genetic materials, such as a patent on a particular cell line. Rather, these assemblages are engaged in the publicly funded production of scientific knowledge as part of a commons ("upstream") that can then be subject to privatization in the forms of commodities ("downstream") (Reynolds and Szerszynski 2012, 36–40). Examples include the incorporation of the Karitiana, Surui, and Maya data by ancestral DNA genealogy companies like 23andMe or forensic genetic technologies like the Illumina FGx DNA sequencing system that I analyze in later chapters (Macpherson et al. 2008, 6). Hence, arguments by the Coriell Cell Repository representatives that no one has made any money from the Karitiana and Surui cell lines and DNA extracts use a very narrow conception of commodification (Rohter 2007). This 1991 article's data and findings became part of the corpus of genetic research. However, a series of developments in the US judicial system would give the Karitiana and Surui and to a lesser extent the Campeche Maya important roles in the DNA Wars as conflicts emerged over the introduction of forensic genetic testing as evidence in the US and Canada, to which we will now turn.

2

Judicial Review, Not Peer Review: The Karitiana and Surui in the DNA Wars

The preceding chapter considered how the Ticuna, Karitiana, Surui, and Maya became resources and objects of exchange in the scientific commons, routinely used by Kenneth Kidd and his colleagues, shared with other scientists, and sold internationally by Coriell Cell Repositories (until 2015 for the Karitiana and Surui, though the Maya are still for sale). However, this chapter will show how these Indigenous peoples' introduction into forensic genetics was not by such a conventional route. Kidd's entry in a pamphlet entitled *2006 National Institute of Justice* DNA *Grantees Meeting Presenter Biographies* summarizes this introduction: "His molecular lab was also active in the discovery of new polymorphisms and especially in determining their frequencies in different human populations. His comments in testimony in the [1990] Yee case led to his laboratory's unpublished data on South American Indians being introduced into the courts. The isolated Karitiana Indian tribe became famous in forensic circles" (National Institute of Justice 2006, 12). This brief mention of Kidd's testimony about the Karitiana is not explained; only that they are an isolated tribe that became famous. As I will show in this chapter, this mention of the Karitiana is something of a backhanded compliment. For what occurred was a breach of Kenneth Kidd and his colleagues' genetic diversity research assemblage that contained and used the Karitiana, Surui, Nasioi, and other Indigenous peoples as productive resources. This breach occurred when defence counsels appropriated and used these peoples' data to try to undermine Kidd as an expert witness supporting the prosecution's introduction of genetic evidence into the Canadian and US court systems.

Foucault's concepts of biopolitics and governmentality intertwine through an emphasis and attention to how people and populations are managed in a productive manner rather than through coercive sovereign violence. The courts and prisons however have long had a role in the administration and management of racialized populations, particularly those deemed subservient within or superfluous and/or threatening to capitalist accumulation (Browne 2015, 12–14; Crosby and Monaghan 2016, 40–1; Neocleous 2013, 12, 17–18). Concepts of race have been hierarchically encoded and enforced through both international and domestic law. During the early 1990s, concepts of race were central to the DNA Wars, which were a protracted set of debates and exchanges over the admissibility of genetic testing evidence into the Canadian and US court systems (Humes 1992; Munsterhjelm 2015, 298–302; Thompson 1993, 30–3). A breach between Kidd's diversity research and his role as a prosecution expert witness occurred when US and Canadian defence lawyers appropriated Kidd et al's Karitiana, Surui, and Maya diversity data to challenge this introduction of genetic evidence and try to discredit Kidd and his colleagues as prosecution expert witnesses. These Indigenous peoples went from being used by Kidd and other scientists as productive objects of research in human diversity studies to being disruptive entities used by US and Canadian legal defence teams to undermine prosecution use of genetic evidence. This process was contrary to popular conceptions of how scientific knowledge advances in which research is subjected to peer review by other experts within the field who assess its significance, validity, and reliability, and so on. Instead, the conflicts over the introduction of forensic genetic testing as evidence into the US and Canadian judiciaries demonstrate how state judicial institutions ruled on the admissibility of evidence, including its scientific validity in particular, and the significance of differences in the genetic characteristics among racial populations, known as substructure.

The role of these Indigenous peoples in the DNA Wars involves a striking study of contestations over racially defining normal and abnormal populations, an example of biopolitical governmentality enacted within the US and Canadian judiciary, which is typically thought of as sovereign, yet shows the interpenetration of biopolitics with law. In this way, it demonstrates the biopolitical emphasis upon the genetic traits of racial populations in a debate over probabilities of risk in a conflict between defence counsels' arguments that forensic genetic testing was still not a reliable means to distinguish between guilty and innocent suspects and prosecution arguments that it was reliable, verifiable, and commonly accepted science. The Karitiana's data were used by the defence legal teams to destabilize and question the random match probabilities (RMPs) of the prosecution that linked suspects to crime scenes. In response, the prosecution experts therefore sought to disassociate the Karitiana from US and Canadian populations through emphasizing their high levels of consanguinity as "extreme."

During the DNA Wars, there was a conflict over how random match probabilities, which link a suspect to a crime scene, could be affected by substructure; that is, by whether someone is from a particular racial population in a specific geographic area. In this argument, the possibility that people from the same racial group in a given geographic area would tend to intermarry at a higher rate would lead to greater chances that two people might share the same combination of genetic markers used in genetic testing. If two or more people could share the same combinations, this possibility would compromise the assumptions used in calculating RMPs that link a suspect to a crime scene. In particular, the defence legal teams tried to use the Karitiana and Surui data to argue that Caucasian, Latino, or other racial populations might also have local subpopulations with different general genetic characteristics. *Population substructure* means that a local or regional population of a particular racially or ethnically defined group might have higher or lower percentages of certain forms of a genetic marker (allele) used in the genetic testing and considers whether various populations' frequencies of these forms of genetic markers accorded with a measure called the Hardy-Weinberg equilibrium.[1] The defence legal teams used the Karitiana and Surui to argue that there was a real possibility that such differences existed in the US and Canada and so challenged the prosecutor's use of large nation-wide racial categories like Canadian Caucasian or American Hispanic, which involves the rhetorical technique of association. These large racial population databases were used in calculating estimates of the likelihood of a statistical match, called RMPs, between genetic samples from a crime scene to a suspect or victim (Humes 1992; Thompson 1993, 61–4). This use was part of a broader challenge to the robustness and reliability of forensic genetic testing, which had only been invented in 1984 by Alec Jeffreys and so was then still a new technology.

In the early 1990s, there were intense arguments over whether and how the genetic characteristics of these larger population categories were differentiated by racial categories. In effect, these arguments were about using racial categories in calibrating the violence of law. And it was the judiciary's sovereign decision-making that shaped forensic genetics by assessing the reliability and validity of these technologies in a process of judicial review, not peer-review. Furthermore, the defence lawyers attempted to use the Karitiana and Surui and other Indigenous peoples genetic data and traits to undermine RMPs as a way of disrupting the state's biopolitical pastoral hunts for and persecution of criminals (Chamayou 2012). Central to these disputes was the attempt by Kenneth Kidd and other prosecution expert witnesses to take the Karitiana and Surui out of the American and Canadian courtrooms where defence lawyers were using them to challenge the administration of justice and put them back in their productive role as isolated populations in genetic diversity research, a rhetorical process of dissociation (Cooren and Taylor, 2000, 180–5). In actor network theory terms, Kidd and the other prosecution expert witness eventually put the Karitiana and Surui back into their productive role in diversity research and thereby ended the treason the Karitiana and Surui data had committed by serving defence legal teams to disrupt prosecutions use of forensic genetic testing (Callon 1986).

There are a number of books and numerous articles on the DNA Wars, but none of the authors analyzed the conflict over the significance of the Karitiana and Surui data (see, for example, Aronson 2007; Thompson 1993; Giannelli 2011, 68–9). Barbara Prainsack (2015, 80) comments that science and technology studies scholarship on the DNA Wars "remained to a large extent within the existing boundaries of either legal or existing STS discourse." Whatever these authors' respective reasons may have been, these omissions are typical of the marginalization of Indigenous peoples from academic analyses and histories of forensic genetics. Such omission is important because the disputes over the Karitiana and Surui involved what Bruno Latour (1987) terms trials of strength over the traits and organizing properties of racially defined genetic differences between European settlers in the US and Canada in comparison to Indigenous peoples.

SUMMARY OF DISPUTES OVER EARLY TECHNOLOGIES

An early testing technology of this period called RFLP analysis used a variable number of tandem repeats (VNTRs), a type of so-called "junk DNA" that was thought not to code for proteins.[2] VNTRs are sequences of DNA ranging from about ten to one hundred base pairs long, which then typically repeat anywhere from five to fifty times in a row in a particular location. The loci used in US and Canadian forensic testing in the late 1980s and early 1990s were between nine and thirty-eight base pairs in length (Butler 2010, 51).[3] Because the number of repeats of each VNTR sequence are all highly variable between individuals, theoretically, completing this process over four or five different locations (each location is checked using a "probe") on the genome would be able to differentiate between two individuals to an extremely high degree of accuracy.

These estimates of differentiation between two individuals are called *random match probabilities* and are phrased in terms of the probability of two people sharing a number of matches at the different VNTR loci. However, the distributions of these frequencies of VNTRs can vary depending upon the population, so knowledge of these distributions in a given population was considered important in the calculation of random match probabilities. As we will see in the following analysis, defence expert witnesses showed how the Karitiana and Surui data had very strong differences in the incidence of some VNTR markers, allowing defence legal teams to argue that such differences might also occur among Caucasian populations depending upon the region.

THE CONFLICT BEGINS

During the initial few years following the introduction of forensic genetic testing as evidence, defence attorneys generally did not challenge the veracity and validity of such evidence (Aronson 2007). However, a number of trial lawyers began to have doubts, and they enrolled an impressive array of genetic researchers as part of their defence efforts. This conflict began in 1989 in the *US v. Castro* double murder case in which the victims, a mother and her daughter, had been stabbed to death. An important precedent was the trial judge's decision not to allow the prosecution's submission of an analysis done on a drop of blood on the watch of the accused by the genetic testing company Lifecodes (Lifecodes scientist Ivan Balazs was a coauthor of the Kidd et al. 1991 paper on the Karitiana, Surui, and Maya) due to defence objections that the analysis had been improperly conducted (Aronson 2007, 74–6). However, faced with other evidence, Castro eventually confessed that the blood spot on his watch was from one of the victims and was found guilty (76–7).[4] Nonetheless, the issues raised over forensic genetics were not resolved and in the aftermath of the trial Peter Neufeld and Barry Scheck, two of the defence lawyers involved in the *Castro* case under the aegis of the National Association of Criminal Lawyers, established a DNA Task Force (224). The *Castro* case was significant in that the defence was able to challenge the hitherto unquestioned ability of forensic genetics as a manhunting technology to reliably distinguish between friend and enemy.[5]

The legal rulings in *Castro* and other cases were a dramatic shock to judicial use of these new technologies. For example, from 29

November to 1 December 1988, a conference entitled DNA Technology and Forensic Science was convened at the US government's Cold Spring Harbor Laboratory over the issues that the *Castro* case raised.[6] Here we see an early example of the seamless interaction between genetic researchers like Kenneth Kidd in their various academic and legal advisory capacities, state scientific, judicial, and police agencies' scientists like Bruce Budowle of the FBI, and biotechnology companies, with the organizers stating, "We are particularly grateful to Cellmark Diagnostics (ICI Americas, Inc.), Collaborative Research, Lifecodes Corporation, and the NIJ (US Department of Justice) for underwriting the costs of the conference," along with a list of thirty-one corporate sponsors (Ballantyne et al. 1989, v, xiii). The use of racial categories in forensic genetic testing was discussed. Kenneth Kidd stated "With enough of these systems [of human gene mapping information/databases] and variation in [noncoding VNTR] frequencies among populations, one could begin to look at it overall to infer racial origin" (Pugliese 1999, 439; Track et al. 1989, 344). Jack Ballantyne then responded, "Even conventional genetic markers can occasionally give you precise racial data. For example, in a case last week, peptidase A 2–1 was found in evidence at the scene of the crime. It is highly likely that it originated from someone of Negroid origin" (Pugliese 1999, 439; Track et al. 1989, 344). Kidd responded that "If we looked at D6S2 and found a particular allele, there would be a high probability that the person was of African origin. That allele is so far only in pygmies. We will find ethnic-specific alleles as an automatic by-product of the kind of research we are doing. Once we know them, these alleles can be examined in DNA samples that are sometimes available when conventional serology is not" (Track et al. 1989, 344–5). In this exchange, the use of established racial categories was stated outright.

A logic of emergency over a threat to the functioning of the manhunt informed the epideictic rhetoric of the organizing narratives of both sides. In contrast to scientific research articles, the organizing narratives by the defence and by the prosecution involve what Torronen (2000) terms an *interrupted narrative* that seeks to persuade the jury as representatives of the People towards a particular course of action. Defence legal teams, including their expert witnesses, argued that an unreliable technology not yet accepted by the scientific community was being used to imprison potentially innocent people (see, for example, Mueller 1993; Leslie Roberts 1992a, 1992b). In contrast, prosecution legal teams argued that these were reliable technologies accepted by the scientific community and so defence concerns over population substructure were exaggerated and largely irrelevant, and worse yet, not allowing the submission of forensic testing as evidence would allow murderers and rapists to go free (see, for example, Mueller 1993; Leslie Roberts 1992a, 1992b).

In the aftermath of the success of the defence team in *US v. Castro* in preventing the admission of some forensic genetic evidence, the next high profile conflict was *US v. Yee*. These disputes began to gain mass media attention, with the *New York Times* carrying a 22 June 1990 story entitled "Law: DNA Fingerprinting Showdown Expected in Ohio" on the discovery hearings for the *Yee* murder case (Labaton 1990). This 1988 murder near Sandusky, Ohio, involved a case of mistaken identity in which three Hells Angels gang members killed David Hartlaub, a 28-year old record store manager. Prosecutors theorized Hartlaub drove a van that resembled that of a member of a rival motorcycle gang called the Outlaws. In this account, the shooter, John Ray Bonds, hid in Hartlaub's van and then shot him thirteen times from behind using a Mac-11 submachine gun with a silencer. Bonds then escaped with his two accomplices, Mark Verdi and Steven Yee, in Yee's car. Police investigators obtained blood samples from the murder scene that did not match that of the victim and suspected that Bonds had been injured by a ricocheting bullet or fragment (Coleman and Swensen 1994, 7). Following an extensive investigation, the three suspects were finally arrested and charged. Of particular significance were a set of bloodstains found in the back of Steven Yee's car that the prosecution argued, based on DNA testing, matched the unidentified blood samples found at the murder scene (Coleman and Swensen 1994, 7–9).

Whereas the focus of defence legal efforts in the *US v. Castro* was Lifecodes' testing methods (a private company), their main focus in the *US v. Yee* was the FBI's new forensic genetic testing system, which had been developed under the leadership of Bruce Budowle (whose 2015 conference speech was discussed in the introduction). Both the defence and the prosecution brought in well-known expert witnesses. The defence witnesses included Harvard University's Richard Lewontin, a prominent population geneticist whose oft-cited 1970s genetics research showed that 85 per cent of the differences between individuals occurred within the same racial grouping while only 15 per cent occurred between racial groupings. Daniel Hartl of the Washington University School of Medicine was another prominent defence expert witness. For the prosecution, expert witnesses included Kenneth Kidd from Yale University and Thomas Caskey from the Baylor College of Medicine.

Among other things, the defence accused the FBI of not subjecting their research and analytical methods to peer review. Hence, the defence argued there had been no independent validation of the FBI's forensic genetic methodologies and so it failed the Frye Test standard of the technology being commonly accepted in the scientific community (Giannelli 1993, 106–7).[7] Furthermore, the defence team focused on discrepancies in how the FBI formulated RMPs, and the Karitiana data became a point of contention as exemplars of how population substructure can affect these probabilities; something the FBI's use of broad population categories of Caucasian, Black, or Latino did not sufficiently consider in calculating RMPs. The FBI countered that they had presented papers at conferences and consulted with members of the forensic genetic community and that the conservativeness of their estimates more than compensated for any population differences, and if anything, erred in favour of the defendant (Giannelli 1997, 418–19).

This dispute was over taxonomy, in particular the definition and attendant spatial-temporally defined agency that characterized the various racial and ethnic categories. The defence argued for the need for finer regional subpopulation distinctions within the larger racial categories of White, Black, Hispanic (this category applies mainly to the US), Oriental (called East Asian today), and Native American or Native Canadian (depending on whether it was the American or Canadian settler state jurisdiction). Hence, as American legal scholar Jonathan Kahn (2015, 73) discusses, the overall use of racial categories was never at issue, only the level of resolution of such categories. Lewontin and Hartl in their testimony for the defence during the *Yee* hearings argued that more population data was needed to understand substructure, based on how geographic and social boundaries could influence population genetics, particularly the frequency of particular alleles (Kahn 2015, 73).

During the extensive discovery hearings in the summer of 1990 that preceded the beginning of the *US v. Yee* trial, defence legal teams learned of and requested Kenneth Kidd and his colleagues' data on the Karitiana, Surui, and Campeche Maya. According to Thompson (1993, 77n253), there was a protective order placed on the data "that prevented the experts for the defense from discussing it outside the courtroom," to allow for the publication of the 1991 article by Black, Kidd, Kidd, Balazs, and Weiss discussed in the preceding chapter (Kidd et al. 1991). However, the data eventually circulated between different defence lawyers and became a staple of defence legal strategies during the 1990s and into the 2000s.

The *Yee* case is significant in that it decided on the admissibility of the FBI's new forensic genetic testing that had arguably not been subject to extensive scientific peer review. As Cornell University law scholar Faust

Rossi (1991, 32) discusses, "The court, while purporting to apply the Frye test, stated that 'the circumstances of this case call on the court to use the legal process as a surrogate for scientific procedures to ascertain the reliability of the government's evidence.'" Hence, the legal process assessed the scientific validity of the FBI's forensic genetic testing procedures, not peer review nor other similar processes of scientific knowledge evaluation. The Karitiana are not discussed in most publicly available accounts of the *Yee* case.

Unfortunately, a limitation of this book is that I could not conduct a detailed analysis of the *Yee* court materials. Digital versions are not available, and the costs of obtaining copies of the court transcript and related materials from the National Archives at Chicago are prohibitive.[8] However, the University of New Brunswick Law School has compiled a publicly available website called the Allan Legere Digital Archive on the trial of serial killer Allan Legere. It contains PDF copies of the transcripts of the 1991 *Her Majesty the Queen and Allan Joseph Legere* (hereafter *The Queen v. Legere*) voir dire hearing and court transcripts in which Kenneth Kidd and a number of other experts testified and that also include many references to the *US v. Yee*. The *Queen v. Legere* led to one of the first criminal convictions in Canada based mainly on genetic testing evidence. The public availability of this material makes the case useful because it is not hidden away in archives with prohibitive access fees and so will allow readers the ability to make their own assessments of the court documents.

The following analysis will narrowly focus on the forensic genetic sections of the voir dire and trial transcripts in *The Queen v. Legere* dealing with the Karitiana, Surui, Nasioi, and other Indigenous peoples. The transcripts from the *Queen v. Legere*'s voir dire hearings over the admissibility of genetic evidence and the trial itself reveal conflicts over the significance of the Karitiana, Surui, and to a lesser extent the data of the Nasioi. In the *US. v. Yee* and then the *Queen v. Legere* case, (also written as *R. v. Legere*), the defence argued through analogy that the differences in the Karitiana and Surui data were directly relevant to Caucasian populations as examples of major differences in allele frequency within a regional or continental population category. These differences meant that more research was required on substructure among other populations before these new technologies could be admissible, which involves a rhetorical strategy of association. In sharp contrast, the prosecution expert witnesses argued the defence interpretation of these Indigenous peoples' genetic data could not be universalized. The prosecution expert witnesses did this by asserting that these Indigenous peoples' genetics have been shaped spatially and temporally through close intermarriage in ways that make them distinct from and so incomparable to Caucasian settler populations, which involves a rhetorical strategy of dissociation.

THE QUEEN V. LEGERE

The trial of serial killer Allan Legere in 1991 involved a high profile conflict over the introduction of genetic evidence into the Canadian judiciary in what became one of the first convictions based on the use of DNA evidence. Legere became known as the Monster of the Miramichi, a reference to the river in rural New Brunswick in eastern Canada where his series of murders and the resulting police manhunt occurred in 1989. Legere had already been in prison after being convicted along with two other men for the 1986 murder of sixty-six-year-old John Glendenning and the beating and rape of his sixty-one-year-old spouse Mary Glendenning. In a 2006 article about his experiences in this case of *The Queen v. Legere*, the New Brunswick lead prosecutor Queen's Counsel John Walsh recalled the impact of Legere's escape: "On May 3, 1989, he escaped custody in Moncton N.B., during a medical transfer from the Atlantic Region Penitentiary in Renous, N.B. What ensued for the people of New Brunswick, particularly the people of the Miramichi region, was a nightmare of horrific brutality; a dark time of murder, arson and palpable fear. It was a time of bolted doors, sentinel lights, a night time patrolling helicopter and an army of weapons wielding police officers. It ended on the morning of November 24, 1989 with his capture" (Walsh 2006, 1). Over the seven months between May and November 1989, Legere raped and beat two elderly sisters, Annie Flam, who died, and Nina Flam, who survived; he raped and murdered another two sisters, Donna and Linda Daughney, who were both in their forties; and finally he murdered Father James Smith, a sixty-nine-year-old local Roman Catholic priest. The manhunt for this sadistic rapist/murderer created a state of emergency in which security forces were deployed to track and capture or kill the criminal in an effort to restore law and order.

While the manhunt continued, the prosecution lawyer John Walsh travelled to Ottawa where the Royal Canadian Mounted Police's (RCMP) new DNA identification lab had found matches between sperm samples from the female victims and scalp hair and pubic hair samples that had been taken from Legere during his first murder trial in 1986 (*Her Majesty the Queen v. Legere*, Vol. VIII, 7–8 May 1991, 260). The RCMP calculated a four-probe match from the Flam crime scene with "an estimated random match probability of 1 in 5.2 million in the male Canadian Caucasian population," and the Daughney crime scene "revealed a 5 loci match, with an estimated random match probability of 1 in 310 million in the male Canadian Caucasian population" (Walsh 2006, 5).

The other evidence directly connecting Legere to the multiple murders was weak, so the genetic evidence took on a pivotal role in the prosecution's case. Legere was formally charged with the four murders in December 1989. In August 1990, Legere was found guilty of several crimes related directly to his May 1989 escape for which he was sentenced to another nine years in prison. The serial murder case against Legere then took some time to prepare. A voir dire hearing on the admissibility of genetic evidence began in February 1991. In Canadian jurisprudence, a voir dire is defined as a trial within a trial over the admissibility of evidence; one that is decided by the judge without the presence of a jury.[9] According to Walsh, "In and out of the courtroom, the defence telegraphed the intention to make a frontal assault on all aspects of forensic DNA typing. The Crown having decided to try and prove admissibility by establishing that DNA typing was generally accepted, at least in the forensic scientific community, and/or reasonably reliable, what was left to be decided was how and with what experts" (2006, 5). As lead prosecutor, Walsh was tasked with assembling a group of expert witnesses. Kenneth Kidd was suggested to him and Walsh, along with an RCMP officer, travelled to Yale University where they met with Kidd. In the end, they returned to Canada with an agreement from Kidd to testify as an expert witness on the prosecution's behalf (Walsh 2006, 7).

A major point of contention in the voir dire hearing that ran from February to June 1991 was whether the Caucasian population of Canada had significant regional substructure or not. Following the strategies and tactics of American defence legal teams in the 1990 *US v. Yee* discovery hearings, Legere's defence lawyers utilized the Karitiana and Surui data to challenge prosecution expert witnesses. However, I will first outline the prosecution's arguments then analyze how the defence not only challenged the prosecution but also the orthodox interpretation of genetic research regarding the traits of Indigenous peoples and settler populations. I should reiterate that what follows is a narrowly defined focus on the role of the Karitiana and Surui data (and to a lesser extent that of the Nasioi) in *The Queen v. Legere*; I do not want to overstate their roles in these conflicts over the admissibility of forensic genetic evidence (the trial transcripts total thousands of pages). Rather, I think this focus is significant in what it reveals about the conflicts over and the persistence of established popular racial/ethnic categories in the emerging racial taxonomy of forensic genetics.

THE PROSECUTION'S ARGUMENTS

If we consider the above general organizing narrative of the prosecution in the *Queen v. Legere* case, the Karitiana and Surui data can be viewed as one type of a subnarrative that is part of a larger effort to include genetic evidence. In particular, the prosecution asserted that the defence were misrepresenting the data from these Indigenous peoples in an effort to rhetorically dissociate them from Canadian populations and put them back in their productive place in diversity research assemblages. Therefore, a significant subnarrative was the role of Kenneth Kidd in defining the Karitiana as the "extreme" of known human genetic difference while acknowledging that genetic technologies could still individually distinguish between them. For example, in his May 1991 expert witness testimony during the *Queen v. Legere* voir dire hearing, Kenneth Kidd argued that there was still "lots of variation" among the Karitiana:

> We are looking at very primitive tribal populations in the Amazon basin and even though all children born in the last 15 or 20 years, I don't remember the exact date, are descended from one man three to four generations ago, because he was the tribal chief and had five wives, still at the DNA loci every individual that we studied had a distinct DNA pattern so that the frequencies there might differ but the patterns are – there are still lots of variation lots of patterns so that small variations in frequency become very unimportant. And that, I might add, that's probably the most extreme sample example that I know or that's ever been studied in humans for being a very tight, closed, inbred population. (Her Majesty the Queen v. Allan Joseph Legere, Voir Dire, Vol XI, 15 May 1991, 187)

Kidd in the above extract refers to the late Karitiana leader Antônio Morais and his five wives (discussed in chapter 2) and utilizes a series of common colonial taxonomic temporal-spatial and social distinctions. Describing the Karitiana as "very primitive" places them at a low level of social and technological complexity distinct from modernity, which is then reinforced through mention of the geographic location in the Amazon Basin as a signifier of isolation. His emphasis on the polygamist marital arrangements that led to high levels of consanguinity places the Karitiana outside Western social norms. By arguing that there remains a lot of genetic variation detectable among the Karitiana as the "most extreme" example, Kidd claims a type of universal applicability for these forensic genetic technologies (something that became important in the use of the Karitiana after 9/11, as we will see later). While Kidd mentions the shared descent of most of the Karitiana from Antônio Morais and his five wives to emphasize their genetic difference, he does not mention that this common descent was related to the Karitiana's plummeting numbers due to settler colonial invasion and displacement as discussed in chapter 1.

Under cross-examination by the defence lawyer, Kidd argued that there was plenty of variation among the Karitiana: "Chance is still the primary factor in these situations. Chance can produce a variety of different outcomes. But I have seen chance produce in one of the smallest, most isolated populations I know out of this Amazon basin Indian population, I have seen chance produce everybody having a unique genotype" (Vol XII, 16–17 May 1991, 53). Kidd then made use of the Nasioi of Bougainville Island stating: "I have studied a hundred plus DNA markers of a small tribe from the highlands of a small island off the coast of New Guinea where there are 27 mutually unintelligible languages spoken on that island and very tight inbreeding for generations, and there's tremendous amounts of genetic variation" (Vol XII, 16–17 May 1991, 53–4). Kidd later sought to minimize the magnitude of the overall reduction in genetic variation between Caucasians with the Karitiana and Surui:

> Q: And what were your basic findings in your study about the Amerindians?
> A: We found looking at approximately 30 loci in two Amazon Indian populations and in a population of Mayans in the center of the Yucatán Peninsula that overall the amount of genetic variability was reduced by no more than 27%. (Vol XII, 16–17 Mauy 1991, 75)

Kidd's use of the phrase "no more than" as a qualifier attempts to minimize the significance of differences in allele frequency among different populations.

Later prosecution counsel Walsh questioned Kidd about how to interpret the potential significance of population substructuring: "What, if any, opinion do you have in respect to the effect of that particular data with respect to what you've seen with respect to Amerindians and native Indian populations in Canada, what effect does that have on your opinion with respect to North American Caucasian population, Canadian Caucasian populations and the database in this particular case? Does that in any way, what effect does that have on your opinions that you've previously given?" (Vol XII, 16–17 May 1991, 154).

To this carefully hedged question, Kidd responded:

> The main effect it has in my opinion is that I know the very strong contrast in the nature of the population structures. The Amerindian populations are very subdivided, many different languages and not until fairly recently that much admixture, so that there is a fair degree of differentiation, allele frequency variation among the different subgroups, the different tribes. Whereas in Europe the degree of variation across all of Europe is much smaller and the North American Caucasian population are a very much admixed selection from Europe where we're starting with a fairly comparative more homogeneous population to start with. (Vol XII, 16–17 May 1991, 154–5)

Kidd is reiterating the common view among population geneticists on the differences between Caucasian populations as a more homogeneous group and Indigenous peoples of the Americas having more genetic differences between populations due to geographic isolation and lack of admixture.

Following a series of questions to which Kidd responded that there were minimal differences across European populations, referring to the Karitiana, he again emphasized the high variability of these VNTR genetic markers: "Because if there is a high variability it's very unlikely that two people will have the same pattern by chance. The coincidence factor, I mentioned the Amazon tribe where in fact no two people did have the same pattern even though they were closely related" (Vol XII, 16–17 May 1991, 199).

CROSS-EXAMINATION

These attempts to reduce the significance of genetic differences and the attendant questions these differences raised about forensic genetic technologies were challenged by the defence counsel Walden Furlotte. A common cross-examination strategy involves asking a series of questions that then leads to an inconsistency in the witness's understanding of the issues. Furlotte appears to use this strategy when questioning Kidd about the Karitiana, Surui, and Maya data. He challenged Kidd's assertions about subpopulation substructure by citing a major difference in the frequency of alleles that Daniel Hartl found in his analysis of the Karitiana and Surui data for the *US v. Yee* case: "Q. Do you recall anybody telling you that if you had a combined probability pattern across three loci, the MS estimate is – this is for Village A, without naming it, the estimate is one in three hundred and seventy thousand and if one relies on Village B data base with the same patterns the combined probability is one in five hundred and seventy, could it vary that much, doctor?" (Vol XII, 16–17 May 1991, 135–136). Kidd responded that he that he did not "know the basis for that calculation but of course, the villages were different, they are small basically individual families" (Vol XII, 16–17 May 1991, 136). The defence lawyer then asked Kidd about his reaction to Hartl's analysis, as they had been friends in university:

> Q. And you mean to tell me that he never presented you with his findings and you weren't curious enough to ask?
> A. It was months later that I found out that he had done something and I don't particularly care what he did, one day I may find out, he certainly did not send them to me nor did the defence lawyer.
> I must say one of the reasons I am stopping testifying is the fact that these sorts of things are being circulated among lawyers without my knowledge and that I find that not a very proper approach. And I am not terribly pleased with the way the legal profession is handling this. (Vol XII, 16–17 May 1991, 136–7)

Kidd was clearly upset with the line of questioning and the use of his own data against him. The defence counsel tried to use Hartl's analysis of the Karitiana and Surui data to show that Kidd did not fully understand his own data in an attempt to undermine Kidd's credibility. In response, Kidd denied the relevance of the Karitiana and Surui data by describing them as "basically individual families."

Later, the defence tried to use an inductive argument that the differences between the Karitiana and Surui data show that substructure is potentially a general property of human populations, particularly geographically isolated ones, and that this property was relevant to Caucasian populations in rural New Brunswick where Legere was from. Dr William Shields, a defence expert witness from the State University of New York, entered a graph as an exhibit (VD-132) based on his own analysis of Kidd's Karitiana, Surui, and Maya data (such use also shows how the data was circulating among defence expert witnesses):

> VD-132 are graphs that I prepared based on raw data provided via the case from Doctor Kidd study, which I have a copy of, of three Indian groups, two in Brazil and one in the Yucatán in Mexico … I did an analysis on the raw data and also present them in figures

> and all of them illustrate that there are statistically significant differences between these populations of Native Americans, even though these two come from tribes that are about 600 km apart in South America. That's at that particular locus. This is at a locus that's not yet used in forensics but it is a VNTR locus, D14S1, and it illustrates one of the potential problems with ignoring substructure. (Vol XIII, 27 May 1991, 40)

Shields tries to generalize from the differences in the Karitiana and Surui data that this substructure means that "you can expect that if you take an individual out of his sub-population and test him in another sub-population that that individual will be biased against by doing that form of analysis" (Vol XIII, 27 May 1991, 40). This bias is because of the other subpopulation having a different frequency of alleles that would exaggerate the uniqueness of an outsider's random match probability.

Later Shields used Hartl's analysis of substructure among the Karitiana and Surui from the *US v. Yee* case. Having noted how it was similar to his analysis, Shields quoted Hartl's strongly worded report on Kidd's errors about the variability of alleles among the Karitiana and Surui:

> Q. Dr. Kidd made that statement, that's what Dr. Hartl is saying?
> A. That's what Dr. Hartl said he said, and they do differ. "Three, that it does not matter what data base you use for forensic calculations because all the numbers are small and any particular band pattern is uncommon. In the particular example used in this report the accumulated error over three VNTR loci caused by using the Karitiana rather than the Surui data base is a factor of at least 500. In my opinion no credible research scientist would ever treat the numbers one in 213,000 and one in 400 as if they were equal on the grounds that both are small." (Shields quoting Hart in Vol XIII, 27 May 1991, 172)

Shields utilizes Hartl's statistics that show significant differences between the Karitiana and Surui in their respective allele frequencies to indicate the potential for major differences between regional subpopulations of a larger regional racial population. Central to the defence strategy was the idea that allele frequencies for racial populations could vary widely so how to calculate them was still not yet settled in the scientific community. By attacking Kidd with his own data, defence counsel sought to undermine the authority and credibility of one of the prosecution's key expert witnesses and to raise doubts about the veracity of the random match probabilities.

In his concluding arguments for the voir dire hearing, prosecuting lawyer Walsh emphasized Kidd's eminence (using the rhetorical form of ethos), including how "His testimony has been accepted over some very prestigious scientists in the most major cases in the United States in DNA typing" (Vol XIV, 6–7 June 1991, 41). Walsh then summarized Kidd's arguments on substructure differences between the Karitiana and Surui as irrelevant to North American Caucasians:

> He has looked at, very tightly – in his opinion the most inbred populations in the world.
>
> He has looked at Mennonites, he's looked at the Mennonite community in Saskatchewan and Alberta. He's looked at Amazon tribes, one particular tribe that is the most tightly inbred population he's ever seen in the world. He's looked at the Middle East. His conclusions are that all of these frequencies have high variability in that they would not have any effect, significant, meaningful – there is no significant meaningful substructure as applies to the Caucasians in North America that would affect frequencies of this particular case. (Vol XIV, 6–7 June 1991, 41)

Ultimately, the prosecution arguments were successful in the *Queen v. Legere* voir dire hearing. The judge ruled in favour of the prosecution and decided that genetic testing was admissible as evidence. The court in effect ruled in favour of keeping the Karitiana and Surui out of the courts and instead in their place in human diversity research.

THE LEGERE TRIAL

During the *Queen v. Legere* trial proceedings in the autumn of 1991, Kidd again testified on behalf of the prosecution. Prosecution counsel Walsh questioned Kidd on the representativeness of the RCMP's Caucasian databases used to assess the frequency of VNTR alleles to estimate random match probabilities, and Kidd answered he thought the database was a "very good representation of the Canadian population as a whole" and contained a "sizable number of individuals who are from New Brunswick" (*Her Majesty the Queen v. Alan Joseph Legere*, University of New Brunswick Law School, Vol XVIII, 21–22 October 1991, 4506). Kidd supported the assertion that sampling of smaller Caucasian populations was not necessary by citing studies of Bougainville Islanders (including the Nasioi) and Amerindians (including the Karitiana) showing high levels of consanguinity while still having significant genetic variation between individual members.

Q. And you've indicated you've studied isolated populations from around the world?
A. Yes.
Q. I take it you're referring to extreme examples of areas that are very to themselves, so to speak?
A. I mentioned the studies on the island of Bougainville that I first got involved in over twenty years ago and now we're going back and doing some studies at the DNA level and in fact have data on some of these same systems for one of the small tribal groups on Bougainville. We're also studying Amerindian tribes and have looked at what is probably the most extreme example of a human genetic isolate that I know of, a single tribe in the Amazon Basin that has over the last couple of decades been reduced by illness and other things to one village. The whole language is only spoken by the people in this one village. Everybody born in the last ten to fifteen years are all descended as fourth or fifth generation descendants from one single chief and one or more of his three wives, so it's basically one big family, and I think a total of maybe 175 individuals, total of whom we sampled 54, and even in that which has to be it's certainly the most extreme example I know of isolation and genetic differentiation, being different from everybody else. (Vol XVIII, 21–22 October 1991, 4507–8)

Kidd, in this iteration of the Karitiana as Other, begins by emphasizing the Karitiana are the "most extreme example." Situating them in the Amazon Basin with its connotations of geographic remoteness, he elaborates further by stressing their population decline to one village, but naturalizes this decline in terms of "illness" and "other things," thereby omitting specific mention of settler colonial violence and displacement. He then again cites the common descent from Antônio Moraes and polygamous social relations. However, he adds that they are "one big family," again concluding they represent the extreme of genetic difference. The prosecution lawyer then asked, "You could differentiate between even those individuals?" to which Kidd responded, "That's right, everybody has a unique pattern." "And that's the most extreme example you know of?" the lawyer asked, to which Kidd responded, "That's correct" (Vol XVIII, 21–22 October 1991, 4508). Kidd's argument is that greater resolution of regional subpopulations was not required because these technologies still functioned reliably even under conditions of extreme genetic difference, which was reiterated shortly after:

Q. Apart from identical twins and without even putting a probability figure – just ignore those probability figures for the time being – to a match, have you in your experience ever seen a four or five probe match between different individuals with these highly polymorphic probes?
A. No, I have never seen it, and that includes these very isolated populations such as the Amazon tribe I was talking about where, really, everybody is very closely related, and that's where you would expect the highest chance of seeing two different people with the same pattern, and I have not seen it. (Vol XVIII, 21–22 October 1991, 4513)

Kidd asserts that an encounter with the Karitiana is an encounter with the limits of human genetic difference, and the technology was still able to differentiate between individual Karitiana, proving its universal applicability; so even if there was substructure, it was not relevant because the technology had proven itself at the limits of human genetic difference.

Over defence objections, in anticipation of Shields's testimony later in the trial, prosecution counsel Walsh decided to ask Kenneth Kidd to "comment on his understanding of Dr. Shields's opinions" about population substructures (Vol XVIII, 21–22 October 1991, 4530). Kidd gave a brief history of population flows into Europe stating, "All of Europe was settled out of the Middle East within the last six to ten thousand years, so these are not populations that are remarkably different to start with, and we simply do not see the kind of subdivision and substructuring in Caucasians that's relevant to this issue" (Vol XVIII, 21–22 OCtober 1991, 4533–4). Kidd then continued:

Now, Dr. Shields used as proof of substructuring in his testimony here and in other testimony that I am aware of in other courts, he used my study of the Amazon Indians, but that's a study of tribes that speak different languages that are known to be highly inbred, that are known to be greatly isolated. I studied them because they are so different from Europeans and North American Caucasians. They bear no relationship on substructuring within Caucasians. Of course, substructuring does exist in the Amazon Basin and yet even there in [Francisco M.] Salzano's book of about ten years ago where they summarized literally over a hundred studies of something like nearly two hundred populations in the Amazon Basin, different tribes in the Amazon Basin, they concluded that there was so much gene flow that even though you could see this structure it was a very transitory thing and had no long term significance and that in terms of long term genetic change the whole Amazon Basin was acting as a unit. (Vol XVIII, 21–22 October 1991: 4534)

Kidd rejects comparisons of substructuring between the Karitiana and Surui with Caucasians. He then minimizes the significance of substructuring by situating the Karitiana and Surui as part of a larger regional Amazon Basin population.

In contrast, the defence did not emphasize the Karitiana and Surui as extreme liminal figures. Rather, like the voir dire proceedings, the defence again attempted to normalize the Karitiana and Surui as evidence that indicated more research on substructure among different Caucasian populations was required. The defence lawyers and expert witnesses argued that the differences in allele frequency between the Karitiana and Surui meant there was a real possibility of substructure occurring among Caucasians, and so these differences raised doubts over Legere's RMPs since he was from a rural area that might have considerable population substructure. Defence counsel Walden Furlotte cross-examined Kidd on the fact that there were significant differences in allele frequency between the Karitiana and Surui as Amerindians and whether this raised the possibility of similar differences among Caucasians:

Q. And a second argument would be that as a result of difference between ethnic groups that there's likely also that same degree of difference within the Caucasians?
A. And I think that's an absolutely fallacious argument for which there is extensive data arguing the other way. As I mentioned earlier, I know that Dr. Shields has used the study that my wife and I did of Amerindians in several courts around the country, around the US, arguing that that study proves there is substructure in the white Caucasian population of the United States, and that's just nonsense. (Vol XVIII, 21–22 October 1991, 4563)

Furlotte rejected Kidd's summary of Shields's argument:

Q. Well, that's not his argument. To be fair, Doctor, that's not his argument, is it?
A. Well, that's the way it's been relayed to me because I have not read those transcripts.
Q. Isn't his argument that it's because there's sufficient substructure within the Amerindians is that therefore there's no proof that there

> isn't that same degree of substructure within Caucasians; isn't that his basic argument?
> A. Well, that's equally nonsense.
> Q. That's equally nonsense, in your opinion?
> A. Yes, in my opinion. (Vol XVIII, 21–22 October 1991, 4563)

Furlotte continued by questioning whether there were many experts in the field who would disagree with Kidd. Kidd rejected the idea that his views about forensic genetics were not widely accepted in the scientific community:

> I am not aware of that. I know of many scientists who definitely agree with me. I know of people who have misinterpreted the data we got, who have treated the Karitiana as though they were like Poles in their evaluation of it and saying – which is certainly not the case.
>
> The Karitiana are basically one family, one tiny, tiny group of people, and as I said, the extreme I know of in the human race for a tiny isolated inbred group, and they were studied for that reason, to look at the extremes and there with six of these loci every one was polymorphic. One had only a small amount of variation but every one showed variation among individuals and every individual we sampled had a unique DNA pattern, and some of those individuals are more closely related than full siblings. (Vol XVIII, 21–22 October 1991, 4564)

Kidd maintained that these technologies could still be used to identify individuals among the Karitiana: "There is a lot of genetic variation even in these isolates which are far more extreme than any possible subdivisions within the Caucasian population" (Vol XVIII, 21–22 October 1991, 4565).

In his cross-examination, Furlotte continued to question Kidd over the possibility the substructure of the Karitiana and Surui was relevant to Canadian Caucasian populations. In the following section, they argue over the significance of bin frequencies, which refers to a way of estimating the length of the VNTR segments. There is a high level of variability in the length of these segments, and the RFLP technology could distinguish between ten and twenty repeats but could not readily distinguish between whether there were nineteen repeats or twenty, an important limitation of this technology. Therefore, the RCMP, like the FBI, utilized a set of categories called fixed bins to sort the VNTR segments depending on their length.[10] The possibility of a match at a loci occurring was calculated based on the known distribution of alleles (variants) among the population, in this case Canadian Caucasians. Therefore, a central point of contention was the representativeness of Caucasian databases with the defence arguing that New Brunswick rural areas were isolated leading to higher levels of interrelatedness and as a result had to be treated differently from the RCMP's Caucasian databases. The following extended extract shows how Furlotte tried to argue for this point against Kidd:

> Q. But just based on the bin frequencies you can conclude that it wouldn't be proper to use a data-base for one tribe when the accused belongs to a different tribe. Or if you were trying – being an accused or somebody you were just trying to identify?
> A. I would say based on my looking at the frequencies I would be much happier if the crime were committed by an Amazon Indian to have more data than I have now, and certainly if the crime were committed by a Karitiana I would not use a Mayan or Surui data base. On the other hand, I don't expect there to be any two Karitiana with the same DNA pattern, so if I got a match I wouldn't be terribly concerned about the probabilities. I could actually go in and type all 150 Karitiana. I can't do all however million Canadians there are.
> Q. So in the Amazon if you didn't do a sample database from each different tribe, had you just went and treated them all as one general population, it wouldn't be fair to them, would it, to use the general population database for those people?
> A. I'm not sure what fair is in this case. One can ask many different questions and if I don't know what tribe the criminal came from or what tribe, even, the accused came from because maybe he's somebody who moved into the city and his four grandparents came from four different tribes, then the best database would be a general database of Amazon tribes pooling the data from all of the different tribes, and that would be probably the fairest. Again I would probably want to in that situation, because I know there is some reduction in variability at some of these loci, some of them only have three or four alleles within a tribe, I would probably want to do a couple of more loci because the individual loci aren't quite as good, but it would be very easy to construct a database and to assure oneself that the probabilities were low enough that one was approaching very, very low chance that any two people had the same probability.
> Q. O.K., Doctor, you mentioned that you don't know what fair is but the position is representing Mr. Legere I want the expert

testimony to be as fair as possible to Mr. Legere. That would be understandable, right, so the evidence as I understand as the R.C.M.P. are bringing in is the likelihood of somebody other than Mr. Legere being a male Caucasian, and these are the figures; would that be right?
A. We cannot ever know precisely what the likelihood is of someone else having the same pattern as Mr. Legere without typing everyone around. We can make estimates and these are the estimates that have been made based on the Canadian population for these loci. We can also place around those estimates some indication of our confidence, of how accurate we think these estimates are, and those are the confidence intervals that I mentioned in my testimony yesterday, and even going overboard with the quick and dirty method that I've used to combine confidence intervals across loci, one that's not mathematically or statistically correct but is simple and more favourable, more fair to the defendant, I come up with an upper limit of one in 66 million, so I know the true frequency is somewhere below there. (Vol XVIII, 21–22 October 1991, 4579–41)

In the above exchange, Kidd once again dismisses Furlotte's attempts to use the Karitiana and Surui substructure to support the possibility of substructuring among Caucasians.

Later, near the end of his cross-examination of Kidd, Furlotte once again tried to argue that substructure might be relevant to Legere.

Q. But the fact that substructures exist is not a red herring?
A. Substructures exist in some sections of the human species. They certainly exist in Bougainville and Papua, New Guinea, where we're collecting samples. They certainly exist in the Amazon Basin where we're collecting samples. One can possibly argue some about whether they exist in Europeans but I would say it is in Europeans mostly a red herring because of the bulk of other evidence that we have about DNA variation and about genetic variation among European populations. (Vol XVIII, 21–22 October 1991, 4604)

Kidd once again dismissed Furlotte's arguments by making the distinction between isolated Indigenous peoples and European populations.

The defence counsel continued to try to draw parallels between the substructure of the Karitiana and Surui and Caucasian populations. Later, during the defence's cross-examination of another prosecution expert witness George Carmody, who was then an associate professor of biology at Carleton University in Ottawa, there was a discussion of the significance of the differences between Canadian Indigenous peoples and Canadian Caucasian populations, which included mention of Kidd's work on the Karitiana and Surui:

> Q. So depending on the database for the different sub-populations you can come up with a wide variety of numbers and great differences?
> A. Except that in those native databases they are remarkably unlike Caucasian profiles. They show extreme cases of what we would call founder effect; that is, that these populations were probably the result of a rather small group of people that immigrated to North America across the Bering Straits at some point, that these populations often have been further fractionated during historical time where they form small tribal groups, and it would be similar to the data that Dr. Kidd was referring to in terms of Amazonian tribes and so forth, that there are many instances of smaller aboriginal bands in North America that show quite strong profile differences from Caucasians. The differences that we see between any Caucasian populations that have been looked at are quite, quite smaller than any of the differences we see in the Amerindians.
> Q. O.K., but if you take your Amerindians, for instance, and I believe Dr. Kidd also testified about one small island [Bougainville], he gave the dimensions which I forget just now but I think he said something like there was about 26 different tribes and all different languages on there, and to calculate the frequency for each different one they were, well, quite different?
> A. Yes, they could be. (Vol XIX, 23–24 October 1991, 4908)

Near the end of the day on 24 October, the defendant Allan Legere was present in the prisoner's dock and gave his own chilling interpretations of the expert witnesses' testimony:

> MR. LEGERE: Can I be excused, Your Honour? I am tired of hearing that over and over and over. All the money they're wasting on these fellows here they could have had a database taken on the Miramichi [River], but the only thing is you'd run into the Indian caste system.
> THE COURT: Let's stop here now. How much longer are you going to be, Mr. Furlotte?
> MR. FURLOTTE: My Lord, I believe this doctor [Carmody] wants to get

away on a 4:55 flight or something and I could probably shorten it up.
MR. LEGERE: No, I want to leave now. I don't want to stay and listen to him anymore. The Crown's got a half a dozen of these guys coming up like elephants and telling the same story over and over but all they have to do is take a database of the Miramichi, but they'd have to take one up in Nordin, up in Douglastown, up the Chaplain Island Road.
THE COURT: Sheriff, if the accused doesn't want to stay would you take him out, please?
MR. LEGERE: I don't want to stay because he's missing the factor. He's wrong in the substructure and Mr. Shields is right and that's why they're all screwing things around. Move to the Miramichi and take a database and if I'm wrong I'll eat my shirt. You have fathers making love to their own daughters and vice versa. My own son was raped at five years old and the police did nothing about it, they keep it underground. You don't see this stuff, see? That's your missing factor, substructuring, guaranteed. Go back on Chaplin Island Road, you have cousins marrying cousins, you wouldn't believe it. You never seen it before, it's like the modernized Sodom and Gomorrah of the Biblical era. You've never seen it before. Open, lock. I'll have to make a key.
THE COURT: Well, I would ask the jury to go out. We'll bring you back in a minute.
MR. ALLMAN: I respectfully suggest that we adjourn and leave the officers to deal with their problem. (Vol XIX, 23–23 October 1991, 4930–1)

As Walsh predicted, the defence counsel again called Shields as an expert witness. During his cross-examination of Shields, prosecution counsel Walsh pointed out that Shields had made use of Kidd's Karitiana data on allele frequency differences and cited Kidd's earlier testimony, when Kidd disagreed with Shields' interpretation (Vol XX, 29 October 1991, 5123). There was an extensive discussion of the effects of substructure, including the following exchange between Walsh and Shields that mentions Kidd's testimony on the Karitiana data in a discussion on polygamy and its effects on consanguinity:

Q. Again, for the layman?
A. Polygamy is where there is one male and many females as part of the marriage system. It still happens, for example in some South American Indian tribes.

> Q. Like that tribes that Doctor Kidd looked at, the one that has one king and at least three queens.
> A. Yes.
> Q. Okay.
> A. And when you have that small a level of ancestry that creates high levels of background band sharing. I believe there is evidence that there were only a few thousand French Canadians who settled – French people who settled Canada. That's a small number of founders so you have what we call founder effects there as well, and that will generate higher levels of background band sharing than you would expect randomly. Selection can do it. (Cross Examination, Vol XX, 29 October 1991, 5138)

Background band sharing refers to how the autoradiograph images are like bar codes and that certain bands representing particular alleles may be more common among populations sharing a small number of founding ancestors. What Shields was trying to argue was that Legere, who is of French-Canadian ancestry and from a rural area of New Brunswick, might have extensive substructure due to ancestral endogamy (interrelatedness due to people marrying locally over a long period of time). However, by the end of his questioning on 29 October, prosecution counsel Walsh had led Shields to concede that the semen samples from the victims were those of the accused. In his concluding question, Walsh asked Shields, "And if this semen was from Mister Legere you would expect it to match the DNA patterns in his hair and blood?" to which Shields replied "yes" (Vol XX, 29 October 1991, 5156).

THE VERDICT

On 3 November 1991, Allan Joseph Legere was found guilty on all four counts of murder and sent to a maximum-security prison.[11] In his comments to the jury following their delivery of the guilty verdicts, the judge reflected on the significance of the trial, particularly the future role of DNA evidence and, citing Kidd's testimony made during the voir dire, stated:

> DNA, I think you'll agree, is something that's here to stay and it represents a tremendous forensic instrument for the investigation of crime. I would say perhaps for the benefit of the media more than anybody else but at the voir dire before we finished on June 7th … Kidd was on the stand and he had been on the stand for some days,

> and when he got through his testimony I put a question to him from the Court, what is the future of DNA, where do you see DNA going over the next few years and in the immediate future, because you know it's only six years old now, really, and it's going to make tremendous strides, and he gave an answer which amounted to a chapter in a book. You could just take the answer that he gave. He talked for several minutes on it and it was a precise, articulate exposition on where DNA was heading, and if the media are interested they would be well advised to dig out through the court reporters his answer to that question and any related questions and they have a ready-made article or story, so I think we've all been privileged to take part in this trial where DNA has been sort of – it hasn't been on trial in any sense of the term, necessarily, but we've been exposed to it and it will play, you can be sure, a part in a great many trials in the future, and it will avoid a great many trials in the future. (Charge to the Jury and Verdict, Trial, Vol XXII, 3 November 1991, 135–6)

The judge's commentary reveals how forensic genetics was itself on trial in a way, a judicial review rather than peer review. A common definition of the legal term *voir dire* is a "trial within a trial," which summarizes the *Legere* voir dire hearing on the admissibility of genetic evidence. Furthermore, the genetic evidence in the case was a central point of contention throughout the trial proceedings. And it is here that American legal theorist Robert Cover's (1986, 1623) observation of the defence counsel utilizing any element of the structure of judicial interpretation to try to undermine the prosecution was clearly evident in their use of the Karitiana and Surui data. In effect, the New Brunswick court engaged in a judicial review of forensic genetics and, significantly for our purposes, the court ruled on how racial categories might affect RMPs linking a suspect to a crime scene.

THE 1991 *SCIENCE* ARTICLES

Conflicts over the issues of substructure's potential effects on random match probabilities were discussed in the corridors of conferences and appeared on the pages of prestigious scientific journals. This next phase was alluded to by Shields during his *Legere* voir dire testimony on May 27, 1991. In response to a question whether Hartl had changed his position on substructure, Shields stated, "What I know for sure from having talked to him on the phone is that he and Dick Lewontin are preparing a paper for *Science* and I think it's actually been submitted that makes similar and even stronger criticisms than he made in the *Yee* case, so I would say he has not retracted his position" (Vol XIII, 27 May 1991, 54).

The significance of the Karitiana and Surui data was again debated in this unusual exchange of articles on the reliability of genetic fingerprinting in the 20 December 1991 edition of *Science*. The circumstances of this exchange are unconventional because rather than simply publishing the peer-reviewed article by Richard Lewontin and Daniel Hartl by itself, the journal gave Kenneth Kidd and Ranajit Chakraborty an immediate opportunity to respond to the article in the same issue. A 1992 *Science* article on the DNA Wars describes how "in mid-October Caskey and Kidd, who had both gotten hold of the paper, cornered one of *Science*'s editors at a genetics meeting and urged her not to publish it without a rebuttal. *Science* editor Daniel Koshland agreed, commissioning a rebuttal by Kidd and Ranajit Chakraborty at the University of Texas, which was published in the same issue. Koshland also called Lewontin a few days after the genetics meeting, asking for revisions in the [peer-reviewed and accepted] paper, which was already in galleys" (Leslie Roberts 1992b, 735).

Kidd and Caskey were directly interfering in the peer review process for political and judicial reasons, not scientific ones (Giannelli 1997, 406–7). Kidd was unequivocal in this regard, stating, "I felt publishing the article would create a very serious problem in the legal system, and that that was their intent" (Leslie Roberts 1991b, 1722; Giannelli 1997). Furthermore, Caskey, who was a member of *Science*'s board of reviewing editors, and who licensed a DNA fingerprinting technology he had developed to Cellmark (a private genetic testing company), recommended the rebuttal article, saying that "He was concerned that 'publishing defence testimony in a scientific journal' gives it such weight that courts might reopen, perhaps to overturn convictions obtained on the basis of DNA evidence" (Anderson 1991, 500; Giannelli, 1997). A decade later, Kidd reiterated his reasons for writing the rebuttal. According to Aronson (2007, 139), based on a May 2002 interview with Kenneth Kidd and a May 2003 email, "Kidd said that he was motivated to write the rebuttal because 'some aspects of the Lewontin and Hartl calculations were really out in left field, and in my opinion, the article was clearly motivated by Dick Lewontin's philosophical bent of always being for the defendant and against the state.'"

In their December 1991 *Science* article about DNA fingerprinting, Lewontin and Hartl (1991, 1745) argued random match probability calculations are liable to generate potentially serious errors because ethnic subgroups within major racial categories exhibit genetic differences that have been maintained by endogamy, which is the practice of marrying within clan or ethnic group. Lewontin and Hartl began by first analyzing endogamy effects among Caucasians, then Blacks, and finally Hispanics. In their discussion of Hispanics, they used the Karitiana and Surui data (originally obtained during the *Yee* case) as an example of the problem of substructure.

> Moreover, American Indians differ markedly in allele frequencies of VNTR and other DNA markers. For example, the Karitiana of Brazil are virtually fixed for alleles of VNTR D14S1 in the size range 3 to 4 kilobases, whereas the Surui of Brazil have 62% heterozygosity for this locus (11, 27); and certain three-locus VNTR genotypes differ in frequency by a factor of more than 500 in these populations, even though they are separated by only 420 kilometers (11). Because of the extreme heterogeneity among "Hispanics" and among "native Americans," it is doubtful whether any reference population could be defined that would be reliable in a forensic context. (1749)

Lewontin and Hartl followed this with a segment on "What Is to Be Done," where they state, "Both the theory of population genetics and the available data imply that the probability of a random match of a given VNTR phenotype cannot be estimated reliably for 'Caucasians,' probably not for 'blacks,' and certainly not for 'Hispanics,' if the present method of calculation and the databases presently available are used," and that these affected the estimate calculations, introducing an error "possibly by two or more orders of magnitude" (1749). They argued for a series of measures that would change the assumptions about how racial groupings affect calculations of random matching probabilities. While the Karitiana and Surui occupy only a few sentences, they are nonetheless important rhetorically as Lewontin and Hartl attempt to use the Karitiana and Surui differences in allele frequencies as a type of concluding example that substructure is important, which supports the scientists' advocacy for changes to forensic genetics testing procedures and interpretation.

In their non-peer-reviewed rebuttal, Chakraborty and Kidd criticized Lewontin and Hartl's (LH) use of the Karitiana and Surui data. They first try to disassociate the Karitiana from Hispanics by arguing: "In the discussion of the origin and genetic makeup of the Hispanics, LH refer to the allele frequency differences among the Karitiana and Surui of

Brazil. It is true that Hispanic is a term used for people that have diverse origins. However, the data on Amazon Basin tribes are not the only considerations. Indeed, the Karitiana represent an extreme example of a small isolated inbred population. No Hispanic group in North America is so small or inbred; whatever tribal structure did exist among their Amerindian ancestors has long since been broken down."

Next, they assert that the VNTR technology is able to distinguish even between highly interrelated individuals, such as the Karitiana: "Although the accumulation of gene diversity among the Amazon Basin tribes is significant, the entire Amazonian population appears to have sufficient gene flow among populations, so that an equilibrium has been reached, and these populations behave over the long-term as a unit. Even within the Karitiana sample, which contains many pairs of individuals more closely related than full siblings, there were no two individuals with identical VNTR profiles" (Chakraborty and Kidd 1991, 1738). The above response is similar to Kidd's argument during the *Legere* trial in which he cited Francisco Salzano's study about gene flow in the Amazon, which is also cited in the above section.

However, an 18 April 1992 *New Scientist* article discredited Kidd's assertion that VNTR technology could reliably distinguish between the Karitiana. Entitled "Courtroom Battle over Genetic Fingerprinting," the article illustrates another way the Karitiana data was used to undermine Kidd's credibility as a prosecution expert witness. An evolutionary biologist named Laurence Mueller at the University of California Irvine, who testified in a number of cases in the US, ran a computer analysis of Kidd's data on the 54 Karitiana who were sampled to compare all 1,431 potential pairs.[12] "When the results emerged, Mueller was astonished to find 322 pairs that matched at four loci, 61 matching pairs at five loci, five pairs of Indians that matched at six loci, and two pairs that matched at all seven loci. This was exactly what Kidd had testified did not occur. Kidd was not available for comment, but according to court papers filed in an Ohio case, does not dispute Mueller's analysis" (Charles 1992, 10). Mueller's 1993 paper used Kidd's data and FBI matching criteria to detail one of the Karitiana matches at seven probes, but this match was eventually found to be a handling error, which somewhat diminished the reliability claims of Kidd et al.'s genetic testing procedures, particularly given the small number of cell lines involved. Furthermore, Mueller's paper also showed a six-probe match between samples from a Surui and a Maya.

In the Letters section of a June 1992 issue of the prominent journal *Science*, it was clear that defence counsels were fully aware that Mueller's analysis of Kidd's data found matches between individuals. Citing this analysis, Patrick J. Sullivan of the Hennepin County Public Defender's office in Minneapolis asserted that data on the Karitiana should be considered: "The existence of individuals who match across a number of loci is not unprecedented. Kenneth Kidd's Amerindian (Karitiana) data show a seven-probe match between two individuals, a four-probe match between another two individuals and a number of three-probe matches" (1992, 1743). Sullivan's role is significant because he led another one of the first successful challenges to genetic evidence in the 1989 *State of Minnesota v. Schwartz* murder case. Countering Sullivan's letter in *Science* were two letters, both of which stressed the subjectivity of the Karitiana data as extreme of human genetic difference. The first letter by population geneticists Neil Risch and Bernie Devlin of Yale University (citing personal correspondence with Kenneth Kidd and Lucas Cavalli-Sforza) asserted, "Matching samples from the Karitiana tribe, mentioned by Sullivan, are irrelevant to forensic inference for general populations in the United States. The Karitiana tribe is an extremely inbred group founded by a few individuals. Human leukocyte antigen data show that the Karitiana tribe is an outlier even among the isolated inbred Amazonian tribes" (1992, 1744).

The taxonomy of the Karitiana as outliers is emphasized to differentiate them from the US. In support of their claim, Risch and Devlin in a footnote provide yet another version of Antônio Moraes and his wives as common ancestors shared by the Karitiana: "The Karitiana tribe is dominated by a father, mating with four females (six generations ago), and a son, mating with six females. The females were of unknown relationship to each other or to the males, except that one of the son's wives was also his stepdaughter. Many of the matings involve biological relationships closer than second cousins. In subsequent generations, there was limited mating outside of the family (a total of seven more individuals). The 'population' or family now consists of about 130 individuals in one village" (1992, 1745). In the *Legere* court testimonies, only Antônio Moraes is mentioned but in this iteration his son is added, along with the degree of consanguinity of "second cousins" and the limited external mating. This series of premises lead to a conclusion of sorts as they place population in quotations to emphasize that the Karitiana did not accord with conventional definitions of the term.

In the second letter to *Science*, Bruce Budowle similarly criticized Sullivan's support of Mueller's analysis of the Karitiana data. Budowle, acting in his capacity as a scientist at the "Forensic Science Research and Training Center, Laboratory Division, FBI Academy, Quantico, VA" in the same exchange of letters, wrote a sharp retort to Sullivan: "Finally,

Sullivan makes a puzzling reference to the Karitiana population study. Any reference to the Karitiana should be accompanied by the caveat that the Karitiana are an isolated, inbred kinship living in the Amazon basin of western Brazil. The members of the kinship are much more closely related than family members found in populations in the United States. There is no relevance of data about matching probabilities derived from the Karitiana to that of unrelated individuals in the United States" (Budowle 1992, 1746). In this exchange of letters, Mueller's analysis is used by Sullivan to raise defence criticisms of prosecution estimates of multiple-probe matches and population substructure. In response, those affiliated with the prosecution could no longer say that genetic technologies were definitively able to find differences among the Karitiana. Rather, they shifted their argument to emphasizing the Karitiana as extreme examples of genetic difference and so irrelevant to the US, typical of the rhetorical strategy of dissociation.

Mueller's analysis of Kidd's Karitiana data with its matches at multiple loci eventually appeared in the mainstream press. A 29 November 1992 *Los Angeles Times* article entitled "THE DNA WARS: Touted as an infallible method to identify criminals, DNA matching has mired courts in a vicious battle of expert witness" provided a detailed overview of the ongoing conflicts. Included in the article was a case from a rural Texas town in which David Hicks was accused of raping and killing his grandmother, which raised the issue of false positive matches in "insular" communities:[13]

> Defence attorneys recently sought to embarrass one of the government's most prominent DNA experts, Kenneth Kidd, a professor of genetics, psychiatry and biology at Yale University, for claiming that a four-probe match between two different people was virtually impossible. This claim was refuted – with Kidd's own research on an isolated Amazonian Indian tribe called the Karitiana, in which about a third of the 54 people tested had identical DNA patterns for four different probes (Hicks was sent to Death Row on the basis of two distinct probes.) Once confronted with his own data, Kidd and prosecutors dismissed the inbred Karitiana community as an aberration that does not apply to the much larger gene pool of U.S. populations. (Humes 1992)

Kidd sought to maintain the robustness of the random matching probabilities by dismissing the Karitiana and Surui as destabilizing liminal figures that should be excluded from the courtroom, since they were unlike US populations.

In a 1993 presentation paper, Kidd and his colleagues Judith Kidd and Andrew Pakstis criticize defence legal teams appropriations of the Karitiana data. Implicit in their criticism is that this reduction in genetic diversity was a natural process: "Endogamy prevails among the Karitiana and most marriages are consanguineous in several ways. Because of their linguistic distinctiveness, the general isolation of Amerindian groups and Amazon groups in particular, and the documented family structure that shows this group is really just one family, the Karitiana should represent an extreme example of random genetic drift in humans" (J.R. Kidd, Pakstis, and K.K. Kidd 1993, 24). While endogamy is part of the Karitiana's marital system and this practice has contributed to consanguinity, the Karitiana adopted very close endogamy exemplified in the Antônio Moraes narrative to survive due to their near extinction by settler colonialism. In Kidd et al.'s explanation, the Karitiana are again categorized using the taxonomy of Amerindian as a general population category and Amazonian peoples as a subcategory to explain their consanguinity. Random genetic drift occurs after an external event causes a sudden loss in population, which is called a bottleneck, and the diversity of survivors' alleles will over subsequent generations be reduced and may become fixed at certain loci, as occurred with the Karitiana. The genetic drift is random in the sense that those alleles that disappeared and those that survived were not due to positive selection or other genetic advantage. While trying to defend themselves, Kidd and his colleagues strongly criticize how defence legal teams appropriated their Karitiana data:

> In 1991, we published a study that included several of the VNTR loci used by forensic laboratories as a minor part of the data (Kidd et al. 1991). Mention of those data in court testimony by KKK [Kenneth K. Kidd] prior to publication, the requirement by the court that the data be given to the Defence attorneys, and KKK's technically erroneous statement ("no two Karitiana have the same genotype") (Chakraborty and Kidd, 1991), have propelled the Karitiana into the forefront of the U.S. legal debate on DNA typing. Unfortunately, this prominence is the result of what we consider to be misrepresentation and misuse of the data by defence witnesses. (24)

Kidd and his colleagues sought to maintain the robustness of the random matching probabilities by dismissing the Karitiana and Surui as destabilizing liminal figures that should be excluded from the courtroom since they were unlike US and Canadian populations.

TRACES OF THE KARITIANA AND SURUI AND THE DNA WARS

The Diversity Project announced in 1991 was primarily cast as a potential way of understanding genetic variation in the pursuit of curing disease and understanding human origins. However, given that it coincided with the conflicts of the DNA Wars, numerous mentions of these conflicts in various contexts discuss the potential of the Diversity Project for forensic genetics research.

The DNA Wars' disputes over forensic genetics are referred to directly and indirectly in a number of Diversity Project articles and documents. A September 1991 Human Genome Organization's Committee for Human Genetic Diversity proposal entitled "A Project for the Study of Human Diversity with Special Attention to Vanishing Human Populations" states it will "provide base line information for the investigation of genetically based variation in disease incidence between populations and for identification of individual samples for paternity, forensic and other applications" (HUGO Committee for Human Genetic Diversity 1991, 2). The 1991 public announcement of the Diversity Project in the journal *Genomics* includes an article by Bowcock and Cavalli-Sforza (1991, 491) in which they state, "A study of human variation provides information on our evolutionary history and on fundamental genetic mechanisms such as selection, mutation, and genetic drift and is essential for analyses involving individual DNA fingerprints such as in forensic science." Later in another apparent reference to the disputes over the DNA Wars, they state: "One other important application requiring a study of DNA variation in different human populations is forensic DNA typing. This study requires a knowledge of allele frequencies in different human populations. Evidence of a match between two DNA samples is meaningless if the approximate population frequency of the DNA pattern is not known" (Bowcock and Cavalli-Sforza 1991, 491). A 1992 article by Kidd and his colleagues, written when he was in the midst of the DNA Wars, directly refers to them: "Recent controversies over the value of data for forensic identification purposes, despite public misunderstanding, clearly show the importance of better and more systematic data" (Weiss, Kidd, and Kidd 1992, 81). Kidd and his colleagues Kenneth Weiss and Judith Kidd then reiterate the need to capture disappearing Indigenous peoples: "Some populations are in imminent danger of being submerged into larger gene pools or even passing out of existence. It is important to try to characterize them genetically before the opportunity is lost forever" (81). This article concludes with a call to action similar to the call made by Cavalli-Sforza et al. in 1991: "Once assembled, the *resource* will facilitate many types of studies and thus offer possibilities for increased long-term research support for physical anthropologists as well as archaeologists, ethnographers, linguists, and others. The HGD [Human Genome Diversity] project could be a transforming event for several branches of anthropology. But first, the challenge of organizing this effort must be met" (82, my emphasis). Kidd and his colleagues focus on the technical and organizational challenges that must be overcome to contemplate the bright future their transnational assemblages can achieve in transforming Indigenous peoples' samples into a resource.

1993 US SENATE TESTIMONY

One of the Diversity Project's major proponents, Cavalli-Sforza, did not consider the project to have major forensic potential. A summary of his participation in a discussion during a US Senate hearing states, "The U.S. government is interested in looking at variation within its borders for forensic purposes. That raises concern among some groups. Dr. Cavalli-Sforza pointed out that the Project was not likely to have major value for forensic purposes because most of the populations it will sample are not heavily represented in the U.S. population" (US Senate 1993, 69). In contrast, Kenneth Kidd, Kenneth Weiss, and Judith Kidd's submission to the Senate committee under the heading "Forensic Studies" states, "There is a great deal of interest and controversy these days over the use of DNA data for forensic purposes, another of the traditional areas of interest to anthropologists" (US Senate 1993, 91). After outlining early genetic testing technologies based on variable tandem repeats (VNTRs), they state: "One of the major forensic issues being debated is the degree to which the DNA polymorphisms being used for forensic identification can be interpreted when questions of small isolated populations are relevant to the specific case" (91). This is an implicit reference to the Karitiana's and the Surui's roles in the DNA Wars.

Defence legal teams' appropriation and use of Kidd's data on the Karitiana, Surui, and Maya against Kidd and other prosecution expert witnesses shows the unpredictable agency of scientific data once it is produced and in circulation. The conflict over the significance of the Karitiana and Surui data and the larger issues of racial population substructure helped undermine the credibility of some of the early genetic testing companies including Lifecodes. This conflict likely explains why a Lifecodes researcher coauthored the controversial 1991 paper on the Karitiana, Surui, and Maya with Kidd et al. discussed in chapter 1 (Kidd et al. 1991).

Further issues over the lack of transparency in private companies' often proprietary testing methods and attendant reliability of lab procedures (e.g., mislabeled samples) and disputes over statistical interpretations resulted in further debates and high profile consultations (Aronson 2007, 57–172; Kaye 2010, 68–9; Thompson 1993). These consultations involving the National Academy of Sciences, FBI, and other science and security agencies eventually led to de facto standards and guidelines for forensic genetic testing to ensure its routine acceptance in the courts. These efforts also contributed to the introduction of offender databases, such as the FBI's CODIS system with its thirteen standard short-tandem repeat (STR) loci that was implemented in 1998 (Aronson 2007, 208–9).[14] While the significance of the Karitiana and Surui had been much debated during the peak of the DNA Wars, by 1996, they were relegated to a footnote in a National Research Council report: "Another example that has been mentioned as evidence of multilocus matches is a highly inbred group, the Karitiana, in the Amazon. See Kidd et al. (1993) for a discussion of the lack of relevance of this example to populations in the United States" (National Research Council 1996, 109).

Kenneth Kidd's service to the US government in the development and implementation of forensic genetics was recognized. His Yale University webpage states he received two awards for his efforts: a "'Profile in DNA Courage' award" from the US NIJ in 2000 and an award in "Recognition of Your Efforts During Our Decade of DNA, 1988–1998" from the FBI in 1998 (Yale University 2022).

Meanwhile, the 1991 Kidd et al. data continued to circulate among defence lawyers in the US courts. For example, in a 2002 case *The People of the State of California v. Paul Eugene Robinson*, the defence counsel used Mueller's findings on the Karitiana and Surui. In an admission of an academic violation given under oath, the defence counsel was able to get Chakraborty to admit to accepting Kidd's analysis of the data without actually having done his own analysis despite being a credited as a coauthor of the December 1991 rebuttal of Lewontin and Hartl that appeared in *Science*, including the statement that no two Karitiana matched at all seven loci (The People of the State of California 2002, 2065–6):

> Q. Okay. Did you write the statement or did he?
> A. The statement, I believe, was written by Ken Kidd and since he is the originator of the Karitiana database I agreed with – I relied on his expertise.
> Q. You didn't review the data yourself?

A. No. By then I did not, no.
Q. Okay. Even though you weren't a co-author you were listed as the principal author of this document, correct?
A. Yes. (The People of the State of California 2002, 2065)

Chakraborty's admission in court about authorship is a correction to accounts of the DNA Wars. As well, Chakrabory stated that he found the error several months after the 1991 publication but never published a retraction of it, though he said he discussed Mueller's analysis in a couple of papers.

Later, the prosecution lawyer pointed out that Chakraborty and Kidd's 1991 article with the error about the Karitiana was now twelve years old and the issues around potential substructure in calculation of RMPs had been resolved:

Q. Okay. With respect to the Karitiana tribe, fair to say that that particular tribe does not exist in the United States?
A. As far as I know I have not seen a Karitiana individual in my life. As far as I know there's probably one person who stated that he or she has seen an individual of Karitiana origin somewhere on the East Coast. But in general Karitiana is not a relevant population for U.S. (The People of the State of California 2002, 2104)

The rapid technological advances in biotechnology during the 1990s meant that this line of critique involving the Karitiana and Surui had become irrelevant. This exchange in the 2002 *Robinson* trial was four years after the implementation of the CODIS system with its thirteen STR marker individual identification panel that had replaced the VNTR based systems of the late 1980s and early 1990s (Kaye 2010, 191). These more advanced STR technologies when combined with polymerase chain reaction (PCR) amplification had completely transformed individual identification. In a way then, the technological advances of STR genetic profile displacement of the VNTR meant that defence teams could no longer use the Karitiana and Surui data, thereby finally removing them from Canadian and US courtrooms. This displacement ended the breach between judicial use of forensic genetics in Canada and the US and diversity research involving the Karitiana and Surui. From a governmentality perspective, these Indigenous peoples were now back in their place as resources in human genetic diversity research where they could be used productively, not disruptively (Crosby and Monaghan 2012).

CONCLUSIONS

Critical views of the judiciary in the US and Canada see it as an important system of governance of populations, particularly those racialized populations that are deemed superfluous to capitalist reproduction (Browne 2015; Dorothy Roberts 2013, 155–9). In this way, biopolitics and necropolitics as management of populations function through the judicial system as a form of regulated state-market sovereign violence (Seigel 2018). The judiciary uses racial categories to divide – impose caesurae between – those privileged populations under biopolitical governance (e.g., middle and upper-class White populations) and those disproportionately subjected to necropolitical exposure to death through the violence of law (e.g., African Americans and Indigenous peoples) (Foucault 2003, 254–6; Mbembe 2003, 27–30).

The DNA Wars involved conflicts over RMPs between different racially defined populations. In this way, the sovereign decision-making power exercised by the judiciary helped to shape and coproduce the research and development and implementation of forensic genetic technologies in the US and Canada (cf. Toom 2012, 165). So while the debates in the DNA Wars were couched in terms of the biopolitical collective characteristics of populations, particularly the substructure of geographically defined racial groups like American Caucasians or Latinos, the significance of these with regard to the US and Canadian judiciaries were ruled on by the courts. In the US, the FBI, as the main US federal policing agency, effectively imposed a series of technological standards for genetic testing that were based upon individual liberal legal rights, particularly the clarity of evidence presented against the suspect. The legal procedural rights of the suspects overrode the neoliberal commercial intellectual property rights of the proprietary random match probability processes of Lifecodes, Cellmark, and other forensic genetic testing companies.

Latour (1987) considers how the claims of a scientific assemblage are required to undergo trials or tests of strength against dissenters. The DNA Wars involved legal tests (e.g., the Frye Test) in which the defence teams, including their expert witnesses, attempted to use Kidd et al.'s Karitiana and Surui data against the prosecution. In the *Yee* and *Legere* murder trials, the courts sided with the prosecution and did not agree with the defence that the data was significant enough to undermine the admission of genetic evidence. While the defence was able to make Kidd look sloppy by revealing that he had actually not checked the data for multiprobe matches, Kidd's loss of credibility was not significant enough to sway the court decisions in favour of the defence. The courts in these cases rejected the defence's inductive argument that the Karitiana and Surui findings meant more research on population substructures was needed before forensic genetics could be admitted. Instead the courts in *Yee* and in *Legere* sided with the prosecution argument that the Karitiana in particular were "extreme" and in no way comparable to North American Caucasian populations. The breach between Kidd's diversity research and the court systems was eventually repaired as the courts in effect ruled the Karitiana and Surui findings of multiprobe matches were inapplicable to the Canadian and American Caucasian populations, thereby disassociating these populations and so validating the dominant taxonomy of genetically inscribed spatial-temporal population relationships among Indigenous peoples of the Americas versus European settlers.

The judicial imposition of life sentences or the death penalty (depending on the jurisdiction) as a legal process involves the "ultimate measure of popular sovereignty[;] capital trials are the moment when that sovereignty is most vividly on display" (Sarat 1995, 1105). In this way, the functioning of the legal system itself is an example of sovereign backed violence that represents the authority of the will of the People (1105–9).[15] English political philosopher John Locke (1632–1704) considered the ability to decide on death penalties through the legal system a defining feature of sovereign political power: "Political Power then I take to be a Right of making Laws with Penalties of Death, and consequently all less Penalties, for the Regulating and Preserving of Property, and of employing the force of the Community, in the Execution of such Laws, and in the defence of the Commonwealth from Foreign Injury, and all this only for the Publick Good" (as quoted in Tully 1980, 162). For Locke, sovereign political power involves the legitimate ability to kill through the functioning of the judicial system, something recognized and reiterated in Weber's definition of the state's monopoly on the legitimate use of force.

American legal scholar Robert Cover argued that the law/legal system is fundamentally violent – "Legal interpretive acts signal and occasion the imposition of violence upon others" – and that it strips people of their legal rights and property, incarcerates them, and even kills them (1986, 1601). In the oppositional process of law practised in various jurisdictions, including the US and Canada (even though Canada does not currently have the death penalty), "Capital cases, thus, disclose far more of the structure of judicial interpretation than do other cases. Aiding this disclosure is the agonistic character of law: The defendant and his counsel search for and exploit any part of the structure that may work to their advantage. And they do so to an extreme degree in a matter of life and death" (1986, 1623). The stakes are extremely high in capital cases for they involve the state inflicting civil death and even physical death in the name of the People. Incarceration for extended periods of time is a form of civil death that can, depending on the jurisdiction and the crime, include loss of freedom of movement and association, loss of political rights such as voting or running for office, loss of property, loss of citizenship and deportation after completion of a sentence if someone is an immigrant, and so on. Hence, defence lawyers ideally will use any means available, whether spurious or substantial, to undermine the strength of the prosecution's case.

These attempts to raise doubts were central to how defence experts used reinterpreted Karitiana and Surui data against the prosecution's forensic genetic evidence. In particular, the defence attempted to universalize the Karitiana and Surui as examples of substructure when each of these peoples were sampled just as they were beginning to recover from settler colonial inflicted population collapses. As a result, important points of contention in the DNA Wars were defining not only the characteristics and boundaries of the American and Canadian Caucasian populations for legal purposes, but also their biopolitical taxonomies of racial and ethnic classifications and forms of agency, generally using population categories that followed conventional racial distinctions.

3

Hunting Diverse Humans after 9/11

We're at war in a different kind of war. It's a war that requires us to be on an international manhunt.

US president George Bush

The Bush Administration's launch of the so-called Global War on Terror after the 9/11 attacks ushered in the era of the US security state, Western imperial invasions and long-term counterinsurgencies in Afghanistan and Iraq, and clandestine special operations elsewhere. As Mitchell Dean argues, "The War on Terror thus emerges as a new and frightening kind of just war, a new kind of total war, a new war on annihilation which knows no legal limits because of its vilification of the enemy as an absolute enemy" (2007, 171). While the discourses of international terrorism predate 9/11, it was only after the attacks that the figure of the terrorist as *hostis generis humani*, a total enemy of all Humanity, exemplified by the manhunt of Osama bin Laden, became central to US security state policies (e.g., the Patriot Act) and military doctrine (Thorup 2009, 406–10; Neocleous 2015, 29–30). In the organizing narrative schemas of these security apparatus discourses, the terrorist is semiotically an antisubject that who seeks to kill, disrupt normal life, and create chaos, their unrestrained violence leading to a state of nature-like existence without law or civilization. Moreover, because the terrorist can hide anywhere, the scope of the hunt is necessarily global. Unlike a more conventional uniformed military enemy of international law, they often hide their identity, so anonymity is their greatest strategic asset. The attacks of 9/11 introduced the new enemy antisubject of the terrorist into forensic genetics discourses (e.g., Bruce Budowle's 2015 Human Identification Solutions conference presentation discussed in the introduction).

During the 2000s, scientific assemblages spanning the US, EU, and China began researching the ability to estimate the ancestry and appearance of an unknown individual based on a genetic sample. The figure of the terrorist was not significant in pre-9/11 forensic genetic research discourses. However, following these attacks upon the United States as mass violations of sovereignty, the terrorist immediately became a pivotal enemy antisubject in these discourses' organizing narrative schema. A significant series of changes occurred after the launch of the Global War on Terror in how forensic genetics incorporated Indigenous peoples. In the DNA Wars, prosecution expert witnesses constructed the Karitiana and Surui as genetically "extreme" and the defence counsels that used the data as recklessly trying to undermine the validity and robustness of new forensic genetic testing. In contrast, after 9/11, Kidd and other scientists used these Indigenous peoples' attributes of genetic difference and variation as productive elements in forensic genetic research and development. After 9/11, using new security agency funding and advanced new genetic analysis technologies, Kidd and other scientists transformed these Indigenous peoples into sources of maximal genetic difference that might be encountered by the technologies, thereby improving robustness and providing estimates of geographic/regional ancestral origins. Based on an analysis of mass media coverage, US government funding applications and research reports, scientific research papers, and patent applications, this chapter explains how after 9/11 the Karitiana and Surui and various other Indigenous peoples were transformed into valued and productive sources of genetic difference in forensic genetics. This captured human genetic diversity would now help scientists develop technologies to capture genetically diverse humans.

BIOTECHNOLOGICAL ADVANCES

The escalation of the War on Terror in the early 2000s also coincided with rapid advances in forensic genetics, including the development of so-called next-generation or massively parallel sequencing technologies and mass data computing. These new developments allowed testing for new types of genetic markers, in particular SNPs. Much of the processing and interpretation of genetic data and test results has become increasingly automated and routine, a far cry from the DNA Wars controversies over bins and band interpretation of old VNTR technologies during the early 1990s. In the wake of the 9/11 terrorist attacks, these Indigenous peoples' agency and value within forensic genetics drastically changed when the scope of forensic genetics as a manhunting technology expanded to include ancestry and phenotype. Forensic genetic discourses began to promote the use of racial categories of ancestry and phenotype as intelligence that could be used to guide police, national security investigations, and military operations, particularly counterinsurgency. The US declaration of its so-called Global War on Terror transformed the Karitiana, Surui, and other Indigenous peoples from around the world into valuable sources of forensic genetic difference and as representatives of their respective territories and collectively of their geographic regions. The remainder of this book considers how research networks based in the European Union, the US, and People's Republic of China cooperated in joint forensic genetic research utilizing an organizing narrative schema of revealing the identity of these enemy antisubjects of the criminal and/or the terrorist.

GENETICS AS INTELLIGENCE

Biometrics, including forensic genetic identification, has become central to the disciplinary control of individual bodies and collectively through racial profiling of those deemed as dangerous populations (Aradau and Van Munster 2007; Dillon 2008; Epstein 2007). Racial profiling has become a central strategy of post 9/11 security apparatuses like those of the US, the European Union, and China (Bell 2006, 2013). There is then a seamlessness between this individual identification within a taxonomy of racial categories and attendant levels of scrutiny, as Nikolas Rose (1999, 46) notes, "subjectification is simultaneously individualizing and collectivizing." Anthropologist of science Amade M'charek argues that the development of phenotyping genetic forensic testing would blur the distinctions between individual and collective identities, describing DNA profiles as "collective rather than singular objects" that combine various types of information into a "heterogeneous network of relations" (2008, 521). The effect of this seemingly scientific reification of race in policing is to reinforce well-established racial categories in conjunction with other biometric technologies to create racially coded "risky bodies" subject to disproportionate police surveillance (Epstein 2007, 156; M'charek 2008, 523–4; Rollins 2018, 109–10). Biometrics involve rapidly developing technologies that measured the body in different ways, including retina scanning, fingerprinting, gait recognition, facial recognition, voice printing, and genetic profiling to create a type of composite profile (Epstein 2007, 152–3; Kruger 2013, 246). This ever-growing set of biometric measures involves a process called identity dominance in which more forms of identifying information are viewed as increasing public safety and security (Pugliese 2010). These forms of bodily information are a central focus in the research and development of biometric technologies in which ancestry and phenotype expanded the scope of genetic profiling beyond individual identification (Pugliese 2010, 17–18, 122–3). Under the guise of the Global War on Terror and fighting crime, during the 2000s and 2010s, the security apparatuses of the US, the European Union, and China engaged in extensive international research cooperation to expand the scope of forensic genetic profiling to overt racial categories of ancestry and phenotype (visible appearance).

Forensic genetics has been in part shaped by the larger institutional shifts that led to intelligence-led efforts of police and security agencies in the War on Crime and the War on Terror. In such an operation, forensic genetics can be understood as a series of technologies based on the metonymy of genetics FOR various forms of identity in distinguishing hunted from nonhunted. In the early forms of the technology, VNTR technologies could exonerate or help convict individuals. The early RFLP technologies of the DNA Wars were very costly, taking weeks to complete and requiring high-quality genetic samples due to the long lengths of VNTRs. As a result, they could not process degraded samples, for example old samples or those exposed to the elements. However, over the course of the 1990s, VNTRs were replaced in forensic genetic identification with a new STR-based marker systems that were cheaper, quicker, could be used on degraded samples, and could take advantage of PCR amplification, meaning that much smaller quantities of genetic materials were required. These small sets of STR markers allowed for the development of national DNA databases by various agencies. For example, in 1998, the FBI established the CODIS database using a thirteen STR marker standard, which transformed the investigative usage of genetic profiling by allowing for computerized searching of potential matches among enrolled arrestees, convicted criminals, and unidentified profiles from crime scenes.

Importantly, these databases transformed the use of forensic genetic data into intelligence that could be used routinely to guide investigations, where enrolees whether found guilty or not become "permanent suspects" whose genetic profiles are routinely scanned in criminal investigations (Dorothy Roberts 2011a, 264–88). In her analysis of US forensic genetic databases, Dorothy Roberts (2011b, 569–70) asserts "that the benefits of this genetic surveillance in terms of crime detection, exonerations of innocent inmates, and public safety do not outweigh the unmerited collateral penalty of state invasion of individuals' privacy and the larger harms to democracy. These harms are exacerbated by the disproportionate collection of DNA from African Americans as a result of deep racial biases in law enforcement ... and help to perpetuate a Jim Crow system of criminal justice."

This intelligence function is distinct from the use of genetic materials as evidence that proves guilt or innocence in a court case, though the same sample could readily serve both functions at different stages in a particular case. Forensic geneticists Manfred Kayser and Peter de Knijff (2011, 184) assert that forensic genetic phenotyping "can usefully contribute to criminal investigation, to include or exclude groups for further investigation, fitting the trend in criminal law towards intelligence-led policing." Intelligence can be described as information that informs and guides the conduct of security agency operations, so phenotype and ancestry genetic testing expands the scope from individual identification to providing information to guide investigations (Kruger 2013, 238). This intelligence function is intertwined with the larger development of intelligence-led policing that began in the US and the UK during the 1970s and 1980s. The idea of intelligence-led policing was to utilize information and data to manage risk and thereby more effectively allocate police resources towards crime prevention and intervention rather than only reacting to crime events (James 2013).

Forensic genetic phenotyping and ancestry creates estimates of suspects' potential traits that can then be compared to other forms of available profile data and population data in intelligence to help guide police and national security investigations. Because it takes the form of an estimate, it fits within an overall risk paradigm of security, and, so its proponents argue, if phenotype and ancestry help apprehend suspects more quickly, it can not only punish perpetrators but also prevent them from carrying out future crimes. This involves a process in which "fear is what must be produced and reproduced by governmental agents in order to establish the control, supervision, or enhancement of the social body through multiple mechanisms of measurement, calculation, improvement, and preservation of life" (Debrix and Barder 2009, 400). The fear of terrorist attacks and criminals has been central to the research and development of genetic surveillance technologies that have legitimated categories of race as scientifically valid within security apparatuses and judicial systems that have long been deeply racialized (Dorothy Roberts 2011a, 582–5).

Of particular concern is how racial categories of genetic phenotyping for visible traits and ancestry could be used in security-related sovereign decisions over recognition and suspension of rights, with the technology reducing these decisions to a seemingly technical matter. For example, police could use "DNA dragnets" (mass genetic sampling in criminal investigations) to target particular racialized groups based on a forensic genetic estimate of the probable biogeographic ancestry and visible characteristics of a sample from a crime scene. Already, reified racial categories have been used to guide DNA dragnets in murder and rape investigations by police in countries like the UK and the US (M'charek 2008, 520; Murphy 2015, 175–88; Toom et al. 2016, e2; Will 2003, 136–7). These dragnets have involved the de facto suspension of the rights to not incriminate oneself and protection against unreasonable searches (M'charek 2008, 520; Murphy 2015, 178–80; Toom et al. 2016, e2; Will 2003, 136–7). This is a suspension of rights since refusing to provide samples during a genetic dragnet involves becoming a suspect by virtue of refusing, hence also violating presumptions of innocence until proven guilty (Duster 2006, 296–9; M'charek 2008, 520; M'charek, Toom, and Prainsack 2012, 16; Murphy 2015, 175–88; Toom et al. 2016, e2). These human hunting technologies have begun to guide the suspension of legal rights by police based on racial-type categories of phenotype appearance and biogeographic ancestry. As we will see in this book, the apparent scientific validity of these racial taxonomies has powerful implications because the use of racial taxonomies is even more problematic in authoritarian states like the People's Republic of China, where state security forces are engaged in severe repression of the Uyghurs, Tibetans, and other Indigenous peoples.

RACE IN THE DEFENCE OF THE US

The robustness of US forensic genetic technologies was overwhelmed by the 9/11 terrorist attacks that killed some three thousand people (Biesecker et al. 2005). As a technology involved in guiding sovereign decisions over friend and enemy, the limitations of individual-identification oriented genetic testing were evident in the mass identification efforts that followed the 9/11 attacks. The sheer force of the planes' impact with thousands of gallons of jet fuel exploding and the subsequent collapse of the World Trade Center buildings mixed the fragmented remains of hijackers with those of the victims (Conant 2009). While identification of the victims' remains for return to families was the main priority, an important subnarrative in this return was the problem of the mixing of victims' remains with those of the hijackers and the attendant sovereign requirement to distinguish between friend from enemy in death (Munsterhjelm 2015, 302–3; Conant 2009).[1] A 2009 *Newsweek* article quoted Robert Shaler of the New York City Department of Forensic Biology: "'They did not want the terrorists mixed in with their loved ones,' says Shaler. The families said, 'These people were criminals and did not deserve to be with them.' The families asked for the remains of the hijackers to be separated out and kept someplace else" (Conant 2009). The state's duty to distinguish friend from enemy would ensure that proper reverence was paid to the remains of the victims, not inadvertently those of the hijackers. This duty is based on the metonymy of remains FOR person, which made differentiating the victim or hijacker a central quest that demonstrated the sovereign ability of the US state to differentiate the identity of friend from enemy at the genetic level. Therefore, these efforts involve a distinct sort of intersection of sovereign decision-making with the anatomo-politics of the state surveillance of individual bodies and biopolitical governance of populations.

The post-9/11 identification efforts highlight how state sovereignty claims extend to the dead in distinguishing friend from enemy. According to a 2004 article in *Popular Science* on the possibility of using ancestry as part of racial profiling: "In 1997, when members of the national DNA Advisory Board officially selected the gene markers for DNA evidence matching, they could have included a few markers associated with ancestral geographic origins (European, East Asian, sub-Saharan African) – which are a good indication of race and ethnicity. 'We deliberately chose not to do so,' says Ranajit Chakraborty, director of the University of Cincinnati's Center for Genome Information. Chakraborty says the board skirted the racial-marker issue in part because of the political minefield it represented" (Sachs 2004). However after 9/11, such reluctance to use genetic markers as indicators of race and ethnicity was removed and Chakraborty stated that his lab made use of ancestry markers at the request of a 9/11 victim's family to ensure that a piece of remains was not that of a hijacker (Sachs 2004).

The 9/11 identification effort problems were frequently mentioned in subsequent forensic genetic discourse. For example, launched by the Bush Administration in 2003, the US President's DNA Initiative called Advancing Justice through DNA Technology asserts that new methods should be developed to convict the guilty, exonerate the innocent, and identify the missing. The sacred dead victims of 9/11 are invoked as part of this: "The events of September 11, 2001 demonstrated on a national scale the potential for anguish when the remains of a missing person go unidentified. In the wake of this tragedy, the Department of Justice brought together DNA experts from across the country to develop improved DNA analysis methods identifying the World Trade

Center victims" (US Department of Justice 2003). Among those experts were Bruce Budowle and Kenneth Kidd (Biesecker et al. 2005). In effect, there is a quest outlined in which the United States must never repeat the problems that it had in identifying World Trade Center victims. As such, these victims as sacred dead have crucial forms of agency within this interaction (Munsterhjelm 2015; Anthony Smith 2000). Invoking and mobilizing them creates a type of quest that can then be translated and appropriated by those with the capacity to help (however that is defined).

CO-OPTING DIVERSITY

Kidd Lab exemplifies the post 9/11 US security state co-option and integration of human diversity. Kenneth Kidd's Yale School of Medicine webpage biography mentions his role in the DNA Wars, stating, "Among his other awards, he has been recognized by the U.S. Federal Bureau of Investigation and the National Institute of Justice for his contributions toward acceptance of DNA methodologies in the courts" (Yale University 2022). The biography then mentions his role in the 9/11 and Katrina identification efforts: "He recently served on national advisory panels for DNA identification of victims of the World Trade Center attack and victims of Katrina" (Yale University 2022). These two central sovereign usages of forensic genetics in the courts and in mass casualty events were important in Kidd Lab's post 9/11 US government grants to research and develop individual identification SNP marker panels and ancestry informative SNP marker panels. These indicate how Kidd and other research scientists in forensic genetics function as security professionals who serve their respective security apparatuses but do so in part by exceeding those apparatuses through transnational research assemblages (Bigo 2008).

The shift in the purpose of diversity research is evident in the following list of research grants awarded by the US NSF, which begins with evolutionary and anthropological diversity research but ends with diversity research for forensic genetic purposes.[2]

One of the projects in table 3.1 is entitled the ALFRED database, which is an important example of an assemblage of capture of Indigenous peoples. The Allele Frequency Database (ALFRED) developed by Kidd Lab at Yale University over the last twenty years incorporates data and findings from thousands of articles and many research projects. The original NSF funding grant abstract for the ALFRED states one of its goals was "identifying/developing, testing, and collecting population data on multiple marker loci appropriate for use in the Human Genome

Table 3.1 | Kidd Lab National Science Foundation grants

Award number	*Awarded amount to date*	*National Science Foundation grant title*	*Start date (mm/dd/year)*	*End date (mm/dd/year)*
8619703	$80,000	DNA Markers and Genetic Variation in the Human Species	04/01/1987	09/30/1988
8605114	$258,635	Acquisition of a VAX 8800 Computer	09/15/1987	02/28/1990
8813234	$120,000	DNA Markers and Genetic Variation in the Human Species	07/15/1988	12/31/1990
9208917	$37,500	Anthropology Histories of Selected New World Populations Based on Nuclear DNA Polymorphisms	07/15/1992	12/31/1993
9413152	$50,000	A Pilot Database for Collaborative Studies of Human Genome Diversity	07/15/1994	06/30/1996
9408934	$70,000	A Framework for Genomic Diversity Studies in the New World	07/15/1994	06/30/1997

Award number	*Awarded amount to date*	*National Science Foundation grant title*	*Start date (mm/dd/year)*	*End date (mm/dd/year)*
9632509	$264,000	Use of Haplotypes to Study Population Histories	09/15/1996	08/31/2000
9912028	$315,250	Use of Haplotypes to Study Population Histories	03/15/2000	02/29/2004
0096588	$2,652,755	A Genetic Database for Anthropology	06/15/2001	01/31/2008
0725180	$370,691	ALFRED: A Resource for Research & Teaching Human Evolution	09/01/2007	08/31/2010
0840570	$200,000	ALFRED: Making Very High Throughput Data Accessible	09/15/2008	08/31/2010
0938633	$731,577	ALFRED: Ongoing Growth of an Anthropological Resource	08/01/2009	07/31/2013
1444279	$379,240	Ongoing Development of a Human Population Genetics Resource with Forensic Application	08/15/2014	12/31/2016
Total	$5,529,648			

Diversity Project," while another was "developing a global population database of such loci designed specifically to provide a framework that can be built on by others" (NSF 9632509). The ALFRED was originally intended to assist in the development of the proposed Human Genome Diversity Project databases, though the latter project was shelved due to Indigenous organizing (NSF 9632509).

After 9/11, Kidd Lab began to receive a steady flow of funding from the US Department of Justice for ancestry informative SNP–related (AISNP) research totalling over US$8.5 million, far exceeding NSF funding. Table 3.2 is a summary of grants Kidd Lab received for this research based on a search of the Department of Justice's NIJ funding webpage (http://www.nij.gov/funding/awards/Pages/welcome.aspx). Clearly, this new flow of funding indicates that Kidd Lab's "unique collection" was considered a valuable research asset by US security agencies after 9/11.

The NIJ and NSF funding of US$14,060,842 shows how Kidd Lab's ancestry research embedded its respective project organizing narratives within larger US national security state organizational narrative schema after 9/11. The change in the focus of the NSF funding after 9/11 shows how human diversity research and evolutionary genetics was translated and so co-opted into serving the US security apparatus. This shift is evident because prior to 2001 none of the NSF grant abstracts mentioned forensic usage. In contrast, after 2001 four of the five abstracts mention forensics (0096588, 0725180, 0840570, 1444279).

Another change is the abstracts' inclusion of defensive rhetoric. Perhaps in response to Indigenous organizations' criticism of the Diversity Project (see, for example, Harry 1995), three of the post-2001 NSF abstracts conclude with a statement of how the ALFRED will "prevent" and "combat" racism: "It is also our belief that understanding normal genetic variation is one of the surest ways to prevent racist misuse of genetic data" (0725180). "Additionally, one of the most effective means to combat the misuse of genetic information is to make data regarding genetic variation in our species widely available through facilities like ALFRED" (0840570). "It is also our belief that understanding normal genetic variation is one of the surest ways to prevent misuse of genetic data" (0938633).

Grant abstracts are very short and concise summaries of the research projects' goals and significance. Therefore, the concluding sentence is crucial epideictic rhetoric that seeks to justify and defend the project and persuade the reader about the moral and ethical virtues of the project (Munsterhjelm 2014, 101; Torronen 2000, 84). The above claims are typical expressions of this belief shared by many of the proponents of the Diversity Project (Reardon and TallBear 2012, 234). Such ethical

Table 3.2 | Kidd Lab National Institute of Justice grants

Grant number	*Grant amount*	*National Institute of Justice grant title*	*Fiscal year*
2004-DN-BX-K025 (0)	$824,540	Population Genetics of SNPs for Forensic Purposes	2004
2007-DN-BX-K197 (0)	$680,516	Population Genetics of SNPs for Forensic Purposes	2007
2007-DN-BX-K197 (1)	$471,914	Population Genetics of SNPs for Forensic Purposes (renewal of above grant)	2009
2010-DN-BX-K225 (0)	$1,351,352	Further Development of SNP Panels for Forensics	2010
2010-DN-BX-K226 (0)	$643,771	Developing a Forensic Resource/Reference on Genetics Knowledge Base	2010
2013-DN-BX-K023 (0)	$952,255	High Resolution SNP Panels for Forensic Identification of Ancestry, Family, and Phenotype	2013
2014-DN-BX-K030 (0)	$314,468	Continued Development of FROG-kb: A Forensic Resource/Reference on Genetics Knowledge Base	2014
2015-DN-BX-K023	$1,146,143	Powerful Forensic Markers Optimized for Massively Parallel Sequencing	2015
2016-DN-BX-0162	$1,062,736	Enhancing and Sustaining the ALFRED-FROG-KB Forensic Resource	2016
2018-75-CX-0041	$1,083,499	Better Forensic Markers: Microhaplotypes and Ancestry SNPs	2018
Total	$8,531,194		

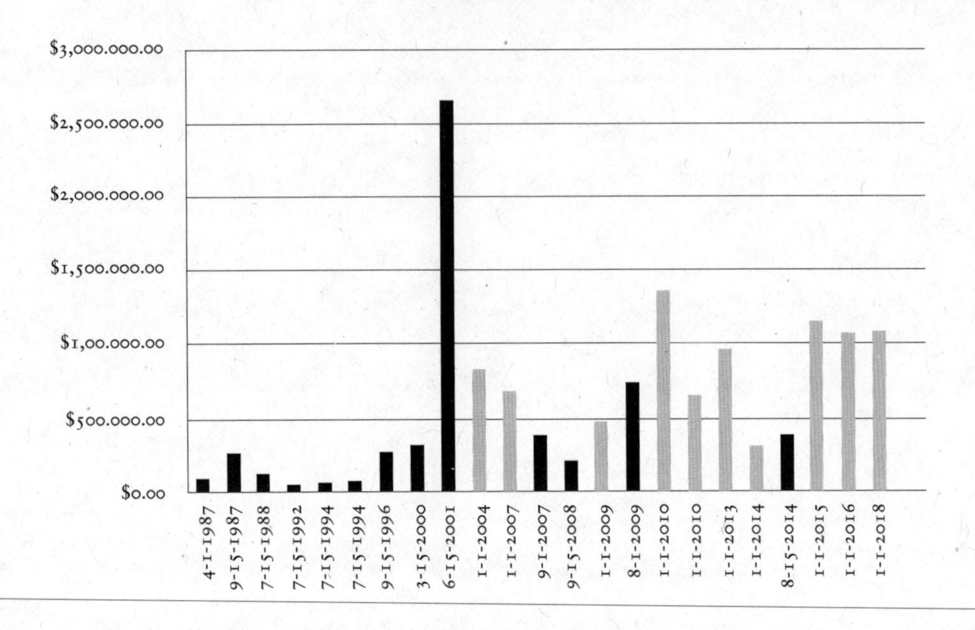

Figure 3.1 | This is a bar chart of Kidd Lab's US National Science Foundation and the US National Institute of Justice grant funding for the years 1987 to 2018.

claims were sufficient for involved ethics review committees and funding agencies, though these claims would come to be sharply questioned in 2019, as we will see in chapter 11.

In figure 3.1, the solid bars are the NSF grants while the patterned ones are NIJ grants. Though this analysis does not include the funding from three US Public Health Service grants (I have not been able to find those abstracts and amounts), even without that data, 2001 is likely still a dividing line in the amounts of funding and the types of research, which shifted the emphasis from diversity and public health to forensic genetics. The US$2.6 million grant in 2001 (0096588) had a 15 June start date – that is, it started pre-9/11 – but its end date was January 2008, meaning most of the research was done after 9/11.[3] In short, after 9/11 NSF funding drops off while NIJ becomes a significant source of funding. As the funding table indicates, human diversity research was co-opted into forensic genetic research and development after 9/11.[4]

Kidd and his colleagues began integrating the ALFRED into their forensic genetic studies of individual identification SNP markers with a listing of forty of the populations and their ALFRED IDs in a 2006 article (Kidd et al. 2006, 22). Kidd and his colleagues used the ALFRED in selecting potential individual identification SNPs: "To determine reasonable screening values we analyzed data we have collected on other projects (in ALFRED and unpublished)" (Kidd et al. 2006, 23). Furthermore, Kidd's first funding report to the NIJ describes how they made use of the ALFRED for tasks such as SNP selection and also for depositing individual identification SNP information in the database (Kidd 2008, 17, 19–21, 37, 41).

Publicly funded and freely available, the ALFRED is a repository that centralizes data from thousands of papers, research projects, and datasets into a readily accessible and searchable format. This Yale-based repository also features a forensic genetics-oriented web interface to use the ALFRED database: "FROG-kb seeks to make allele frequency data for panels of SNPs more useful in a forensic setting. The primary objective of FROG-kb is to provide a web interface that, from a forensic perspective, is useful for teaching and research and can serve as a tool facilitating forensic practice. The underlying data are housed in the already extensively used and referenced ALlele FREquency Database (ALFRED)" (ALFRED 2018). The ALFRED's global human genetic diversity research is co-opted through its translation into forensic genetics research and development. There are a large number of panels for various ancestry and phenotype inference SNP (ALFRED 2019a). ALFRED is an example of how creating centralized bodies of knowledge, such as maps and records, are pivotal in overcoming the problem of "how to act at a distance upon unfamiliar events, places and people? Answer: by somehow bringing home these events, places and people" (Latour 1987, 223). Such bodies of knowledge allow scientists to represent and so gain agency from these distant peoples, a form of governing at a distance (Rose 1999, 49–50). To do so requires research to be *mobile*; *stable*, that is, it does not deteriorate or otherwise change form; and *combinable* with already centralized existing bodies of knowledge (Latour 1987, 223). Figure 3.2 illustrates how the ALFRED has acted as a global repository for the collection of datasets representing hundreds of Indigenous peoples and other populations.[5]

The ALFRED database is a publicly funded part of a larger information commons for biotechnology research and development in which Indigenous peoples are important resources (Faye 2004; Reardon and TallBear 2012, S235). It has a number of AISNP, phenotype,

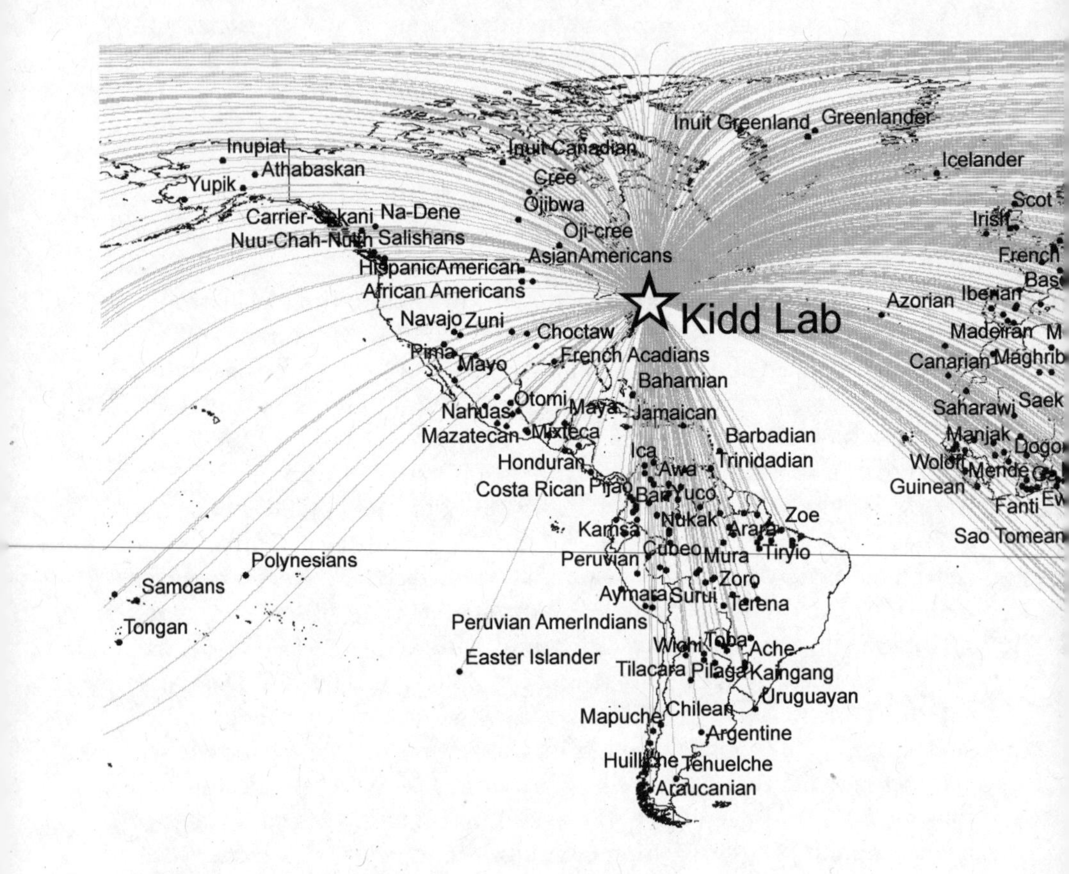

Figure 3.2 | Representing at a distance: This is a flowmap of over seven hundred populations in the ALFRED database run by Kidd Lab at Yale University in New Haven, Connecticut, about ninety miles north of New York City.

and individual identification SNP panels, including controversial twenty-seven SNP and seventy-four SNP ancestry panels developed by Li Caixia and her colleagues at the Ministry of Public Security's Institute of Forensic Science in Beijing, which involved the mass sampling of Uyghurs and other Turkic peoples, as we will see in chapters 5 and 11 (ALFRED 2019a). Also included in the ALFRED are the Kidd

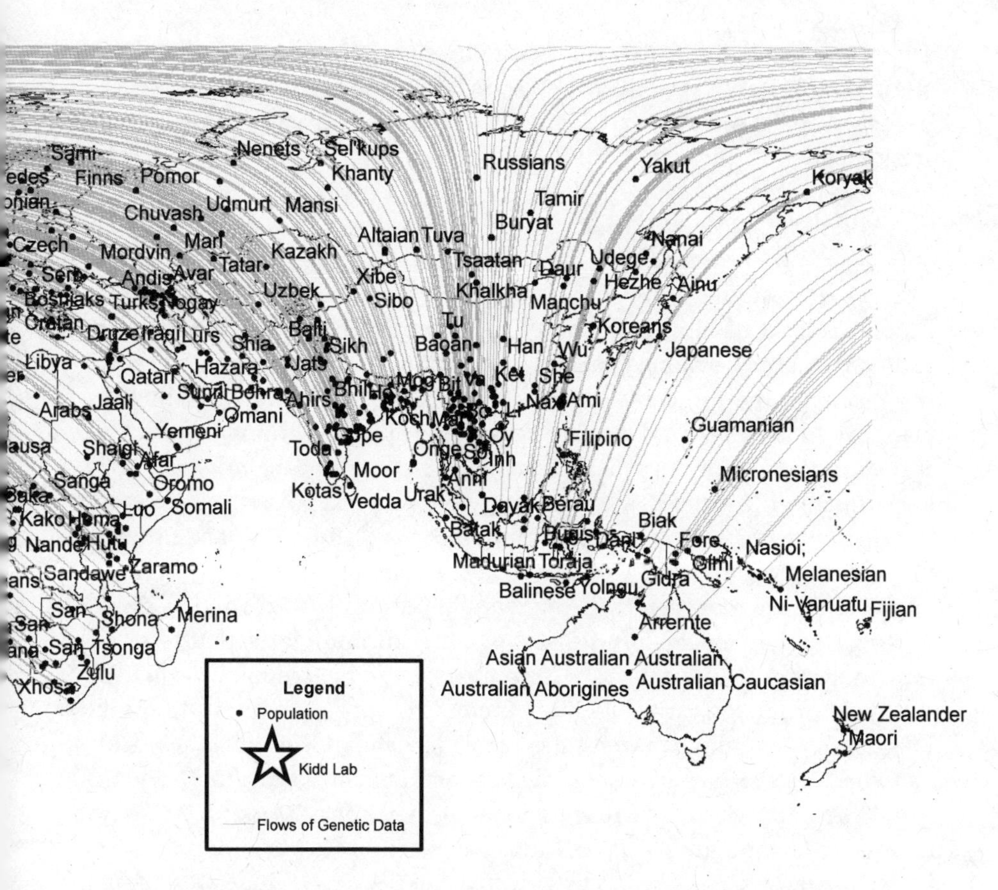

Map by Mark Munsterhjelm 2022. Based on data from ALFRED database and alfredPops.zip (dated October 2013). Locations are approximate.

Lab ancestry inference and individual identification SNP panels. Kidd's funding reports of the 2000s and early 2010s to the NIJ and associated journal articles show the research and development of these panels also involved extensive use of the Lab's collection of Indigenous peoples cell lines and samples.

KIDD'S RESEARCH FUNDING REPORTS

These grants involve Kidd and his colleagues submitting to the NIJ's quest for alternative genetic markers, one that is in turn embedded in the larger quest for security and safety in which the Department of Justice acts on behalf of the American People.

Manipulation Phase

The government acting on behalf of the People and/or Humanity (the People are part of Humanity) needs help against the threat of criminals and/or terrorists or some other manifestation(s) of the state of nature. For example, the original 2003 US President's DNA Initiative defines a mandate to advance DNA technologies "to identify criminals quickly and accurately," "protect the innocent," and "identify the missing" (US Department of Justice 2003). Each of these goals seeks to use technologies to help in sovereign decisions that determine friend and enemy and protect the innocent from the the violence of law. In citing the events of 11 September 2001, it equates identifying missing persons with the sacred dead of 9/11, thereby associating tragic outcomes of criminal activities with those of terrorism; the agency of the antisubject of the criminal is equated with that of the terrorist. Later, the increased scope was detailed on the President's DNA Initiative website, which advocated research on the use of "alternative genetic markers": "While short tandem repeat (STR) typing forms the backbone of current national DNA databases and day-to-day forensic casework, additional DNA marker systems are under exploration in the research arena. These new DNA marker systems include SNPs, Alu-insertion elements, and phenotypic predictors. Alternative genetic markers and assays can potentially provide further information about biological samples under investigation, such as an estimation of ethnic origin, physical characteristics, and skin, hair, or eye color" ("Alternative genetic markers" n.d.). This call to action seeks to expand the scope of forensic genetic testing from individual identification of biological samples to developing technologies that can produce probability-based estimation of an array of racially and ethnically defined characteristics (Kruger 2013). These technologies were now to be shaped by a persuasive ethical and political calculation that the risks of resurrecting once discredited categories of race in forensic genetics were outweighed by their potential ability to improve police investigations and national security applications, a calculation that could be politically defended in the name of protecting the People from violent attacks by terrorists and criminals

Commitment Phase

Scientists and others within the biotechnology time-space accept this quest to help by conducting genetic research on their area of expertise. This acceptance of the quest as a form of submission creates a fiduciary duty that provides a moral authorization since they are serving the macro-actors of Humanity and/or the US People. These scientists commit to advancing these larger quests and situate their respective research projects within them accordingly. On the NIJ webpage list of funded projects is Kenneth Kidd's "Population Genetics of SNPs for Forensic Purposes," which was awarded US$824,540 (UK Research and Innovation n.d.). This 2004 award abstract submits to the quest for alternative genetic markers and assays by stating that its goal is to develop panels of individual identification single nucleotide polymorphisms (IISNPs) and AISNPs. Kidd's collection of cell lines is cited as a major research asset in this quest: "The laboratory of Dr. Kenneth Kidd has samples from over 2000 individuals from 41 populations" (National Institutes of Justice n.d.a).

This shift is evident in how Kidd et al. in their 2008 and 2011 NIJ research grant funding reports and a number of scientific papers cite the problems of identifying the 9/11 victims (Kidd et al. 2008, 17–18; Kidd et al. 2011, 23–4). Each of these reports cover several scientific journal articles and involve an editing together of some of the content of these within an overall narrative of serving the NIJ organizational mandates and the larger call to action to help law enforcement and defend US sovereignty through the development and implementation of forensic genetic technologies:[6]

> We justified our goals of continuing to develop both IISNP and AISNP panels based on *our unique collection* of population samples (Table 4–1), our well-equipped molecular laboratory, our extensive experience in population genetics, and considerable experience testifying during the early use of DNA in forensics. We felt we knew what the Courts would require as scientific support for use of SNP panels and that we were in an ideal position to develop panels meeting those criteria. The necessity for population data for forensic SNPs was especially evident when the need for SNPs in identification of victims on the World Trade Center attacks could not find any with adequate scientific support for use in a multiethnic population. (Kidd 2011, 23–4, my emphasis)

In this way, Kidd and his colleagues situate and embed their research projects within the larger organizing post-9/11 narrative schema of the state security apparatus. By claiming various abilities that qualify them to participate in this quest, they seek to be translated as helpers in the institutions' organizing narratives. Kidd et al. anticipate the potential usage of these technologies in sovereign decisions by the judiciary thereby shaping the research and development processes and showing the importance of the lab in sovereignty and security enforcement as violence work (Bourne, Johnson, and Lisle 2015; Seigel 2018). Of the forty-four populations in his 2011 NIJ report's Table 4–1, seventeen are typically considered Indigenous peoples (Kidd 2011, 26). Together they comprise 794 of the 2,479 cell lines or samples, categorized by "Geographical Region," which included Africa, southwest Asia, northwest Asia, south central Asia, East Asia, northeast Asia, Pacific Islands, North America, and South America (Kidd 2011, 26). Finally, in this passage Kidd brings together his lab's experience as well as his extensive experience as a prosecution expert witness during the DNA Wars.

Kidd and his colleagues used their collection of genetic samples, cell lines, and data to metonymically represent global genetic ancestry: "*Our collection* of population samples also provides a *unique resource* for validating SNPs that can be used in investigations to identify the ethnic ancestry of the individual leaving a DNA sample at a crime scene. As seen in Figure 3–1, SNPs that vary considerably in frequency can carry information on ancestry. *Our populations* provide an excellent global overview of human variation as shown in various publications" (Kidd 2011, 24, my emphasis). This collection is a globally representative resource that will be processed through the various stages of the competence phase and to produce knowledge in the service of Humanity or the US People, including the sacred dead of 9/11.

Competence Phase

This phase occupies most of the narrative and includes subnarratives of all the various steps through which the scientists accomplish their quest to aid Humanity and the US People. These can be summarized as (a) gaining moral and financial authorizations, (b) enrolling donors who represent populations, (c) processing cell lines into various forms of data, (d) statistically analyzing the data using various types of software, and (e) discussion and findings.

(a) The scientists make claims of legal/moral authorizations by gaining institutional ethics approval and funding. There is generally a blanket and uncritically accepted assertion of the scientists having obtained informed consent to use the Indigenous peoples' samples and derived cell lines.

(b) The scientists now utilize cell lines derived from samples taken for other projects long ago to include and so represent those peoples and use them to test potential AISNPs: "In our search for a better panel of AISNPs we have primarily pursued testing candidate AISNPs on our resource of >2500 individuals from >47 populations" (Kidd et al. 2011, 1). In this way, the Indigenous peoples in these assemblages are resources to be processed into further research, which implies a predictable established property relation. Therefore, the scientists do not have to travel to other countries nor do they need to negotiate informed consent with Indigenous peoples, collect and transport samples, and so on (Munsterhjelm 2014). Rather, scientists grow and use these collections of Indigenous peoples' cell lines derived from samples taken long ago to represent and articulate these peoples and their territories within research assemblages (Munsterhjelm 2014, 74; 2015, 296).

(c) This segment is a technical account of how the samples were processed in the time-space of the laboratory or data was downloaded from a database like HapMap.

(d) Statistical analysis of the genetic data is conducted and the findings are explained, summarized with tables, and represented in graphs of various types.

(e) In the discussion, the authors explain the significance of their findings in relation to the field.

In the 2008 and 2011 NIJ funding reports, Kenneth Kidd and his colleagues discuss the significance of their findings and its relevance to forensic genetics. It is here that they elaborate on their reasoning and justification for the inclusion of Indigenous peoples such as the Karitiana and Surui. They discuss how measures of genetic difference between populations called fixation index (Fst) values vary. In particular, Kidd claims that the inclusion of these Indigenous peoples will approximate the maximum possible difference of any other potential populations that might be encountered and still produce reliable robust random match probabilities. In his 2008 report for the NIJ 2004-DN-K025 grant, Kidd discusses the inclusion of a number of Indigenous peoples into a set of forty populations to test candidate individual identification SNPs: "We have deliberately included some isolated populations from various parts of the world as a test of the robustness/generality of the results" (Kidd 2008, 57). However, criticism by reviewers about the inclusion of isolated populations led Kidd and his colleagues to classify nine Indigenous peoples as not "forensically relevant" thereby reducing the original list of forty populations to thirty-one: "Our panel of 40 candidate SNPs meeting those criteria and giving 40-SNP genotype probabilities of $<10^{-16}$ in almost all populations was criticized by some as being too stringent because those studies included several small, isolated groups. Therefore, we re-evaluated our data, as well as other data, after excluding the most isolated populations from consideration, reducing the screening panel from 40 to 31 populations, those most likely to be forensically relevant" (59).

However, in the 2011 NIJ report, Kidd discussed the above 2008 reduction to thirty-one populations and why they eventually rejected it: "We have retained the small isolated populations because we believe they are important for demonstrating the universality of the low match probabilities" (Kidd 2011, 33). Furthermore, Kidd and his colleagues added four additional populations for a total of forty-four and reaffirmed the importance of intentionally including various Indigenous peoples: "We have deliberately included several small isolated and inbred populations from different geographic regions in our studies: Mbuti from Africa, Samaritans from Southwest Asia, Khanty from West Siberia, Nasioi from Melanesia, Ami and Atayal from Taiwan, Surui and Karitiana from the Amazon. While these do show larger match probabilities (Figure 6–6) than the large populations, those probabilities are still $<10^{-15}$. Some of these smaller populations are among the smallest, most isolated in the world making it exceedingly improbable that another small population would be dramatically different" (Kidd 2011, 43–4).

All of these are typically considered Indigenous peoples, and Kidd and his colleagues' research assemblage gain significant forms of agency by being able to represent them.[7] In effect, Kidd is claiming the ability to use these Indigenous peoples as proxies to routinely simulate and approximate potential maximum genetic difference from the norms of the Hardy-Weinberg equilibrium with its assumptions of random mating and infinite population size, the abnormal versus the normal. Therefore, rather than restricting the scope of the technology to those deemed forensically relevant to the US and other large populations (as occurred during the DNA Wars), Kidd and his colleagues argue for a more global scope, stating that even as more small populations are studied, "The 44 populations studied here cover most major regions of the world; the regions not covered are flanked by those that have been studied. We would expect the Fst [fixation index] values to increase as more small, isolated populations are studied for these markers. Even so, the frequencies of the most common genotype and the average probabilities of identity are not likely to greatly exceed the largest seen for the 44 populations that we have studied since we have deliberately included some isolated populations from various parts of the world as a test of the robustness/generality of the results" (Kidd 2011, 57–8).[8]

In this way, by including these populations, they are anticipating such genetic difference and so inoculating or immunizing the panel to be able to deal with it.

In the organizing narrative schema of forensic genetics, this inclusion of these Indigenous peoples acts as a sort of immunization (to use philosopher Roberto Esposito's [2010] terminology) due to their ascribed genetic difference of these assemblages in which they function as proxies for maximal genetic difference likely to be encountered in potential future usage (Esposito 2010; Opitz 2010, 100). Liminality is a concept from anthropology that involves the transformation of subjects through ritualized encounters with extreme or rare states that are outside the domain of normal and routine experience (Turner 1969). This transformative encounter with the liminal abnormal state is woven into Kidd's reports through the emphasis that he places upon the concept of robustness in part through reference to Indigenous peoples. In these NIJ research funding reports and forensic genetic research articles, Kidd and his colleagues narrate carefully controlled encounters with Indigenous peoples whose liminal genetic difference is productively channelled into the creation of new knowledge, an instance of colonial governmentality. This channelling keeps the Karitiana and Surui and other Indigenous peoples productively in their place in the research assemblages and prevents destabilization like that caused by defence teams' appropriation of Kidd's Karitiana, Surui, and Maya data in the DNA Wars of the early 1990s.

Performance Phase

Having demonstrated the necessary competences, they claim to have completed their quest when they restate the contribution they have made to advancing the quest for national security and protecting the US People and so on. For example, in the 2011 NIJ funding report, Kidd restates the advances this project has made towards a panel of IISNPs. While noting some success towards AISNPs, the report states that much work must be done and includes a section on the dissemination of findings, including publications and presentations on the IISNP and AISNP panels, their inclusion within genetic databases, and so on. Kidd et al. reiterate their ownership and resource status of the cell lines, as they state their future research plans for improving the AISNP panel by increasing the number of populations: "We have already tested the best SNPs from several of the multiple ancestry informative panels on the set of 55 populations on which we have *unlimited amounts* of DNA (from *our cell lines*) and we will be bringing more of these SNPs up to 57 populations" (Kidd 2011, 78, my emphasis). Kidd identifies the importance of possessing "unlimited amounts of DNA," which allows him to metonymically represent peoples, and highlights the resource status of Indigenous peoples and other populations as research objects. The co-constitution of the subjectivities of scientists as those who transform the world and objectivities of Indigenous peoples as resources in a property relation is inherent in the assertion of ownership in phrases like "our unique collection of population samples" and "our cell lines," which can be grown to produce "unlimited amounts of DNA." This objectified resource status allows Kidd and his colleagues to metonymically constitute and simulate unlimited encounters between more heterogeneous populations and those they deem as liminal Indigenous peoples. This performance itself involves the exercise of graduated sovereignty and discretionary power within the time-spaces of the assemblage where the scientists routinely utilize many Indigenous peoples' cell lines and data in defiance of contemporary secondary usage and international legal norms requiring ongoing informed consent of involved Indigenous peoples (Munsterhjelm 2014; Ong 2006).

Sanction Phase

Sanction is the reward or punishment depending on the success or failure of the research. Positive sanction, the belief that the researchers have succeeded in their quest, is implicit in a journal publication.[9] In the case of Kidd's research funding reports, that the NIJ published his reports and gave him millions of dollars in further funding indicates a positive sanction.

ROBUSTNESS THROUGH LIMINAL OTHERS

As shown above, an important theme in Kidd's research reports and associated articles is improving the robustness of the technologies through the inclusion of various Indigenous peoples as isolated populations. In May 1990, in his *Legere* voir dire testimony, Kidd provided a brief explanation of robustness in response to a question by the prosecution lawyer:

> Q. You use the term "robust"; what does that mean? Does it have any particular meaning in the science world?
> A. I am using it in the English language sense there that one can have slightly different concentrations than you think you have; slightly

> different pH than you think you have; slightly different buffers; different salt concentrations; and it will still work. (Vol. XI, 14 and 15 May 1991, 174)

Though Kidd was using the concept of robustness to explain how a particular enzyme was used in cutting segments of VNTR, his definition is also applicable to his inclusion of the Karitiana, Surui, and various other Indigenous peoples in his post 9/11 research. This ability to deal with a range of conditions while still producing valid results underpins Kidd's conceptualization of robustness as he utilizes the term *isolated* in reference to various populations no less than twenty times in the 102 page 2011 NIJ report (Kidd et al. 2011). In terms of robustness theory, Kidd's inclusion of these "isolated populations" fits with mathematician Erica Jen's (2005, 1) concept, emerging from her studies of complex adaptive systems, that the "study of robustness focuses on the ability of a system to maintain specified features when subject to assemblages of perturbations either internal or external." It also accords with complexity researchers Jean M. Carlson and John Doyle's (2002, 2539) definition: "By robustness we mean the maintenance of some desired system characteristics despite fluctuations in the behavior of its component parts or its environment." Carlson and Doyle argue that robustness is a direct function of complexity, such that more complex systems are able to deal with a wider range of phenomena and perturbations than simpler systems.

From the perspective of actor network theory, robustness might be considered the ability of a scientific assemblage to withstand what Latour (1987) calls trials of strength, including peer review, attacks by external critics, withdrawal or betrayal by involved actants, and other destabilizations. Therefore, in terms of robustness, Kidd and his colleagues' inclusion of these Indigenous peoples as isolated populations involves constituting them as perturbations that provide sufficient genetic difference and hence complexity within the IISNP panel that it will likely be able to deal with any other potential populations encountered. This transformative encounter with the liminal state is woven into his reports through the emphasis that Kidd places upon the concept of robustness in part through reference to Indigenous peoples.

THE FIFTY-FIVE AISNP PANEL

A major issue with AISNPs is how to strike a balance between robustness and minimizing the size of the panel: "The initial and primary emphasis was on an individual identification panel because the optimization criteria for such a panel were clear. Less clear were the procedures and criteria optimizing an ancestry informative panel and indeed, our progress in that area has focused on developing criteria for optimization and a dataset of candidates from which to select a robust set of AISNPs" (Kidd 2011, 24). In his 2011 NIJ funding report, Kidd states that progress towards an AISNP panel was still tentative and ongoing compared to the IISNPs discussed above. However, in 2012, Kidd et al. presented a panel of fifty-five AISNPs in a poster presentation at a NIJ conference (Kidd et al. 2012).

In his December 2015 NIJ grant report, Kidd explains in more detail the development of this panel of fifty-five AISNPs, framing it as important to next generation forensic technologies' commercialization and adaptation into routine use (Kidd 2015a). In another 2015 NIJ report, he considers the implications of the panel for policy: "No highly differentiating set of AISNPs that is both extensively validated and replicated is currently available in the public domain. Investigators can use commercial ancestry companies, but their markers and statistics are often proprietary and the underlying science unavailable. Some past uses of such a commercial company have been controversial" (Kidd 2015b, 22). Kidd may be referring in part to the legal controversies over the proprietary testing processes of companies like Cellmark and Lifecodes during the DNA Wars discussed in chapter 2. In contrast, Kidd argues forensic labs will be much more likely to use his AISNPs: "Our extensively validated and documented data and their analyses are either published or are being prepared for publication to place them in the public domain" and so are open to scrutiny (Kidd 2015b, 22). In effect, this research has involved processing Indigenous peoples' genetics to establish a validated publicly available resource that can then be integrated into forensic genetic research assemblages, developed into commercial products, and eventually used in routine policing and security applications.[10] Private biotech firms also provided various resources-in-kind like test kits and reagents, funding, and personnel to cooperate in research. This research has been supported by state security agencies and science funding agencies in the form of funding and resources-in-kind, including the cooperation and participation of US government scientists at the famed FBI Labs in Quantico, Virginia. In this research, there is a seamless cooperation between Kidd and other academic researchers and US security agencies and biotechnology companies, typical of the state-market security apparatus involved in violence work, which we will see repeatedly in this book (Seigel 2018).

In 2014, Kidd and his colleagues published an article on the panel of fifty-five AISNPs entitled "Progress toward an Efficient Panel of SNPs for Ancestry Inference" in the prominent journal *Forensic Science*

International: Genetics. This paper on the testing of a panel of fifty-five AISNPs on seventy-three populations had ten coauthors including Kenneth Kidd and other researchers at Yale University, along with Fang Rixun and Manohar R. Furtado of Life Technologies, which was acquired by Thermo Fisher in 2013. A large number of the seventy-three populations are Indigenous peoples, including the Karitiana and Surui: "Our interest in AISNPs is forensics: we wish to identify a small number of SNPs that will be good for identifying the geographic/ethnic origin of an unknown sample. The origin estimated must have a high enough probability of being correct that the SNPs will provide a useful investigative tool. In a forensic context a small number of SNPs can mean lower costs and possibly faster turnaround" (Kidd et al. 2014, 23). This argument for a relatively small number of markers for ancestry inference is part of an effort to select a panel of genetic markers that are low cost and quick to process for use in routine policing and security agency operations, a translation of scientific research into practice. This article is an initial stage in the validation process for these fifty-five markers. Indigenous peoples, including the Karitiana, Surui, Ticuna, Mbuti, Nasioi, Ami, and Atayal, all play an important role as test subjects both statistically and graphically in a heat map diagram "of the clustering of the 73 populations and the 55 AISNPs" (Kidd et al. 2014, 26).

The article concludes with the next step in the validation process, which is increasing the robustness of the panel: "Future tests of the robustness of this panel will require that additional populations be tested for these SNPs to determine how well the panel resolves ancestries for individuals from populations that are in poorly represented biogeographic regions and populations intermediate to the existing 73 population samples. Future improvement in resolution of ancestry among populations poorly differentiated by these 55 AISNPs will require searching for appropriate additional SNPs" (Kidd et al. 2014, 31). This sort of robustness is important to commercialization and adaptation of next generation forensic technologies by police and security agencies, which we will now discuss.

COMMERCIALIZING THE AISNP AND IISNP PANELS

Kidd and his colleagues' publicly funded panel of fifty-five AISNPs and over ninety IISNPs has been integrated into Illumina's new benchtop MiSeq FGx next generation sequencer system that targets the forensic genetics market including police departments and security agencies. Kidd is quoted in a January 2015 *GenomeWeb* report on the launch of the FGx: "'The forensic community, especially when it comes to analyzing data that may enter the court, is very concerned that a methodology, machinery, everything is validated and strongly supported,' he said. Kidd added that he anticipates that other companies and organizations will launch NGS-based forensics platforms and protocols" (Heger 2015). A 2015 Illumina marketing brochure cites Kidd et al. (2014) as the source of the "biogeographic ancestry-informative SNPs" used in its new MiSeq FGx next generation sequencing system (Illumina 2015a, 5, 11). The fifty-five AISNPs and over ninety IISNPs were included in Illumina's US patent application No 61/940,942 entitled "Methods and Compositions for DNA Profiling," which was filed in February 2014 and granted in August 2019 (Stephens et al. 2019, 52–3).

Violence work research and development cooperation between university research institutes and corporate sector and security agencies involves a mutual synergy in the upstream research and midstream development but also in the downstream commercial promotion of these technologies. Illumina has long held a dominant position in the global sequencer market with a 70 to 75 per cent market share and in the early 2010s began to target forensic genetics in its business planning, considering it important to future growth at a time of budgetary cutbacks for US government research funding agencies like the NIH (Petrone 2011). As part of its efforts in the market for forensic genetics, in 2011 Illumina announced a collaboration agreement with Bruce Budowle at the University of North Texas. According to a company press release, Illumina's director of applied markets stated, "We are pleased to announce this strategic agreement with UNT and we look forward to the integration of next-generation sequencing into forensic genetics … Dr. Budowle and his team have demonstrated a unique ability to bring advanced technologies into practical and routine use in crime labs around the world" (Illumina 2011). This comment likely refers in part to Budowle's role in the development of the FBI's forensic genetics systems during the 1980s and 1990s (as discussed in chapter 2). This Illumina press release states that Budowle was going to help its new next generation sequencing efforts by streamlining sample processing and other steps so that they could be used routinely by police. By 2015, these and other development efforts met with success. Illumina issued a press release for the 2015 International Symposium on Human Identification, a major academic conference/trade show, entitled "Towards Validation and Implementation of the MiSeq FGx Forensic Genomic System" for a seminar in which "Dr. Bruce Budowle will cover illuminating validation and applications of the MiSeq FGx Forensic Genomics System from his lab, in the host state of Texas!" (Illumina 2015b).

In scientific method, validation involves experimental demonstration of the predictions of a hypothesis or model or in this case technological system. Budowle and his colleagues tested this US$100,000 benchtop sequencer system in a 2016 paper entitled "Genetic Analysis of the Yavapai Native Americans from West-Central Arizona using the Illumina MiSeq FGx™ Forensic Genomics System" (Wendt et al. 2016). This 2016 article has been part of the larger validation process of the Illumina FGx system on specific populations (England and Harbison 2020). This research is significant in that it was one of the first to test the Illumina MiSeq FGx forensic genetic system, in this case on blood samples taken decades earlier from the Yavapai Indigenous people of central Arizona in the southwest United States. The research project was funded in part by a US$253,315 NIJ grant to the University of California Davis's David G. Smith and Sreetharan Kanthaswamy for a project entitled "The Enhancement of the Native American CODIS STR Database for Use in Forensic Casework," which specifically targets Indigenous peoples in the US (National Institute of Justice n.d.b.). In short, the coproduction of what Seigel (2018) refers to as the state-market of capitalist social relations is evident in how this research study was funded by the Department of Justice, an agency of the US federal government, to help in the validation of a private US company's forensic genetic testing system using an Indigenous people originally sampled decades earlier by the NIH in Phoenix, Arizona. First, I will analyze the article using the semiotic narrative schema, then I will discuss the severe political and ethical problems of using the Yavapai samples and the parallels with Kidd's usage of Indigenous peoples as a form of de facto property in the name of security of the American People.

Manipulation Phase

The manipulation phase is a forensic manhunt narrative schema that cites a shortage of genetic data about differences between Native American peoples as a reason to do research to improve forensic identity testing: "There is a paucity of population genetic data on Native American populations. Since genetic differentiation among Native American tribes is likely to exceed that among subdivisions within other major populations residing in the US, population genetic data for various applications including forensic identity testing would be invaluable" (Wendt et al. 2016, 18). The paper's quest is to improve the ability of security agencies to use these new technologies to estimate the appearance and ancestry of Native Americans.

Commitment Phase

The authors begin by first identifying the population they will use to carry out this quest: "The Yavapai are a Native American tribe with a semi-nomadic and hunter-gatherer history and a geographic distribution that spanned the Verde Valley in west-central Arizona and continued westward to the Colorado River" (Wendt et al. 2016, 18). They explain how another group of researchers had used the system to analyze a range of forensic genetic markers at the same time, including STR's X chromosome and Y chromosome as well as IISNPs and AISNPs. They then outline how their project will contribute to this quest: "Such a genetic marker kit enables simultaneous typing of a large number of forensically relevant markers. Herein, 62 Yavapai Native Americans were typed with the broad collection of target markers in the ForenSeq TM panel and STR and SNP allele frequencies and typical population statistics were generated" (19).

Competence Phase

In the materials and methods section on samples, they first identify that they have received permission to use the Yavapai samples from the University of California Davis's internal review board and cite a case number: ID 430207–2. They add that the analysis of the data was "performed in accordance with both UC Davis and University of North Texas Health Science IRB approval." "DNA was extracted from serum, buffy coat, and blood samples obtained from 64 anonymized Yavapai individuals of self-declared ancestry using the QIAamp DNA Blood Mini Kit (QIAGEN; Redwood City, CA) following the manufacturer's protocol" (one of several instances of Qiagen products in this book). This DNA sample is metonymically identified as Yavapai. The authors then provide more information on the source of the Yavapai samples: "These samples are a subset of those reported by Smith et al. and Monroe et al. that had sufficient DNA quantity to be provided for this study" (Wendt et al. 2016, 19). The 2000 paper by David G. Smith et al. lists samples from some 3,800 Indigenous peoples including 115 Yavapai. The source of the Yavapai (along with 1,295 samples from other Indigenous peoples) is stated as "Peter Bennett of the National Institute of Arthritis and Metabolic Diseases, Phoenix Arizona" (561).

In the next subnarrative of the competence phase called "library preparation," the small quantities of DNA that had been extracted above were amplified. Amplification is based on an important technology developed in the early 1990s called polymerase chain reaction (PCR), which allows for the copying of a small amount of DNA into an endless quantity. Furthermore, during this copying process, the copies are "tagged" with a fluorescent dye. This tagging process allows the copied DNA to be processed and read by the FGx sequencer, hence, the metaphor of "library preparation." During the sequencing, the samples were processed and the genetic sequences of 165 markers read using the Illumina FGx benchtop sequencer. Next, the data was processed using various software including the ForenSeq Universal Analysis Software, which is Illumina's own proprietary genetic analysis software package.

The analysis tested for individual identification STR and SNP markers plus ancestry and phenotype inference SNP markers. The STR markers had a high-level individual differentiation in RMPs (Wendt et al. 2016, 20). The authors note some of the limitations with the current set of Y-STRs. In their analysis of Y-STR markers (passed from father to son) among the Yavapai male subjects, they also did a haplogroup analysis. Y-STR haplogroups are sets of Y-STR alleles (variants) on the Y-chromosome, which are passed from father to son, that are relatively stable (low mutation rate) and tend to be inherited together across generations, so they can be used to categorize ancestral origin: "The majority of Y-STR profiles (17 of the 26) were assigned to haplogroup Q, a branch of which, Q1a2a1a1 (or M3), is the predominant Y-chromosome haplogroup of Native America to which it is exclusive" (Wendt et al. 2016, 21). They also note the occurrence of several other haplogroups in nine other Yavapai men may be "non-Native American admixture" (21). Hence, the Y-STR haplogroup analysis is coded within conventional racial categories. Similarly, the fifty-five AISNPs also classified the subjects using the same categories: "biogeographic ancestry analyses indicated that 21 samples clustered with East Asians, 22 individuals clustered with Admixed Americans, 18 individuals resided between East Asian and Admixed American, and one fell between Admixed American and European" (22). The authors use Y-STR haplogroups to describe this final case of western European paternal descent: "For example, Yavapai #73 fell between these populations. This is a male sample who was predicted with 100% probability to carry an R1b Y-haplogroup, the predominant Y-chromosome haplogroup among western Europeans. Concordance between autosomal aSNPs and the Y-chromosomal haplogroup information suggests that this observation is due to potential admixture" (22). Hence, there is a series of genetic estimates creating a type of composite profile that is both individual and collective in scope in which individual identity genetics are affected by and defined in relation to racial categories (M'charek 2008). This usage and the results point towards a problem of these technologies using genetically essentialist criteria to determine Indigenous peoples' group identity, something that Kim TallBear (2013) has also shown in her studies of genetic tests used in tribal membership assessments. This findings paragraph concludes by stating that phenotype informative SNP profiles for hair and eye colour "were obtained for 59 samples, all of which, not surprisingly, were predicted to have black hair and brown eyes with average probabilities of 0.89 ± 0.086 and 0.99 ± 0.040, respectively" (Wendt et al. 2016, 22).

The conclusion refers to the long-running debates over how racial substructure affects random match probabilities, which began with the DNA Wars of the early 1990s as universal proposition: "While correcting for population substructure has been advocated for assigning random match probabilities, there is still a need to understand the genetic diversity of forensic genetic markers in various populations, particularly for more isolated populations such as Native Americans" (Wendt et al. 2016, 22). The authors assert their paper addresses this gap in the literature with their specific proposition: "This study serves as the first investigation into Yavapai Native American population genetics with respect to forensically-relevant loci as well as the first set of population data reported using the ForenSeq TM DNA Signature Prep Kit" (22). In effect, the authors state that they have enrolled the Yavapai within the circuits of forensic genetics and improved the capacity to identify them and other Indigenous peoples in the region. This validation paper documents how the FGx system testing creates a multifaceted composite profile of different types of genetic data and racially coded interpretations involving individual level identification STRs and SNPs (which sample STR loci) with the group level ancestry and basic phenotype inference SNPs (M'charek 2008).

ETHICAL VIOLATIONS

There is a historically repugnant aspect to testing the Illumina FGX system as a human hunting technology on the Yavapai. An influx of white prospectors followed by settlers in the 1860s involved a sustained series of campaigns of raiding against the Yavapai, and nearby peoples including the Apache, with the goal of displacing them from their lands: "Many of the miners who in the 1860s first disrupted Yavapai life arrived from the exhausted California goldfields, where they had developed a practice of hunting down and killing Native peoples, often for sport, a practice they continued in Yavapai country" (Braatz 2003, 226).

Later, in the early 1870s, the US Army began a sustained systematic conquest against the Yavapai. Under the leadership of General Crook, it mounted a series of winter campaigns that involved a sustained pursuit of different Yavapai bands:

> Sometimes called the "Yavapai Wars," Crook's campaigns were one-sided, murderous onslaughts, carried out by well-armed and organized soldiers against scattered bands of malnourished and poorly armed families; the campaigns were not heroic, romantic, or admirable. Where previous campaigns out of military posts in central Arizona had been sporadic and uncoordinated, Crooks plan for the winter was brutally thorough. He sent out nine self-contained expeditions, each with its own pack train and Indian scouts, to attack camps, shoot down the inhabitants, destroy food stores, and keep the surviving Yavapai and Western Apache on the run until they accepted US authority or succumb to hunger, exposure, and exhaustion. (137–8)

These manhunts included the Skull Cave Massacre of 28 December 1872, in which some seventy-six to one hundred Yavapai were killed by the US Army (with the loss of only one Pima scout) and a raid at Turret Butte in which some fifty or more Yavapai perished (138). The historical violence of the US Cavalry's settler colonial manhunts that imposed US sovereignty over the Yavapai during the 1860s and 1870s are not just a memory because the hierarchies this founding violence imposed continue today.

A research paper by Wendt et al. (2016) illustrates this hierarchy of rights and onerous obligations between settler institutions and Indigenous peoples in which UC Davis scientists' possession of samples taken thirty or more years ago is sufficient for such usage in violation of contemporary norms of informed consent and community consent for secondary usage. Dr David Smith and his colleagues have used the Yavapai as their own resources and objects of exchange since the early 1990s (see, for example, Lorenz and Smith 1994). Dr Smith did not respond to my email requests. I filed an information request (UCDPRA19315) with the University of California Davis over the origins of the samples and the response read, "After a search for responsive records, it has been determined no records exist" (email to author, 7 August 2019). Dr Peter Bennett also stated he has no recollection of ever transferring the Yavapai or other samples to Dr Smith (email to author, 7 August 2017). Bennett has not published any papers on the Yavapai. Budowle, Smith, and their colleagues used the Yavapai samples to test the FGx system, in yet another instance of secondary usage that is far removed from the original sampling, which likely occurred as part of US NIH research in Arizona before the late 1980s.

This forensic genetic research fails several ethical tests. It has no informed consent approval for this secondary usage from the Yavapai, a violation of their sovereignty and rights and international declarations like the UN *Declaration of the Rights of Indigenous Peoples*. The research brings no benefit to them and the genetic data is disseminated without their consent, for example being used in the 2017 paper and entered into the Kidd Lab's ALFRED database (ALFRED 2019g). This unauthorized usage fails the nonmaleficence criteria since testing the Yavapai seeks to aid the US state security agencies in hunting Indigenous peoples. Given the overrepresentation of Indigenous peoples in the US prison industrial complex, the use of the Yavapai samples to refine forensic genetic individual, ancestry, and phenotype identification is ethically deeply problematic in the United States. In Arizona, according to the Prison Policy Initiative (2014), based on an analysis of 2010 US census data, Native Americans were incarcerated in some form of correctional facility at a rate 3.59 times higher than white people, 2,267 per 100,000 people versus 633 per 100,000. What has been termed the *school to prison pipeline* is intertwined with the high levels of impoverishment, social marginalization, and systemic racism, including poor educational provision and police harassment, that govern many Indigenous communities in the US and other settler colonial states. Colonization through repressive policing and mass incarceration has been a fundamental violence work of settler colonialism in governing Indigenous peoples as largely surplus populations (Coulthard 2014; Crosby and Monaghan 2012). This research by Budowle and his colleagues helps not only Illumina but also the larger US settler colonial state's security apparatus.

CONCLUSION

Typically, sovereign decisions are associated with the state acting in the name of the People in deciding whether an individual or entity in a given situation is a friend, potential threat, or enemy. However, recent theorization based on empirical work has shown how sovereign-type decisions over allocation of rights, obligations, and roles are dispersed with state entities delegating such decision-making to nonstate actors like scientists and corporations, or these nonstate entities arrogating these decision-making powers (Munsterhjelm 2014; Ong 2005). Crucially, nonstate actors such as Kenneth Kidd and other scientists in this assemblage acting in concert with corporate interests and state security agencies and funding agencies anticipated potential sovereign decision-making situations. Through their repeated concerns over the robustness of the panel, Kenneth Kidd and his colleagues engaged in a series of performative sovereign decisions that included anticipation of a global scope and liminal encounters with consanguineous populations such that the laboratory research, development, and initial testing seeks to translate and simulate conditions in the field (Bourne, Johnson, and Lisle 2015, 309–11; Epstein 2007, 161). That is, the development of the technology included anticipation of various liminal situations that might be encountered by state security agencies exercising their sovereign function of identifying friend, enemy, or potential threat. This liminal role was once again filled by various Indigenous peoples, including the Karitiana and Surui.

There is within the research assemblages of the fifty-five AISNP panel a complex graduated sovereignty enacted with the research scientists within their respective laboratory time-spaces exchanging Indigenous peoples' cell lines and data as research objects. These assemblages are in turn shaped by Kidd and others by including Indigenous peoples such as the Karitiana and Surui in anticipation of the conditions under which potential sovereign decision-making might occur, including geographically isolated consanguineous populations, which are in turn collectively distinct from more heterogeneous large settler populations.

4

Testing Ancestry SNPs on Uyghurs and Other Turkic Peoples in Xinjiang

KIDD ET AL.'S CALL TO ACTION

Kenneth Kidd coauthored a 2009 paper on the Uyghurs, without any Chinese institutional cooperation, which challenged the findings of two papers (Xu and Li 2008; Xu et al. 2008) coauthored by Xu Shuhua and Jin Li of the Chinese Academy of Sciences (CAS) and Max Planck Society Partner Institute for Computational Biology in Shanghai that Uyghurs were 50 per cent European and 50 per cent East Asian ancestry (we will deal with the CAS and Max Planck Society Partner Institute in detail in later chapters). In their article entitled "Genetic Landscape of Eurasia and 'Admixture' in Uyghurs," Kidd et al. typed sixty-eight ancestry informative markers (AIMs) "on 1766 individuals from 34 populations representing all subdivisions of Eurasia" and estimated that the Uyghurs had "31.2% assignment with the western cluster" (Li Hui et al. 2009, 934). Kidd and his colleagues asserted the PRC researchers lacked enough samples from different populations in the region. They wrote, "Our set of populations contained several population samples for those regions around Uyghurs, such as Kazakhs, Mongols, Tibetan Khams, etc. to the east of Uyghurs and Hazaras, Khanty, etc. to the west. This comprehensive population coverage allowed more reliable estimates of relationships among the populations" (934). The numbers from each of the various populations were still quite small.[1]

Following the July 2009 riots between Uyghur migrants and Han Chinese settlers in the Xinjiang provincial capital of Urumqi, Kidd's expertise on the Uyghurs and other peoples of central Asia seems to have found a ready audience among the Chinese security agencies including the Ministry of Public Security. Kidd began direct cooperation with PRC researchers in 2010–11. The 2009 riots along with various local uprisings and terrorist attacks had an important impact on senior Chinese leaders' political thinking, with an increasingly harsh line that viewed the Uyghurs and ethnic separatism as an existential threat to the PRC regime, one that drew parallels with the breakup of the USSR following increased autonomy under Gorbachev's leadership (Byler 2018, 334–5; Sean Roberts 2020). Some analysts have seen this increasing suspicion and transformation of the Uyghurs into an existential threat to the Chinese state as a form of securitization, something that appears to be reflected in the increasing attention to forensic genetic research targeting Uyghurs by the Chinese Ministry of Security and other Chinese security agencies (Kam and Clarke 2021).

HISTORICAL BACKGROUND

The non-Han peoples under PRC control only account for about 7 or 8 per cent of the total population, but their home territories account for some 60 per cent of the PRC's geographic territory. Xinjiang is over 1.6 million km^2 in area (slightly smaller than Alaska), which is about one-sixth of the PRC.[2] Today, these regions play a significant role in the Chinese economy as resource peripheries that supply oil, gas, uranium, and other minerals, as well as regions for settlement of Han Chinese populations from eastern China and the investment of excess capital and industrial overcapacity, a spatial fix typical of accumulation by dispossession (Clarke 2021, 132; Harvey 2003, 145–52; Sean Roberts 2020, 174–5). Xinjiang has a central strategic role in President Xi Jinping's Belt and Road Initiative ensuring Chinese access to central and south Asia and Russia through major investments in transportation infrastructure like highways and rail, pipelines and natural resources projects, as well as providing markets for PRC goods and services, particularly with the goal of developing the inner provinces (Chacko and Jayasuriya 2018, 97–8; Sean Roberts 2020, 174–5).

While the PRC claims Xinjiang as an ancient part of China, it is not, as indicated by the name Xinjiang, which translates as new territory or new frontier. The region was conquered by the Qing Dynasty in the 1750s, which culminated in the Qing's intentional genocide of the Dzungar Khanate in which some 400,000 to 600,000 people died of disease and famine and massacres in northern Xinjiang's Dzungar Basin (Perdue 2005, 284–91). However, during the 1800s, with the declining power of the Qing Dynasty, the region once again became largely independent. It was then reconquered in the 1870s and made a province of the Qing Empire in 1884. After the Chinese Revolution of 1911, control of the Xinjiang region during the so-called Warlord Era was a complex mix of local political factions, variously aligned with the Chinese Nationalist party under Cheng Kai-shek or the USSR. During this period, influenced by rising Pan-Turkish identity and nationalism, there were two brief republics formed in the region: one in 1933–34 and another backed by the USSR from 1944 to 1949 (Millward 2003, 201–6, 215–30). These republics have since become vital to contemporary East Turkestan nationalism (Brophy 2016). Various authors have shown how concepts of collective identity in the Tarim Basin region tended to be local with particular oases and cities (39–40). Like nationalist movements elsewhere, including Chinese nationalism, its concepts are largely a product of the twentieth century (4–6). In this book, I consider the Uyghurs as an Indigenous people in that they originate from land-based cultures that once had their own political autonomy and independence (Byler 2018; Sean Roberts 2020).

Anthropologist Darren Byler (2018, 2022) argues the PRC's ethnonationalism has predominantly involved forms of ethno-racism. During the 1950s and early 1960s, in the early Maoist phases of the PRC settler colonization of Xinjiang, incorporation of local Uyghur and other Indigenous elites and cadre involved a sort of multiculturalism in which some Han Chinese settlers learned Uyghur. Dominant metaphors were of the Chinese nation as a family in which the Han Chinese were the "big brother" (*dage*) and the Uyghurs the "little brother" (*xiaodi*) involved a hierarchy of dominance and expectation of submission on the part of the Uyghurs and other Indigenous peoples (Byler 2018, 32).

These Maoist revolutionary hierarchies were central to the ideological justification of the 1949–50 invasion and subsequent settler colonization of Xinjiang by the People's Liberation Army, which was the founding violence of the socialist state order, its original expropriation, though it also built on earlier Ching Dynasty conquests (Ince 2018, 889–92). In 1954, Mao founded the Xinjiang Production and Construction Corps (XPCC), known colloquially as the Bingtuan ("soldier corps"). In the 1950s, this paramilitary organization was composed mainly of demobilized soldiers and charged with the mass settlement of Xinjiang, following a practice of establishing military settlements previously used by the Chinese Empire to control newly conquered regions. Most of the expropriation and distribution of the land and its resources and subsequent production have been carried out by this militarized corporate entity. Today it comprises some 2.7 million members (11.9 per cent of Xinjiang's population), over 90 per cent of whom are Han Chinese (Uyghur Human Rights Project 2018, 3). It demonstrates the seamlessness between military and police in the imposition of a capitalist social order in Xinjiang (Neocleous 2013). The Bingtuan involves the absolute fusing of military, police, and state institutions and the corporate sector. It has extensive business interests in Xinjiang and operates as a parallel government structure outside the control of the Xinjiang provincial government. The roles and scope of its operations are well described in a 2014 PRC government paper celebrating the sixtieth anniversary of the Bingtuan and its early pioneering achievements, beneficence, and role in economic development:

> Over the past 60 years, in fulfilling its mission the XPCC [Bingtuan] has adhered to the principle of "not competing for benefits with the local people." The XPCC reclaimed farmland and successively built regimental agricultural and stock raising farms in the Gobi desert to the north and south of the Tianshan Mountains, and in the harsh natural environment of the desolate border areas. The XPCC has gradually established a multi-sector industrial system encompassing food processing, light industry, textiles, iron and steel, coal, building materials, electricity, chemicals, and machinery. It has also achieved significant progress in education, science and technology, culture, health, and other public sectors. By the end of 2013, the XPCC had 176 regiments, 14 divisions, an area of 70,600 square kilometers under its administration, including 1,244,770 hectares of farmland, and a population of 2,701,400, accounting for 11.9% of Xinjiang's total population. (Information Office of the State Council 2014)

The Bingtuan and other programs have driven intensive settler colonization since the 1950s. Today Xinjiang's 20 million people are over 40 per cent Han Chinese, outnumbering the Uyghur who had once been the majority in the region (Byler 2018, 22). Several districts of the Dzungar basin, which makes up most of northern Xinjiang, are majority Han Chinese, including most major industrial centres and the capital of Urumqi. In contrast, the south Tarim Basin region, including the ancient city of Kashgar, remains a Uyghur majority region (Leibold and Deng 2015, 125). Chinese settler colonization has inflicted a severe environmental toll including air and water pollution, extensive nuclear weapons testing, deforestation, and desertification (Alexis-Martin 2019, 157–60; Millward 2007, 312–17).[3]

From the 1980s onwards, the expansion of Deng Xiaoping's economic reforms led to mass Han Chinese settlement programs, transforming Xinjiang into an economic periphery providing natural resources and agricultural products. These programs dominated by Han Chinese included expansion of industrialized agriculture with water intensive crops of cotton (becoming a major source of China's cotton) and then later tomatoes (Byler 2018, 128–9). As well, oil and gas development and other industrial activity expanded (128–9). Settlers often viewed Uyghurs during this time as "backwards" and "troublesome," however, many young Uyghurs began migrating to rapidly growing cities, including Urumqi (Byler 2018, 32; 2021, 108–9).

Increased marginalization by settler colonization over the last thirty years along with grievances over resource disputes, land confiscations, house demolitions, and maltreatment by police and officials has led to localized violence against Chinese government facilities and police (Sean Roberts 2018b, 238–40). Joanne Smith Finley (2019) and Sean Roberts (2018a) have argued convincingly that direct terrorist attacks against Han settlers began only in the 2010s as a result of the crackdown on Uyghur religious and cultural practices during the 2000s. The PRC government intensified repression following the 5 July 2009 Urumqi Riots between Han Chinese settlers and Uyghurs. These communal riots began after peaceful protests by Uyghurs over the murder of two Uyghurs by their Han Chinese coworkers at a factory in southern China were violently broke up by police (Byler 2018, 334–5; Sean Roberts 2020). Over two hundred people were killed during the riots and many more arrested and an unknown number were executed after the riots (Leibold 2020, 46–7; Sean Roberts 2020). However, despite claims by the Chinese government that China has been the victim of various terrorist organizations such as the East Turkestan Islamic Movement, the Chinese have not experienced the kind of insurgency US occupation forces encountered in Iraq or Afghanistan (Sean Roberts 2018a, 2018b). Rather, these attacks and riots have been local and have been met with massive repression, especially given the huge asymmetries between the Chinese occupation forces and those of the Uyghurs, who have no sustained military capability. Nonetheless, the US war on terror and its discourses have been translated into Chinese state security and related sovereignty security discourses. As a result, since the early 2000s China has claimed to be a victim of Islamic terrorist groups and extremism (Smith Finley 2019, 86; Sean Roberts 2018a, 237–8). A 2019 PRC White Paper entitled "The Fight Against Terrorism and Extremism and Human Rights Protection in Xinjiang" defended PRC actions against increasing international criticism: "China's fight against terrorism and extremism is an important part of the same battle being waged by the international community; it is in keeping with the purposes and principles of the United Nations to combat terrorism and safeguard basic human rights" (State Council

Information Office of the People's Republic of China 2019a). The PRC has substituted the antisubject of the terrorist for its old counterrevolutionary and updated the threats to its security apparatus (Byler 2018, 48–9; Rodríguez-Merino 2019). After 2009, the escalating ethno-racism of the PRC increasingly advocated the forceful training of Uyghurs as proper Chinese subjects to protect the sovereignty and future of the Chinese People.

KIDD BEGINS COOPERATION

It is in this context of increasing securitization of the Uyghurs as existential threats in Chinese state security discourses that Kenneth Kidd began research cooperation and exchanges. In 2011, Kidd began his cooperation with a visit at the invitation of the Chinese Ministry of Public Security (Wee 2019). According to a 2015 US NIJ funding report by Kidd: "Invited seminars variously entitled, including 'Better SNPs for Better Forensics,' were presented at five forensic venues in China on three separate invited trips in 2011, 2012, and 2013" (Kidd 2015c, 43). An October 2012 Chinese news story covered how Kidd visited and made a presentation on forensic genetic markers to members of the Public Security Bureaus at China's Southwest University of Political Science and Law in Yubei in Chongqing in Sichuan Province (Wang 2012). The story features a photo of Kidd holding a Yale University banner with a Chinese official at the event. Kidd also made a trip to the Huazhong University of Science and Technology in Wuhan, Hubei Province during early March 2013 where he gave lectures on forensic genetic markers for phenotyping estimation and ancestry inference (Huazhong University 2013).

In October 2015, Bruce Budowle and Kidd both presented papers at the International Conference of Genetics at Xi'an Jiaotong University, organized by BIG and Xi'an Jiaotong University (the latter is an elite C9 League member, sometimes called China's Ivy League). Among the corporate sponsors were the US genetic sequencer makers Illumina and Thermo Fisher (International Conference on Genetics 2015a).[4] Budowle gave a presentation entitled "Molecular markers and advanced methodologies for forensic genetic applications," while Kidd presented on "New forensic markers for the near future." Other presenters included researchers from the Ministry of Public Security's Institute of Forensic Science, Thermo Fisher scientists, and the well-known Chinese-American forensic scientist Henry Lee (International Conference on Genetics 2015b). Ye Jian, who is a deputy director of the Ministry of Public Security Institute of Forensic Science, copresented a paper with her colleague Wang Le entitled "Forensic DNA Databases and Advances of Emerging Forensic DNA Technologies" (International Conference on Genetics 2015b, 2015c).[5] Ye and Bruce Budowle are pictured sitting side by side (second row from the front, towards the right) in the conference group photo (International Conference on Genetics 2015d).

In the 2014 paper by Kidd et al. discussed in the previous chapter, the authors stated that they hoped to expand the number of populations and hence robustness of the ancestry inference SNP panel (Kidd et al. 2014, 31). Kidd was able to realize this goal in part through cooperation with Chinese scientists, including Li Caixia, a senior forensic genetic researcher at the Ministry of Public Security's Institute of Forensic Science in Beijing, who in 2015 spent eleven months at Kidd Lab as a visiting researcher (Wee 2019). This research cooperation led to the publication of "52 additional reference population samples for the 55 AISNP panel" in *Forensic Science International: Genetics* (Pakstis et al. 2015a, 269). The authors define their manipulation phase, stating, "Ancestry inference for a person using a panel of SNPs depends on the variation of frequencies of those SNPs around the world and the amount of reference data available for calculation/comparison." They provide a brief summary of Kidd et al.'s 2014 findings that "seven to eight biogeographic regions could be distinguished using these markers on 3884 individuals from 73 populations" (269). In this opening line of the abstract, the authors define their quest to broaden the global representativeness of the fifty-five ancestry inference SNP panel. Kidd and his colleagues state that one of their reasons for testing the fifty-five ancestry inference SNPs on fifty-two more populations was that these are included in products by the sequencer makers Illumina and Life Technologies (part of Thermo Fisher): "We note that this panel of 55 AISNPs is now implemented for massively parallel sequencing (MPS) in the sequencing products offered by Illumina and by Life Technologies. Since there are now commercial kits using these 55 SNPs for ancestry inference, we have now added the allele frequencies for 52 more population samples for these 55 SNPs to ALFRED and FROG-kb [both run by Kidd Labs], making a more comprehensive reference database available for forensic inferences" (269).

In effect, publicly funded ancestry inference SNP research is being utilized to provide and support the implementation of these emerging new commercial forensic genetic technologies. Furthermore, this ancestry inference SNP project involved developing panels using data or cell lines derived from 7,021 individuals' samples from 125 populations, which includes (by my count) 972 Indigenous individuals representing twenty-five different peoples (Pakstis et al. 2015b). Li Caixia and Wei Yi-liang's lab and Li Hui also contributed genotype data from a total of 244 individuals including Uyghurs, Tajiks, Kazaks, and Kirghiz, which are all predominantly Muslim Turkic Indigenous peoples who live in Xinjiang (those tested by Li Caixia are identified as being from Xinjiang; Li Hui's entries do not). In total, together Li Caixia and Li Hui contributed data based on testing samples from 707 individuals representing seventeen Indigenous and minority peoples in the PRC (Pakstis et al. 2015b, 3). This testing increased the fifty-five ancestry inference SNP panel's coverage of China and Central Asia. Kidd and his colleagues thereby greatly expanded the pool of populations they had access to.

The actual geographic sites of laboratory testing in the 2015 paper reflected the Chinese government's assertion and exercise of sovereign rights over the genetics of the Chinese people, including the requirement that Chinese researchers be directly involved, with Li Caixia not only coauthoring a paper but also being a visiting scholar at Kidd Lab (Wee 2019). "For most new population samples studied collaborators sent DNA samples to Yale and the genotyping was done at Yale using the standard [Thermo Fisher] TaqMan assay system used for the original study," an arrangement that applied to the samples gathered by Turkish Cypriot, Tunisian, and Turkish coauthors (Pakstis 2015a, 270; 2015b, 3). However, the effects of Chinese sovereign legal restrictions on the export of physical genetic samples are evident: "In two cases, the SNP genotypings were carried out in labs in China. Dr. Caixia Li's group in China employed the custom Golden Gate genotyping assay procedure from Illumina, Inc.; Dr. Hui Li used the same TaqMan assays and protocols used at Yale" (2015a, 270). The Chinese samples were processed in PRC labs in accordance with PRC government sovereignty claims, but the use of standardized protocols, including American-made Illumina and Thermo Fisher equipment and assays to do the genotyping, ensured international coordination between the research groups across different national jurisdictions.

The 2015 article adapts a strong epideictic rhetorical conclusion in which it attempts to persuade readers with "unique populations" to join in these growing research efforts: "The ideal forensic ancestry inference resource will consist of a large number of highly informative AISNPs with full data on a large number of population samples representing all regions of the world ... We encourage other researchers to consider adding their unique populations to this growing dataset of population samples which are all tested for the same set of ancestry informative SNPs"

(Pakstis et al. 2015a, 271). This is a call for international cooperation to expand the number of populations and subjects in an assemblage of capture and production that will accumulate and share data and coordinate research.

Kidd restates the need for further cooperation in ancestry informative marker research in conference presentations and publications throughout 2015 and 2016. In his 27 July 2015 presentation at a conference hosted by the Vermont State Department of Public Safety in the US, there is a slide entitled "Conclusion on Ancestry SNPs," in which he argues:

> We do NOT NEED another panel of "continental level" ancestry SNPs
> We DO-NEED more populations tested for the best existing panels of SNPs
> International collaboration is essential if progress is to be made. (Kidd 2015a, emphasis in original)

Kidd and a group of coauthors elaborated on these points in their 2016 paper entitled "Minimal SNP overlap among multiple panels of AIMs argues for more international collaboration" (Soundararajan et al. 2016). The authors argue for the need to choose the "best integrated panel" drawn from the AISNPs that overlap between the many different panels of SNPs currently available "and build toward better (more refined and more robust) resolution of ancestry" (30). Their concluding paragraph breaks typical scientific article genre conventions because it involves a *pending narrative* that seeks to persuade others towards a particular course of action (Torronen 2000; see figure 4.1). They begin their concluding paragraph by stating the limitations of their current situation: "We recognize that expansion of the population coverage is extremely difficult. The Diversity Project panel of roughly 1000 individuals is the *primary population resource* that has been used and both the population distribution and sample sizes make it insufficient" (Soundararajan et al. 2016, 31, my emphasis). They also state that other sources like the 1000 Genomes Project only focus on a few populations and so lack diversity. Having set out these limitations, they next shift to how "progress will require" either:

- Coordinated testing among different research assemblages to "test on their populations the SNPs already identified" (31). However, they note that this first possibility is blocked by different teams having different projects and priorities.
- Researchers to "share DNA among labs to facilitate all populations

> being typed for the same markers" (31). However, this second possibility is blocked by the "small amounts of DNA that many groups possess" combined with losses of "precious samples in the process of sharing" (31). Sharing is also blocked by national governments: "We note, for example, that China, India, and other countries will not allow DNA samples to leave their country, precluding sharing of samples to the detriment of science" (31). Furthermore, "In other cases individual populations will not allow their DNA and genotype data to be shared" (31). "Not allowing" involves noncooperation or opposition that destabilizes attempts by assemblages to enroll and so represent these peoples, which seems to be, at least in part, an implied reference to Indigenous peoples' resistance.

The scientists conclude that these problems can be overcome, but to do so, "Governments and funding agencies need to understand these issues and work to facilitate international collaboration for the benefit of all" (Soundararajan et al. 2016, 31). Implicit here is the demand for positive and negative exceptions for research assemblages that benefiting "all" will require. There is, however, no mention of or appeal to "populations," including Indigenous peoples, to cooperate. Implicit then is the potential for negative exceptions being imposed upon these populations to gain access to genetic samples from them. Kidd et al. implicitly advocate international cooperation that would require restricting or otherwise overriding Indigenous peoples' rights and self-determination. While the conclusions of formal scientific articles typically include suggestions for potential lines of future research, this article is different in that its overt call to action is well beyond the typical discursive boundaries of forensic genetic research articles. In this way, it has some similarities to the 1991 Human Genome Diversity Project call to action (Cavalli-Sforza et al. 1991). Kidd and his colleagues are asking for positive exceptions (a superior set of rights) for these AISNP research assemblages that will allow them to develop more refined coverage of the world's populations for forensic genetics as a sovereignty enforcement technology (Ong 2006, 101). They criticize the exercise of sovereignty by countries like China and India over peoples' genetic materials within their respective boundaries as a type of nationalist exercise of sovereignty that impedes international forensic genetic research cooperation and thereby impedes international security. Progress towards more robust AISNP panels requires scientists coordinate their efforts, which means they need states to grant a new series of positive exceptions to allow more seamless international cooperation such as sharing of samples currently blocked

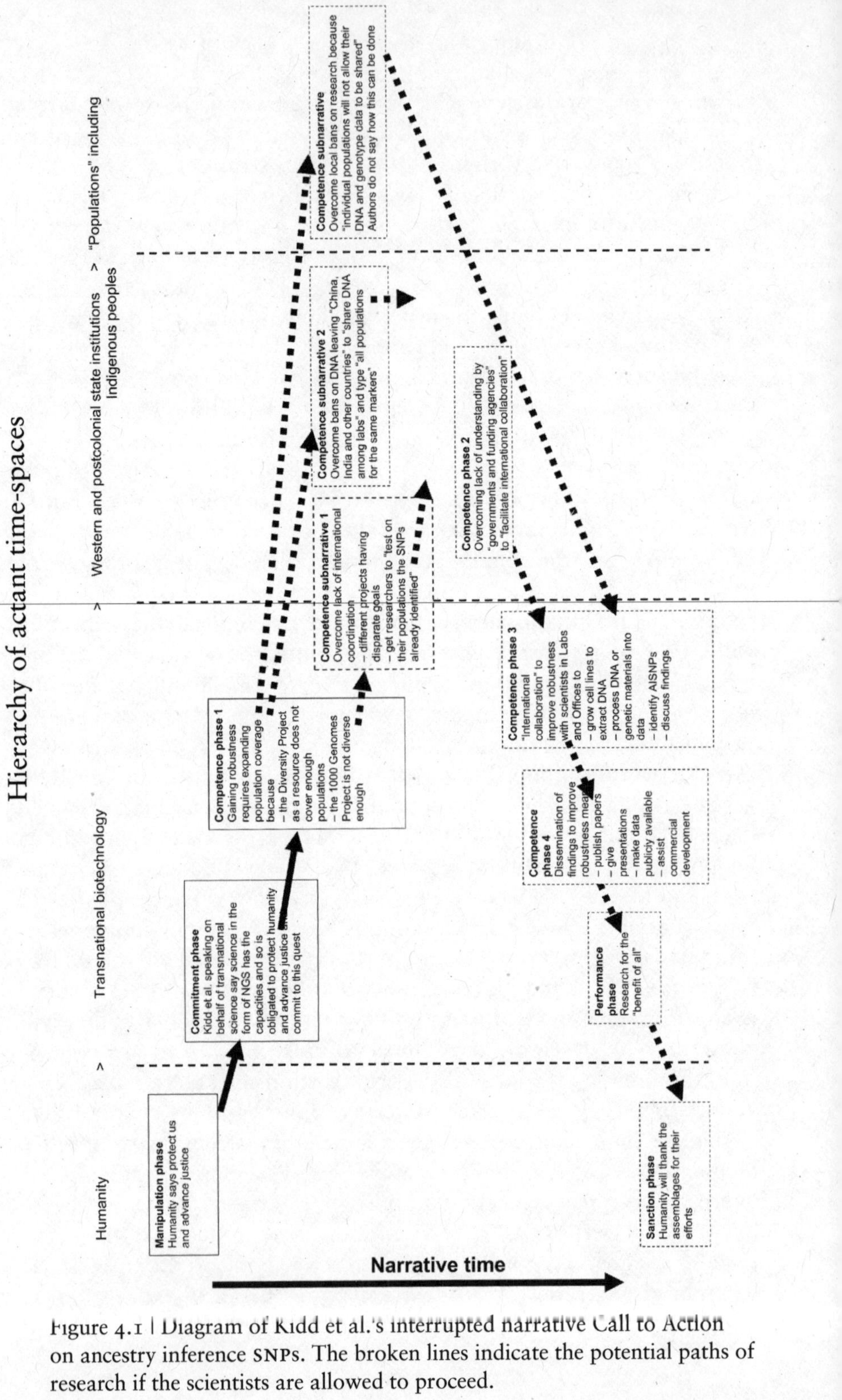

Figure 4.1 | Diagram of Kidd et al.'s interrupted narrative Call to Action on ancestry inference SNPs. The broken lines indicate the potential paths of research if the scientists are allowed to proceed.

by national export bans. As well, there is potentially an implicit call for negative exceptions; that is, reductions in the recognition of rights of Indigenous peoples and minorities whose noncooperation limits the activities of forensic genetic research assemblages (Soundararajan et al. 2016, 31).

Such overt criticisms are absent from a later article in which Kidd Lab again cooperated with Chinese researchers. The January 2017 article "Increasing the Reference Populations for the 55 AISNP Panel: The Need and Benefits," published in the *International Journal of Legal Medicine*, is coauthored by sixteen researchers, including Kenneth Kidd and Bruce Budowle. It includes by my count 1,222 individuals from twenty-seven Indigenous peoples including Kidd's Karitiana, Surui, and Ticuna cell lines and the Yavapai data contributed by Budowle from Wendt et al. (2016). This paper adds a further fourteen populations bringing the total number of individuals tested using the fifty-five AISNPs to 8,105 individuals from 139 populations (see figure 4.2). Most of the additional samples came from Kang Longli of Xizang Minzu (Tibetan Minority) University, who provided eight hundred Chinese samples, including one hundred each from six minority and Indigenous groups: Uyghur, Mongolian, Hui, Tibetan, Miao, and Li. Kang and his colleagues processed these at their lab at Xizang Minzu University in Shaanxi Province.[6] In the test pool of 139 populations totalling 8105 people, there are 1373 people from PRC non-Han minority and Indigenous groups.

This paper continues the earlier 2016 call to action with its title, particularly the "need and benefits," which alludes to the issues of public security. Again, the conclusion is a bit of deliberative rhetoric, though it is more subdued in its request that government agencies cooperate in this research: "We continue to encourage other researchers to consider adding their unique populations to this growing dataset of population samples which are all tested for the same set of AISNPs. Similarly, we encourage others with excellent candidate AISNPs to request that we test them on our population samples" (Pakstis et al. 2017, 916). This is another call to expand the assemblages of capture both in terms of the numbers of populations and subjects and in the coordination of research efforts.

These three papers on the fifty-five AISNP panel are organized around the quest for public security. The shifting assemblages involve scientists at US, European, and Chinese universities, US security agencies, and the Chinese Ministry of Public Security. We see how Kidd initially worked with Thermo Fisher-affiliated researchers and was aided in part by US government grants to develop the fifty-five AISNP panel. This AISNP

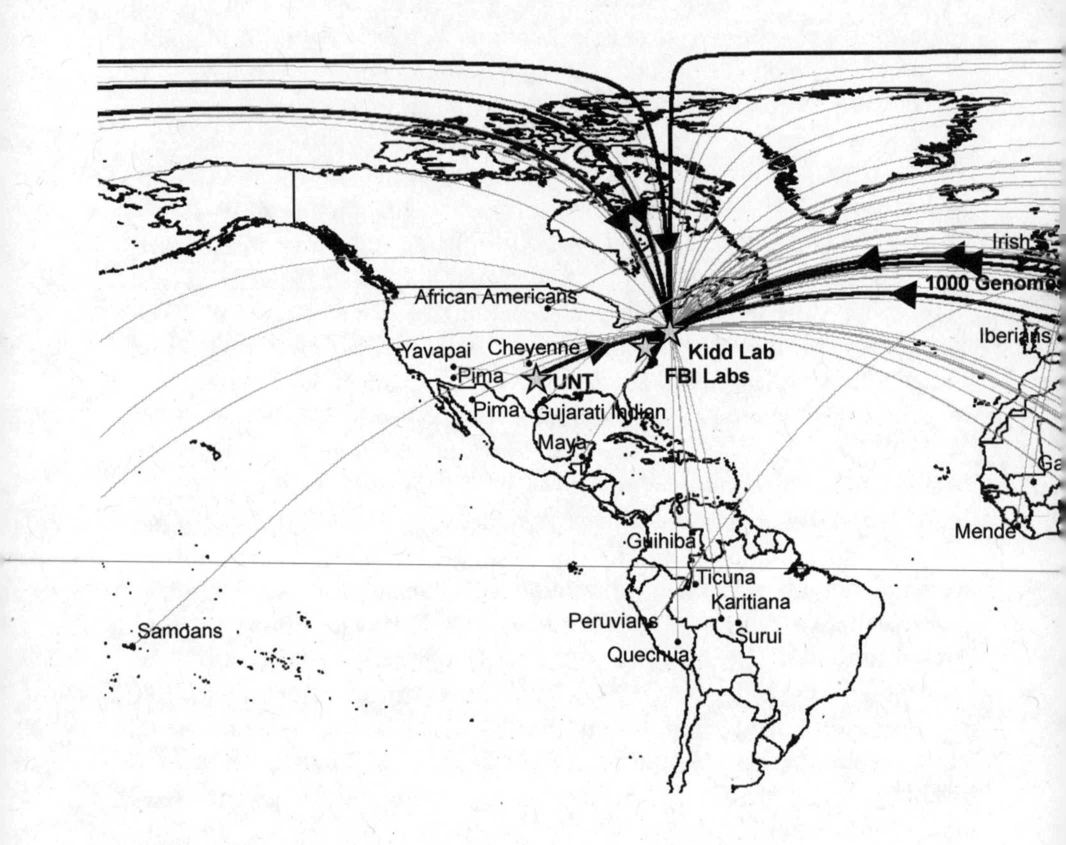

Figure 4.2 | 139 populations tested on Kidd Lab's Ancestry Panel. Map of assemblage of research institutions (stars) in Pakstis et al. (2017) and their respective networks of genetic materials and data accumulated from Indigenous and non-Indigenous peoples.

research in turn rested on research funded by Kidd's US Department of Justice grants going back to 2004 and on grants from the NSF for research on human genetic diversity and evolution dating back to the 1980s and for forensic genetics to the 2000s, and Kidd Lab's collection of Indigenous peoples as resources.

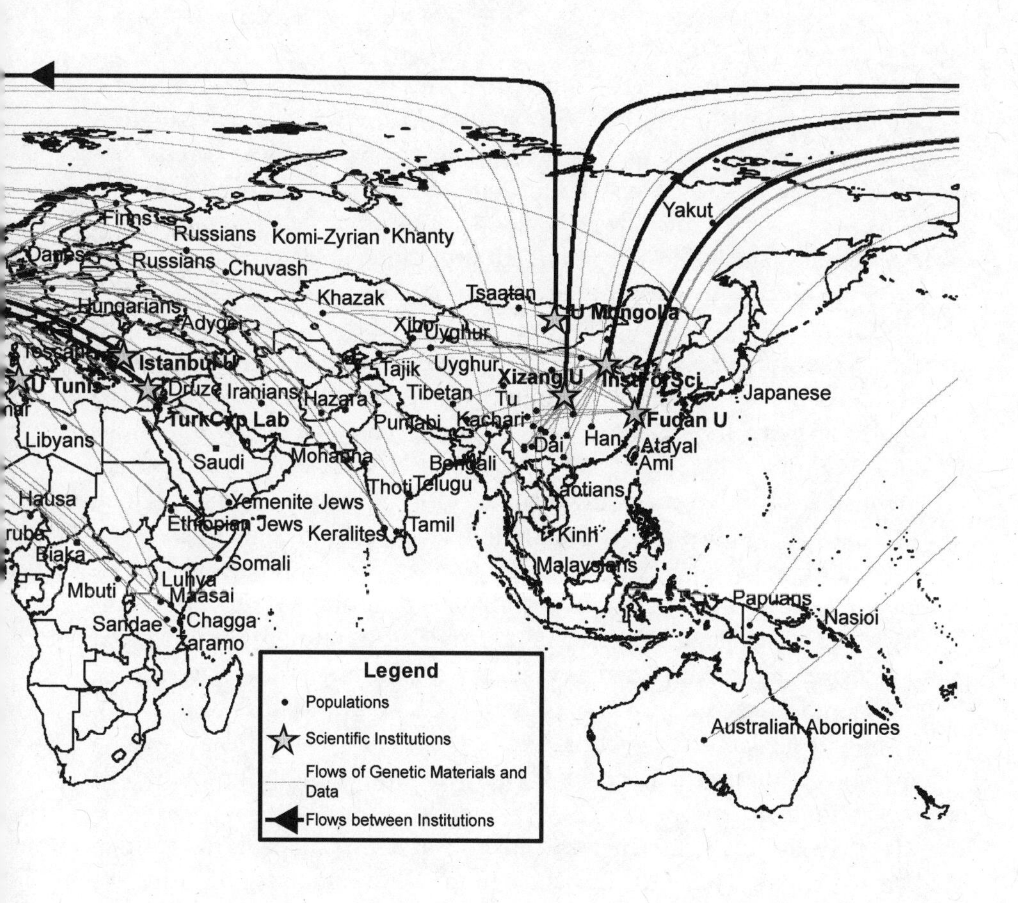

Map by Mark Munsterhjelm 2022. Based on data from Pakstis et al. 2017, Pakstis et al 2015b, and ALFRED. Note: Not all populations are listed.

KIDD AND BUDOWLE'S DNA EXTRACT

Research cooperation takes different forms, including the preceding coauthored papers on the fifty-five AISNP panel, visits, and participation in conferences. A close reading of a number of papers by Li Caixia and her colleagues at the Ministry of Public Security's Institute of Forensic Sciences reveals another form of cooperation in which Kidd and to a lesser extent Budowle each provided large number of samples of DNA extract from cell lines from their respective collections. The first paper, published in 2013, is small in terms of the number of PRC research subjects, for example, only forty Uyghurs (and thirty-eight Tibetans) were included, but these numbers would escalate significantly by 2018 to involve over ten thousand research subjects, including nearly one thousand Uyghurs and some twenty-five hundred samples of DNA extract provided by Kidd and Budowle (Jiang Li et al. 2018a, 2018b). These DNA extract samples would eventually greatly expand the global reach and quality of Li Caixia and her colleagues' efforts, including help developing a twenty-seven marker AISNP panel for routine use by PRC security forces in Xinjiang and elsewhere. As well, attention to the advanced sequencing equipment used reveals how Thermo Fisher Scientific of Waltham, Massachusetts, in the US played a central role as well, which as we will see later in the book would eventually garner this American company, along with Kidd and Budowle, considerable unfavourable international attention.

In January 2014, a paper by Jia Jing and Li Caixia et al. entitled "Developing a Novel Panel of Genome-Wide Ancestry Informative Markers for Bio-geographical Ancestry Estimates" appeared in *Forensic Science International: Genetics*. The authors' epideictic rhetoric in the manipulation phase involves "Inferring the ancestral origin of DNA samples can be helpful in correcting population stratification in disease association studies or guiding crime investigations" (Jia et al. 2014a, 187). Thirty-five SNPs that have high allele frequency differences between the three continental groups "Africans, Europeans and East Asians were selected and validated" (187). The authors used eleven populations from the International HapMap databases and included samples collected by Li Caixia and her colleagues from forty Uyghurs and thirty-eight Tibetans (2014b, 1).[7] A supplemental file source note states, "DNA samples isolated from cell lines provided by Kidd's Laboratory of Yale University" for four populations: forty-four Chagga, an Indigenous people in Tanzania; forty-eight Khanty, an Indigenous people in Western Siberia, forty-seven Danes, and twenty-eight Indians, so Kidd supplied

167 samples from four populations (1).[8] The article's publication history states, "Received 22 April 2013; Received in revised form 6 September 2013; Accepted 9 September 2013," which indicates that these DNA samples extracted from Kidd Lab cell lines were transferred around the time of Kenneth Kidd's trips to China in 2012–13.

These samples were important in developing and expanding the global scope of these Ministry of Public Security researchers. Importantly, this article seems like a trial run based on a very small number of only forty Uyghur samples. As well, the submission and acceptance dates of the paper indicate that it was done before the crackdown that began in late 2013–14. However, the numbers of Uyghurs in subsequent papers would increase dramatically as the Ministry of Public Security began to focus on Xinjiang.

This increase can be seen in a 2016 article (published online in April 2015) entitled "A Single-Tube 27-Plex SNP Assay for Estimating Individual Ancestry and Admixture from Three Continents" (Wei et al. 2016a). Li Caixia, Ye Jian, and researchers from the Xinjiang Public Security Bureau and Bingtuan Public Security Bureau (based in Urumqi) also used DNA extract samples provided by Kidd and Budowle. This article follows the narrative schema of gathering intelligence on unknown DNA samples with its epideictic rhetoric in the manipulation phase stating: "A single-tube multiplex assay of a small set of ancestry-informative markers (AIMs) for effectively estimating individual ancestry and admixture is an ideal forensic tool to trace the population origin of an unknown DNA sample" (27). In response, in the commitment phase, the authors state how they will contribute to this quest: "We present a newly developed 27-plex single nucleotide polymorphism (SNP) panel with highly robust and balanced differential power to perfectly assign individuals to African, European, and East Asian ancestries" (27). Based on a review of the literature on AISNPs, the authors "assembled 968 of the most informative SNPs for inferring continental ancestries and predicting pigment phenotypes" (28). They then analyzed 906 of these markers on HapMap data for eleven populations using a software program called ADMIXTURE and various statistical tests to select a set of twenty-seven markers to differentiate between African, East Asian, and European ancestry.

The primary advantage of a small set of twenty-seven markers is its ability to be tested using widely available capillary electrophoresis sequencers like the Thermo Fisher 3130 XL used by the researchers, which can readily fit within established forensic genetic work flows. In addition to HapMap data, the authors tested the set of twenty-seven markers on a now larger set of samples including sixty-nine Han Chinese, thirty-three Africans sampled in Guangzhou, and five Xinjiang peoples: 58 Tajiks, 238 Uyghurs, 98 Xibe, 100 Kazakhs, and 50 Kirgiz.[9] The article also states Budowle was the source of 257 samples from three populations while Kidd added 187 samples from six populations (Wei et al. 2016a, 29).

The authors' clearly stated research goal is to differentiate Xinjiang minorities: "our 27-plex set has fewer markers but is efficient enough to accomplish the goal of differentiation of three major population groups, providing satisfactory performance in analyzing Eurasians of Northwest of China" (Wei et al. 2016a, 33). For example, the authors claim that their twenty-seven AISNP panel estimated Uyghurs had 49 per cent European and 46 per cent East Asian ancestry, which they state is consistent with the 50 per cent European and 50 per cent East Asian findings of a 2008 paper by Chinese researchers, the same one that Kenneth Kidd and his colleagues responded to (Wei et al. 2016, 34–6; Li Hui et al. 2009; Xu and Jin 2008). The article was received on 29 December 2014, so the samples of 238 Uyghurs used in the article were taken before December 2014, which coincides with the escalating repression of Uyghurs and other Xinjiang peoples that began in late 2013–early 2014 and included President Xi Jinping's May 2014 declaration of the People's War on Terror.

Kidd Lab's large contributions of genetic materials were increasingly combined with larger sets of samples from Uyghurs and other Xinjiang minorities. The mass sampling of Uyghurs is outlined in another paper published in March 2016 by Li Caixia of the Institute of Forensic Science in Beijing and other Ministry of Public Security researchers entitled "Genetic Structure and Differentiation Analysis of a Eurasian Uyghur Population by Use of 27 Continental Ancestry-Informative SNPs," published in the *International Journal of Legal Medicine*. This 2016 paper tested 979 Uyghurs, stating it used the same samples from 238 Uyghurs as the above paper by Wei et al. (2016a) and that the researchers sampled a further 741 Uyghurs from Urumqi for a total of 979 Uyghurs (Wei et al. 2016b, 899).[10] "Overall, 741 Uyghur samples were collected from Ürümqi in Xinjiang (CUX). Samples from all subjects were obtained with written informed consent and self-declared ancestry information. The study was approved by the Ethics Committee of the Institute of Forensic Science, Ministry of Public Security, People's Republic of China, and the experiment was conducted according to the approved guidelines" (898). The scientists genotyped the 979 Uyghurs using a Thermo Fisher 3130xl capillary electrophoresis sequencer with a SNaPshot test kit. The resulting data was processed and interpreted using STRUCTURE software. This software processing involved a reference dataset of twenty-six populations from the Thousand Genomes database and seventeen populations from Wei et al. (2016a), including Budowle's contribution of 257 samples from three populations and Kidd's of 187 samples from six populations as "reference population groups for comparative analysis with the Uyghur population" (2016a, 29; 2016b, 898; 2016c, Supplemental Table S1).[11] The scientists state, "The discrimination effect of the 27 AIMs [ancestry inference markers] was estimated between CUX [Uyghurs] and three Han populations (CHB, CHS, and CHT) in China" (2016b, 899). They found that, "According to match probability results, no individuals of Han ancestry were misclassified (100% accuracy), whereas seven Uyghur individuals were inferred into the Han population with an accuracy of classification of 99.285 % (972/979)" (900). The paper concludes by explaining how the twenty-seven SNP panel was beginning to be implemented into police practice: "This 27-SNP panel is currently used in our lab and [is] going through validation procedures with the cooperation of five local forensic DNA labs from Beijing, Guangdong, Shaanxi, Xinjiang, and Yunnan" (902).

Validation, the next major stage in putting the twenty-seven-plex AISNP panels into security agency practice, is discussed in a January 2017 paper in the Chinese-language journal *Hereditas* (Beijing). Entitled "Optimization and Validation of Analysis Method Based on 27-Plex SNP Panel for Ancestry Inference," the paper was written by Institute of Forensic Science scientists including Jiang Li, Sun Qifan, Zhao Wenting, and Li Caixia. The paper makes repeated use of the term *rén zhǒng* (race) and begins by defining its significance in genetic terms:

> Race refers to people with certain common genetic characteristics that are different from other people. These common genetic constitutional characteristics are gradually formed by the group in a certain area during the long physical evolution and cultural development process, and are the result of the group's long-term adaptation to the natural environment. The triad is a relatively common ethnographic classification method in anthropology: East Asian yellow race (Mongolian race), European white race (European race) and African black race (Negro race).[12] (Jiang, Sun et al. 2017, 167; **种族是指具有区别于其他人群的某些共同遗传体质特征的人群。这些共同的遗传体质特征是群体在一定的地域内，在漫长的体质进化和文化发展过程中逐渐形成的，是群体对自然环境长期适应的结果。三分法是人类学中比较常见的人种分类方法：东亚黄种人(蒙古人种)、欧洲白种人(欧罗巴人种)和非洲黑种人(尼格罗人种)。**)

This introduction sentence functions as a type of epideictic rhetoric that defines the overall quest in the commitment phase in which the scientists say they will build on their previous research, which successfully used four racial categories to classify individuals: East Asian, European, African, and mixed Eurasian. "Based on previous research results, this study optimized and validated an algorithm to infer the ethnic origin of an unknown individual" (基于前期的研究成果，本研究针对未知个体的种族来源推断算法进行了优化及验证)(Jiang, Sun et al. 2017, 167).

The competence phase begins with the materials and methods section where the authors describe the "basic reference database" and the "test sample population information" in which they divide forty-one populations into four categories of East Asian, European, African, and mixed. They use DNA extract from four populations (Danish, Chagga, Japanese, and Korean) provided by Kidd Labs totalling 139 individuals and sixty-six Caucasians provided by Bruce Budowle (Jiang, Sun et al. 2017, 168). They also use genetic data for twenty-two populations from the 1000 Genomes Project database and one from the HapMap project. The Institute of Forensic Science used venous blood samples for twelve populations totalling 1,634 individuals, including twenty-five Africans from Guangdong, forty-nine Tibetans, and forty-nine Yi (who are classified as East Asian). Mixed Eurasian groups from the PRC are made up of 939 Uyghurs (divided into a discovery group of 227 and test group of 712), 96 Tajiks, 86 Kazakhs, and 43 Kirgiz. As well, other mixed Eurasian groups include Indians, Sri Lankan Tamil, Pakistani Punjabi, and Bangladeshi from the 1000 Genomes Project. The 939 Uyghurs referred to as CUX are the same as in Wei et al. (2016b). They make up nearly a quarter of the 3942 individuals in the study, which again places a disproportionate emphasis on Uyghurs. The paper claims that "All sample subjects signed an informed consent form. This study has passed the ethics review of the Ethics Committee of the Material Evidence Appraisal Center of the Ministry of Public Security" (Jiang, Sun et al. 2017, 167; 所有样本对象均签署知情同意书。本研究已通过公安部物证鉴定中心伦理委员会的伦理审查。).

They then utilized the same twenty-seven-plex SNP typing amplification and detection from Wei et al. (2016a, 2016b) in conjunction with QIAamp DNA extraction kits by Qiagen (a Dutch company) and an Eppendorf Mastercycler thermal cycler (a Thermo Fisher brand) for PCR amplification. Capillary electrophoresis was done using an Applied Biosystems 3130 XL genetic analyzer (a Thermo Fisher brand) (Jiang, Sun et al. 2017, 167). The authors then calculated the Population Inference Probability and likelihood ratio (possibility of true positive divided by the possibility of false positive) with Institute of Forensic Science developed software called Forensic Intelligence used for ethnicity inference (167–9).

The thirty-three groups were processed using the STRUCTURE software clustering program (Jiang, Sun et al. 2017, 169). They were visually represented in part through a bar chart coded in red for Africa, green for Europe, and blue for East Asian, though the top of the bar chart is broken up into four categories: African, European, mixed Eurasian, and East Asian (169). Four samples, one for each racial/geographic category, were chosen at random and tested using the model and their respective proportion estimated. The Uyghur test subject had a STRUCTURE estimation of 48 per cent East Asian and 50 per cent European and 2 per cent African components (170). The scientists also completed population tests using a large number of samples in which the East Asian samples were inferred at 99.19 per cent accuracy, European samples at 78.79 per cent, and African samples at 99.08 per cent, while the 712 Uyghurs in the test group were classed unambiguously as mixed at 85.67 per cent but did not exclude 13.62 per cent, and the error rate was 0.70 per cent (four misclassified as East Asia, one as European) (171).[13]

The paper concludes with the claim that the twenty-seven-plex SNP panel can "differentiate between the three major populations of Europe, East Asia, Africa and Eurasian mixed populations and can provide ancestry information for samples of unknown origin" (总体而言，27-plex SNP 种族推断体系能够实现欧、东亚、非三大人群及欧亚混合人群的区分，可为未知来源样本提供祖先信息) and that it can be easily adapted into genetics labs or forensic DNA labs using existing established technologies including PCR amplification and capillary electrophoresis (Jiang, Sun et al. 2017, 172).

The twenty-seven-plex panel of ancestry markers would next benefit from the large number of DNA extract samples that Li Caixia obtained from Kidd Lab following her 2015 stay there as a visiting scientist (Wee 2019). In a May 2018 letter to the editor published in *Forensic Science International: Genetics* entitled "Global Analysis of Population Stratification Using a Smart Panel of 27 Continental Ancestry-Informative SNPs," Jiang et al. return to the twenty-seven-marker panel (Jiang Li et al. 2018a). However, in this short article, they were able to extend the global scope of the testing of the twenty-seven-plex AISNP panel to 10,350 individuals from 110 populations, thanks in large part to Kenneth Kidd's generosity. Its manipulation phase's quest claims that population stratification, the tendency of regional populations to have higher levels of particular alleles and lower levels of others, is a "serious problem that can lead to biased or spurious results in population-based genotype-phenotype association studies" (Jiang Li et al. 2018a, e10). In the commitment phase, they cite their earlier development of the twenty-seven-plex panel, "to analyze three major continental components, namely African, European, and East Asian, for use in forensic individual ancestry inference" (e10). In the competence phase's methodology section, they gathered data from the 1000 Genomes Project and the HapMap phase 3 database along with data from their earlier studies and genotyped an additional 5031 samples representing thirty-two populations from the National Infrastructure of Chinese Genetic Resources, 196 DNA extract samples representing three populations from Bruce Budowle, and 2,266 DNA extract samples representing forty-six populations from Kidd Lab, "yielding a final dataset of 27 SNPs in 10,350 individuals from 110 populations" (2018a, e10; 2018b). The authors did a series of statistical tests called principal components to test the panel (2018a, e10). The authors include a map of the world in which they divide up the 110 populations into fourteen regions such as West Africa, North Europe, East Asia, and Central Asia. They then briefly discuss how to improve the panel, such as adding a couple of SNPs to improve its ability to differentiate between East Asians and Indigenous Americans. In their performance phase, the authors conclude, "The 27 AISNPs panel is highly stratified among four continental ancestries: African, European, East Asian, and Indigenous American. These 27 autosomal SNPs achieve high-resolution performance in distinguishing ancestries and admixture estimates. The reference database, gathered from over 10,000 individuals, provides integrated and confident genetic population backgrounds in the global distribution" (e12). As they state in the abstract, "we propose extensive usage in biomedical studies and forensics."

In the acknowledgments section, their sources of funding include the Ministry of Public Security. They also recognize Kidd's contribution: "The authors would like to thank Kenneth K. Kidd (Yale University) for providing population-based samples" (Jiang Li et al. 2018a, e12). Figure 4.3 is based on the population data spreadsheet from the Jiang Li et al. (2018a) paper to illustrate how this cooperation with Chinese Ministry of Public Security researchers using genetic materials (DNA extract) given to them by Kidd Lab and former FBI Labs scientist Bruce Budowle, along with data from the 1000 Genomes Project, has expanded global coverage. Through these efforts, this team, which includes scientists from the Ministry of Public Security, Bingtuan Public Security Bureau, and People's Liberation Army, constructed a 110-population reference database covering 10,350 people classed into fourteen regions of the world to test a panel

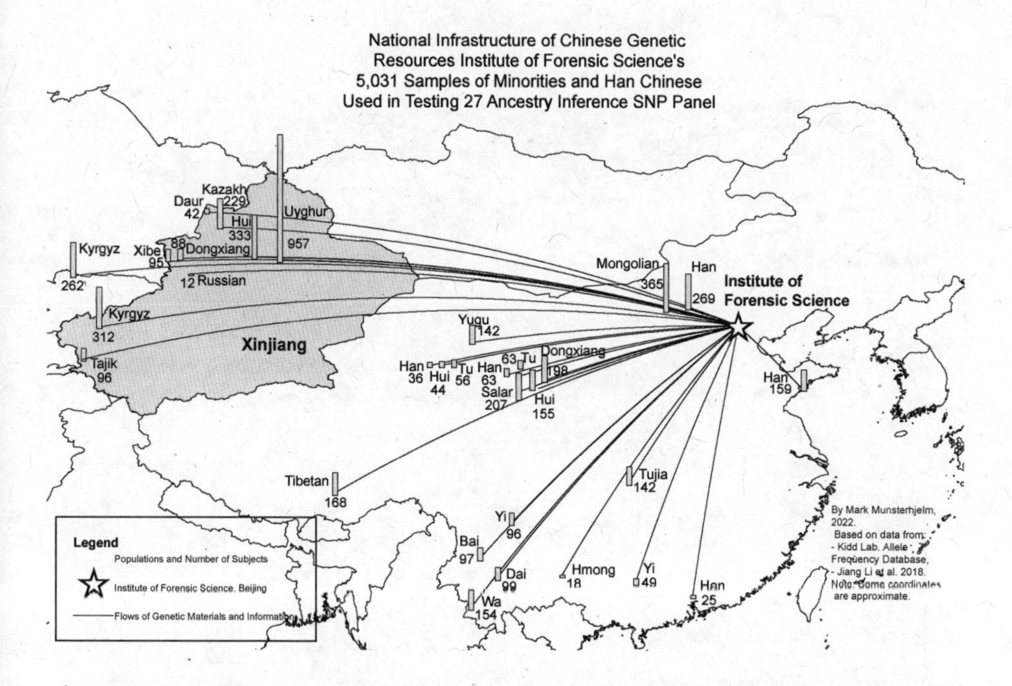

Figure 4.3 | This map shows how Jiang et al. (2018) use only five Han Chinese populations, while the other twenty-seven are minorities and Indigenous peoples. The name of each population is accompanied by the number of research subjects from that population and bar charts to allow comparison between populations, with the 957 Uyghurs towering above the others.

of twenty-seven AISNP markers. In this database are over 2,300 people from various Indigenous groups in Xinjiang and Tibet, including 957 Uyghurs from the National Infrastructure of Chinese Genetic Resources (NICGR). The samples from Uyghurs and other minorities are now part of the infrastructure of China's forensic genetic development efforts, which raises serious ethical questions about informed consent for such inclusion, the ability to withdraw, secondary use, and so on. The Chinese agencies contributed samples for only two of the fourteen geographic regions of the world they set out: Central Asia and East Asia. The samples representing the other twelve regions of the world are from Kidd and Budowle, along with data from the 1000 Genomes Project (in which China is copartner) and HapMap, which are mapped in Figure 4.3.[14]

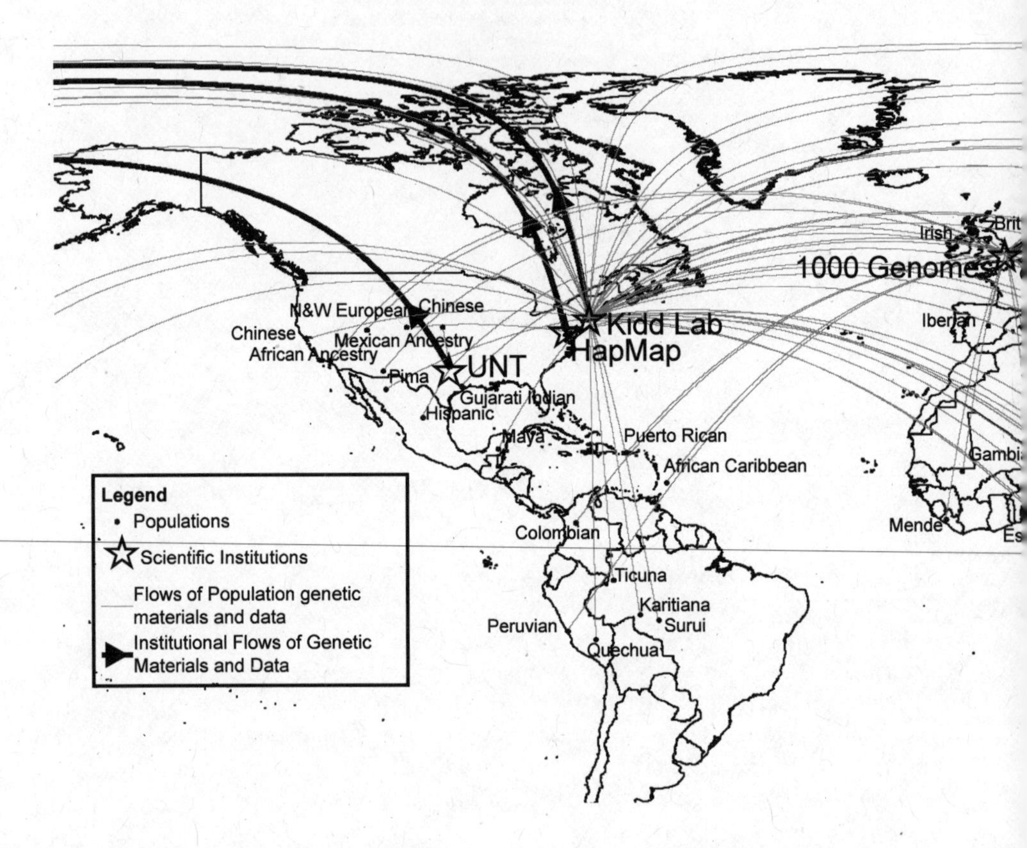

Figure 4.4 | Map of the scientific institutions and 10,350 people from the 110 populations used by Jiang et al. (2018) to test their twenty-seven AISNP panel.

In Jiang et al. (2018a, 2018b), of the 5,031 samples identified as from China, only 552 are Han Chinese, though they represent some 93 per cent of China's population. In contrast, including the Uyghurs, there are over 2,200 individuals from Xinjiang, which is extremely overrepresented. For example, the 957 Uyghurs (CUX) are 19 per cent of the 5031 Chinese samples used in Jiang Li et al. (2018a, 2018b), but Uyghurs are

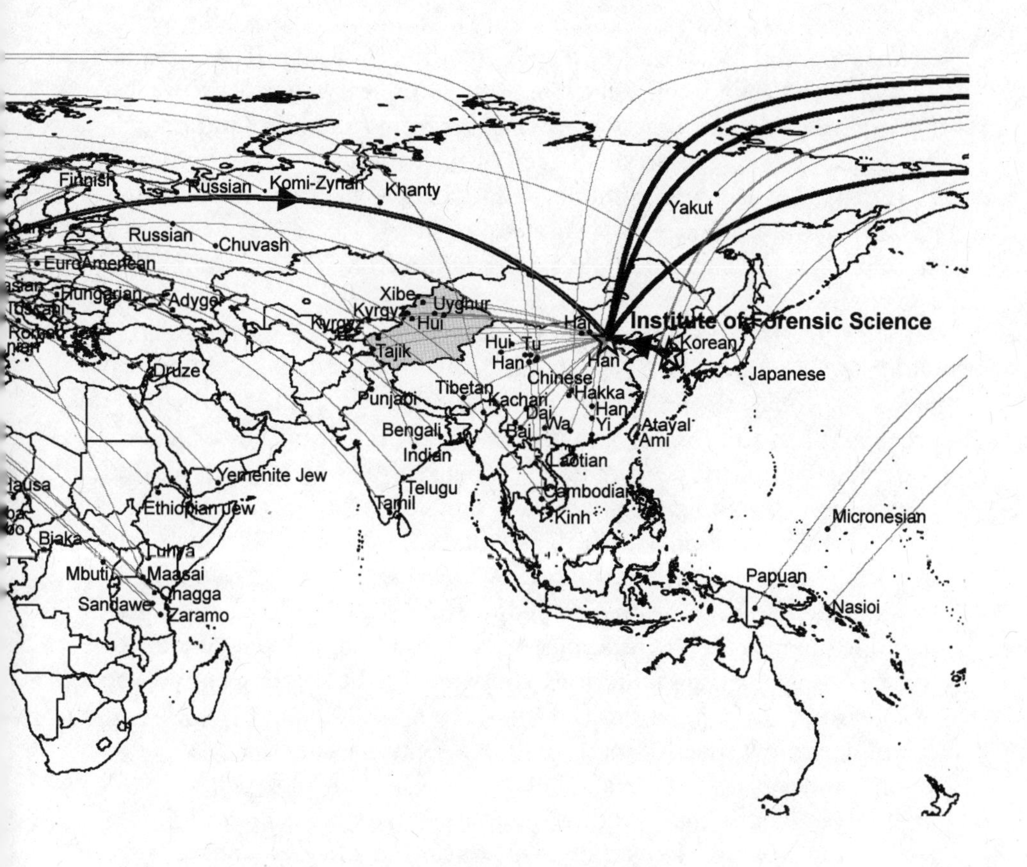

Map by Mark Munsterhjelm 2022. Based on data and population names from Kidd Lab, Frequency Database and Jiang Li et al. 2018.
Note: All coordinates are approximate.

only about 0.75 per cent of China's population, which is a twenty-five-fold overrepresentation. However, when we shift to a global scale, Jiang et al.'s samples are limited to China and Kyrgyzstan.

To create a global data set, Jiang Li et al. used their own collection of cell lines and DNA extracts from Kidd and Budowle and public databases. Sequence data was downloaded for twenty-six populations from the Wellcome Trust–affiliated 1000 Genomes database located near Cambridge, UK, and data for two populations from the US government-funded HapMap project database located in Bethesda, Maryland. As well, Kim Lab in Seoul, Korea, provided samples from South Koreans. The addition of the samples of DNA extract from Kidd Lab and Budowle Lab greatly expands the global reach of the network used in Jiang Li et al. (2018a, 2018b). For example, Kidd Labs provided the Ticuna sampled in 1976, Karitiana and Surui sampled in 1987, and Nasioi sampled in 1985, and the Ami and Atayal Aborigines from Taiwan sampled in 1993–94 (Munsterhjelm 2014).

Kidd's, and to a lesser extent, Bruce Budowle's, cooperation since 2011 with the Chinese Ministry of Public Security helped build the global scope of peoples that ministry researchers have access to. Particularly problematic is Kidd Lab's transfers of significant quantities of genetic materials to Ministry researchers in violation of contemporary ethical norms requiring Indigenous peoples' ongoing informed consent over such secondary usages. The problems raised by Kidd Lab's transfers to ministry researchers are exemplified by how Jiang et al. (2018a, 2018b) integrated these with over 2,000 people from Xinjiang Indigenous peoples, including 957 Uyghurs, plus over 2,000 people from other PRC Indigenous peoples and minorities to create this global data set of 10,350 people from 110 populations (see figure 4.4).

Not only did Kenneth Kidd provide DNA extract samples to Li Caixia and her colleagues for the Jiang Li et al. (2018a, 2018b) paper, but Kidd Lab also uploaded the resulting heterozygosity data Jiang et al. generated to the ALFRED website. For example, the ALFRED entry for the Karitiana reads, "Sample Description: This sample consists of Karitiana individuals from South America. Written informed consent was obtained from all individuals. This sample is part of Caixia Li's group in Institute of Forensic Science, Beijing, China. It was provided by the Kenneth K. Kidd's laboratory to Caixia lab" (ALFRED 2019a). The Uyghur ALFRED contained heterozygosity data for the 957 Uyghurs from Jiang Li et al. (2018a, 2018b):

> Sample Name: Uyghur in Xinjiang, China (CUX)
> Sample UID: SA004656V
> Sample Description: This sample consists of Uyghur individuals from Xinjiang Province, China. Written informed consent was obtained from all individuals. This sample is superset of sample SA004407P. This sample is part of Caixia Li's group in Institute of Forensic Science, Beijing, China.

Number of Chromosomes: 1914
Relation to Other Samples: Superset of SA004407P.
(ALFRED 2019f)

"Number of Chromosomes" refers to the 957 Uyghurs, who each have two chromosomes. This entry would eventually be removed when external events intervened and disrupted these forensic genetic research assemblages, as we will see in chapter 11. Kidd would soon face considerable international scrutiny for his willingness to share Indigenous peoples' genetic materials with the Ministry.

Like the above journal articles, Kidd Lab's ALFRED entry reiterates the Ministry of Public Security's claims of informed consent. Yet the validity of such claims pivots on the idea that there will be no fear of negative repercussions should someone refuse. If we look at the current situation in Xinjiang, this basic assumption of informed consent is untenable because the ethics committee is part of the Ministry of Public Security, which is central to the settler colonization of the Xinjiang region. In particular, the city of Urumqi has been under a very heavy police presence because it was the site of a number of terrorist attacks and riots between Uyghurs and Han Chinese in 2009 that left hundreds dead. Furthermore, the Chinese state has constructed the Turkic peoples of Xinjiang as existential threats to the Chinese nation. Like the hierarchies of power in the sampling of Indigenous peoples discussed earlier, this type of underlying conflict is effectively black boxed in forensic genetics research articles. The goal of the paper by Jiang et al. (2018a) is to distinguish Uyghur and Han Chinese and the authors represent a core set of institutions that have an overriding settler colonial imperative of maintaining Chinese territorial integrity and hence sovereignty over Xinjiang. It includes Li Caixia and Wei Yi-Liang of the Ministry of Public Security's Institute of Forensic Science in Beijing along with researchers from the Bingtuan Public Security Bureau's Institution of Forensic Science in Urumqi, Xinjiang, and the Xinjiang Public Security Department's Science and Technology Institute, also in Urumqi.

USE IN BINGTUAN BOMBING CASE

An important question then is how this AISNP technology will actually be used in Xinjiang. Some evidence is available in a March 2019 article from the *Journal of Forensic Science and Technology*, which, according to the journal website, is "run by China Institute of Forensic Science (CIFS), and supervised by Ministry of Public Security (MPS)." The paper by Zhang et al. (2019) is entitled "Application of 27-plex SNP Race Inference Assay into Tracing a Suspect's Ethnogenesis: Case Report." The paper discusses the application of the twenty-seven-plex ancestry panel developed by Li Caixia et al. by authors from the Bingtuan in the investigation of the bombing of a tractor at a Bingtuan facility in Xinjiang. The authors are Zhang Tao, Feng Baoqiang, Zhou Hao, and Liu Haibo of the Bingtuan Public Security Bureau in Urumqi, Ma Mi of the Bingtuan Seventh Division Public Security Bureau in Kuitan, Xinjiang, and Sun Qifan of the Institute of Forensic Science in Beijing. Zhang Tao, Liu Haibo, and Sun Qifan were also coauthors of the paper by Wei et al. (2016b) on the twenty-seven-plex panel. Due to the involvement of the Kuitan Bingtuan Security Bureau, the likely site of the bombing is near Kuitun, which is located in a major oil production area in the northwest of Xinjiang, near Karamay, in the Han Chinese settler majority dominated northern part of the province.

Manipulation Phase

The article begins by stating that the twenty-seven-plex ancestry inference assay developed by the Institute of Forensic Science can differentiate between three major populations, East Asia, Europe, and Africa to "preliminary assess information about the ethnicity of the source of the biological samples" (Zhang et al. 2019, 276; 从而可初步判断现场生物检材来源的人种族信息).[15] They then use epideictic rhetoric in describing the general utility of the twenty-seven-plex assay, stating it does not require special equipment and "The system can be used on genetic sequencers commonly used in forensic laboratories," and "in general DNA laboratories can meet the application conditions of the system" (276; 该体系可在法医实验室常用遗传测序仪上使用，一般DNA 实验室就能满足体系应用条件). They then state the system is very sensitive and highly accurate in differentiating between racial populations, "reaching 99.37% in 316 blind test samples" (277; 准确性高，在 316 个盲测样本中准确性达到99.37%), citing an earlier 2015 article on the twenty-seven-plex assay.

Commitment Phase

"Since 2016, our laboratory has applied 27 SNP racial inference systems to test case materials, providing important information for investigating and solving many cases. The following is a typical case" (277; 自 2016 年起，本实验室应用 27 重 SNP 种族推断体系检验案件检材，为多起案

件的侦破提供了重要信息。以下为一起典型案例。). They then give an outline of the case study:

> In March 2016, a tractor was damaged with explosives in front of a building of a division of the Xinjiang Production and Construction Corps. Samples of explosives and the explosive remote control were extracted from the scene. After DNA testing, the STR profiles of three suspects were detected. There was one female among them, but all failed to match anyone in a database. The investigation could only be carried out one by one among many suspects, including Han and minority personnel. In order to confirm the race, a single-type DNA was inferred using the 27 SNP system. (277; 2016年3月, 新疆生产建设兵团某师某人家院门前拖拉机被用爆炸物引爆致损。现场提取到爆炸物碎片与爆炸遥控器等检材, 经DNA检验, 检出3名嫌疑人的STR分型。其中有一名单一分型的女性, 但未能在数据库中比中。侦查只能在众多嫌疑人中逐一摸排, 包括有汉族与少数民族人员。为确认种族而用27重SNP体系对单一分型的DNA进行推断。)

The enrolment process is distinct from laboratory studies because it is an actual investigation, making informed consent unlikely. They extracted DNA from the explosion site and a remote controlled detonator found there. This DNA was then processed using a DNA extraction kit made by the Dutch company Qiagen. PCR amplification was done for the twenty-seven-plex SNP markers, which was then processed using a Thermo Fisher SNaPshot testing kit and then genotyped using a Thermo Fisher 3130-XL genetic analyzer. The results were then analyzed using the Thermo Fisher Genemapper IDX data analysis software to determine the twenty-seven SNP alleles (e.g., SNP1 was AG, SNP2 TT, SNP3 AC, etc.). The SNP data was processed using a software developed by the Institute of Forensic Science called Forensic Intelligence (v1.2) and STRUCTURE to calculate the ancestral components and principal components analysis (PCA) to do a sample clustering diagram:

> Cluster analysis of this typed component using Structure 2.3.4 showed that the European and East Asian components of the test materials accounted for 82% and 11%, respectively (Figure 2), and the positive control European and East Asian components accounted for 56% and 36% respectively, all belong to the European-East Asian mixed population [4]; a principal component analysis was performed on the sample using PCA for individual v1.2 (Figure 3). (Figure 3), it can be seen that both the test material (Sample) and

> positive controls (EUR/EAS) were classified as mixed European-East Asian population. (277; 使用Structure2.3.4对该分型成分聚类分析，结果显示检材的欧洲和东亚成分占比分别为82%和11%（图2），阳性对照欧洲和东亚成分占比分别为56%和36%，均属欧洲-东亚混合人群；使用PCA for individual v1.2对该样本进行主成分分析（图3），可见检材（Sample）和阳性对照（EUR/EAS）均被归类于欧洲-东亚混合人群中。)

The suspect's result is overwhelmingly European, which seems to imply they were likely Uyghur. The positive control was a sample from a known European–East Asian mixed population, which is included to assess the validity of the testing procedure; that is, if the control test result was something unexpected, then the test would be invalid: "The above conclusions suggest that the samples were from a mixed European-East Asian population rather than an East Asian population. Based on this analysis and clues, the case was finally solved and the suspect was confirmed to belong to a mixed European-East Asian ethnic group" (Zhang et al. 2019, 277; 综合上述结论，可判定现场检材来源于欧洲-东亚混合人群而非东亚人群。循此分析和线索，案件最终侦破并证实嫌疑人确属于欧洲-东亚混合人群的少数民族。).

In the discussion subnarrative, the authors, citing various earlier papers including Wei et al. (2016b), state that most ethnic groups in the Xinjiang region "have a mixture of Eastern and Western appearances" with the Uyghur as 50 per cent European and 50 per cent East Asian, whereas the Kazakh and Kirgiz populations in Xinjiang have more East Asian characteristics (Zhang et al. 2019). They then discuss how, in utilizing this testing kit, "In cases of multi-ethnic areas, when STR typing does not match the person, the system can be used to analyze the race, combined with other technical means to further characterize the suspect, narrowing the scope of investigation and provide a basis for case identification" (279; 在多民族混居地区的案件中，当STR分型未比中人员时，应用该体系分析种族，结合其他技术手段可进一步刻画嫌疑人，缩小侦查排查范围，为案件定性提供依据。). They warn, "However, in practice, we have encountered some cases where the racial composition of individual samples is at the threshold of East Asian and mixed European-East Asian populations, and it is difficult to determine" (Zhang et al. 2019, 279; 但在实际应用中也遇到个别检材种族成分处于东亚和欧洲-东亚混合人群临界值而难以判定的情况。). Therefore, they warn that investigators should not confuse population results from the genetic testing with ethnicity in the household registration because the latter is culturally defined and "so when applying the results of ethnic inference, we cannot simply rely on ethnicity information, but it should be combined with a variety of factors such as a person's physical characteristics for investigation" (279; 所以在应用种族推断结果时，不能单纯依据民族信息，而应结合人员外貌特征等多种要素进行排查。).

In the paper's concluding paragraph, the authors assert the usefulness of this test in Bingtuan controlled areas:

> In conclusion, in the multi-ethnic Xinjiang Corps area, the 27 SNP race inference system has been applied in various cases, which can be complemented with DNA laboratory STR testing, and can help dig deeper into the genetic information contained in DNA samples and provide more comprehensive information for investigation. At the same time, through more and broader applications in the future, the racial inference system will become more accurate and reliable as more data is accumulated and the database is perfected in field tests. (279; 综上，在多民族混合居住的新疆兵团辖区内，27重SNP种族推断体系在各类案件的应用中初见成效，可与DNA实验室STR检测互补应用，能帮助深入挖掘DNA样本蕴含的遗传信息，为侦查提供更全面的信息。同时，通过今后更多更广的应用，随着实战检验积累数据越多、数据库越完善，该种族推断体系将会更加准确可靠。)

Blowing up the tractor was an act of sabotage, an attack on the Bingtuan's property, and an affront to Chinese settler colonial sovereignty claims. This article shows how the Institute of Forensic Science's research and development of the AISNP was used against a local act of violent resistance in Xinjiang.

FIRST REPORTS OF GENETIC PROFILING

The ethical and political problems with Kidd's long-term forensic genetic cooperation with the PRC's security services became more apparent as the repression in Xinjiang intensified after 2016. In June 2016, US government–sponsored Radio Free Asia reported on the genetic profiling of Uyghurs in Yili Prefecture, who were being required to provide biometric data for passport applications including 3D facial, voiceprint, and DNA samples (Gao, Shan, and Kashgari 2016).[16] In December 2017, a report by Human Rights Watch, a well-known human rights organization, detailed the mandatory genetic profiling of Uyghurs and other Turkic peoples in a number of prefectures in Xinjiang and also elsewhere in China. This report in turn was covered in an article in *Nature News* and received some mainstream press coverage as well. Xinjiang

Public Security Bureau made a public tender with a value of 60 million renminbi (about US$10 million) for twelve high-throughput sequencers and supporting hardware (Brown 2017). Thermo Fisher sold eight capillary electrophoresis genetic sequencers to the Xinjiang Public Security Bureau while another four were supplied by a Chinese company. These machines would be capable of processing some ten thousand samples per day according to the Belgian computational biologist and ethicist Yves Moreau, who was the first to bring attention to Thermo Fisher's sales to security agencies in Xinjiang (Brown 2017). Thermo Fisher denied it has a responsibility over end uses of its products (Kirkwood 2017).

On 13 December 2017, Human Rights Watch issued a report about mandatory genetic profiling of everyone in Xinjiang aged twelve to sixty-five under the guise of a health checkup. During a 13 December 2017 press conference, Chinese Foreign Ministry spokesperson Lu Kang responded to a question about this report:

> Q: According to reports, the Human Rights Watch today issued a report expressing concern about the Chinese government's moves in Xinjiang to obtain DNA from every single resident. What's your response to this?
> A: This organization you mentioned has kept making false allegations on China-related issues all along. That's why I would like to say that such allegation is not even worth your time.
>
> In terms of the situation in Xinjiang, we have said from this podium many times that with economic development, people in Xinjiang are living a peaceful and happy life, and the situation there is sound. Indeed, certain people overseas may be unwilling to see such a situation. I would like to tell them that the Chinese government will continue to uphold the unity of people of all ethnic groups in Xinjiang, safeguard their happy life and promote progress in various fields of Xinjiang. (Ministry of Foreign Affairs 2017)

After mid-2017, all international ethics reviews and related decisions on forensic genetic research cooperation involving Uyghurs and other Turkic peoples should have considered this very significant evidence of Chinese security agencies using mass racial genetic profiling in Xinjiang. However, as these papers demonstrate, international networks of scientists, government institutions, journals, and genetic equipment manufacturers continued to cooperate with the Ministry of Public Security in research on Uyghurs. These conferences promoting the use of forensic genetic testing in the surveillance assemblages and the seamless cooperation between Chinese settler state security agencies and the private sector, which Seigel (2018) calls the state-market, in violence work are instances of what has been termed *surveillance capitalism* and *terror capitalism* (Byler, 2022, 16–18; Sean Roberts 2020, 249).

Thermo Fisher's 2017 Annual Report (2018a, 6) celebrates its rapid growth in China: "Our greatest success story in emerging markets continues to be China, which represents 10 percent of our company's total revenues. Our China strategy is aligned with the country's 5-year plan, placing an emphasis on precision medicine, environmental protection and food safety." The company's global revenues were US$20.9 billion in 2017 with the PRC accounting for over US$2 billion (1). Though forensic genetics in the PRC is only a small portion of its overall sales, during the period of escalating repression after Chen Quanguo's appointment in 2016 as Communist Party Secretary of Xinjiang, Thermo Fisher intensified its sales efforts to Chinese security agencies, promoting large conferences in 2016 and again in 2017. The 2016 Forensic Science New Technology Application Summit Forum in Foshan in Guangdong Province featured Sheree Hughes-Stamm of Sam Houston State University who presented on identifying bombmakers from improvised explosive device fragments (Hughes-Stam 2016; Thermo Fisher Conference 2016). Other presenters included Li Haiyan of the Chinese Ministry of Public Security, whose presentation included testing the Thermo Fisher HID-Ion AmpliSeq Ancestry Panel on Uyghur and Manchu samples using Kidd's fifty-five AISNP panel (Li Haiyan 2016a, 2016b; Thermo Fisher Conference 2016). Later, in November 2017, Bruce Budowle was the keynote speaker at the Thermo Fisher sponsored conference entitled "Revolutionary Forensics: Answers From CE [capillary electrophoresis] to NGS [next generation sequencing]" in Chengdu, the capital of Sichuan Province, which again included Li Haiyan and a number of other Chinese security agency affiliated researchers (Thermo Fisher Conference 2017a). According to the conference website, "A total of more than 300 representatives from public security agencies and related industries from all provinces, autonomous regions, and municipalities across the country participated in this summit forum" (Thermo Fisher Conference 2017a; 共有来自全国各省、自治区、直辖市公安机关和相关行业300余名代表参加本次高峰论坛，会议期间,他们将参加主题演讲, 新技术应用交流及产品演示。). A two minute and forty second promotional video for the conference shows it was attended by senior officials, including Ye Jian (who appears briefly at 0:22, 1:08, and 2:15), who is a deputy director of the Ministry of Public Security's Institute of Forensic Sciences and cooperated with Li Caixia, for example, in Wei et al. (2016a) and Liu et al. (2017), which I will discuss in chapter 10 (Thermo Fisher Conference 2017b).

There was a relative seamlessness in the how Western corporations and researchers readily cooperated with the Chinese security apparatus in such corporate events. These visits and conference presentations occurred during a time in 2016–17 when many senior scientists in the Chinese security apparatus were planning how to integrate advanced new genetic phenotyping and ancestry technologies into China's rapidly growing surveillance infrastructure, which I will discuss in more detail later. And unknown to Western scientists like Kidd, Li Caixia and her colleagues included elements of this research cooperation in a series of patents and applications that were part of their long-term project to differentiate global racial categories, or more specifically, to differentiate between Han Chinese, Uyghurs, and Tibetans.

5

Chinese Patents and Applications

A pattern of racialized/ethnicized patenting that is intended to differentiate racial and ethnic categories is evident in three granted patents and six patent applications filed by the Ministry of Public Security's Institute of Forensic Science in Beijing. In recent years, Chinese government policy has supported the filing of patents as part of a drive to make China a more innovation-based economy (Xu et al. 2021). According to the World Intellectual Property Organization, the Chinese patent office received 241,435 applications in 2009 rising rapidly to 1,460,244 in 2018 (World Intellectual Property Organization 2020). A patent application is a set of legal claims over a specific invention that is filed with the patent office of a country. The patent office has examiners who analyze the application and assess whether it meets the necessary criteria for an invention, which according to Article 22 of China's national patent law are "novel, creative and of practical use" (Chinese National Intellectual Property Administration 2020). A patent itself is a legal right in which the state backs the exclusion of others from using a particular patent for the duration of fifteen to twenty-one years, depending on the jurisdiction (contingent on paying various fees and so on).

This chapter analyzes the use of terrorism discourses and racial categories in a number of Chinese patents and patent applications filed by the Ministry of Public Security's Institute of Forensic Science, which also occurs elsewhere, with the US and European Union accepting such patent applications (see, for example, Kayser, Liu, and Hoffman 2011). The list of patents and applications below reveals a decade-long quest for AISNP panels to racially differentiate minorities as risky or destructive bodies from Han Chinese, a form of biopolitical security technology for sorting populations (Dillon 2008; Epstein 2007).

THE INSTITUTE'S APPLICATIONS AND PATENTS

The epideictic rhetoric of these patents and patent applications occurs mainly in the "Background of the Invention," a section in which the inventors argue for the significance of the invention to solve some problem. The application is a type of submission to the larger organizing narrative of the patent office exercising its sovereign power as a grantor of legal rights that encourage and secure technological innovation. Several of these patents and patent applications' Background of the Invention sections involve the construction of mobile ethnic minorities as potential threats, to which the researchers provide a potential solution. These patents and patent applications variously seek to identify the enemy antisubjects of the criminal and the terrorist through the use of racial categories of race, ancestry, and phenotype.

Chinese Patent CN101956006B (granted 16 October 2013; see table 5.1) is entitled "Method for Obtaining Race Specific Loci and Race Inference System and Application Thereof." It uses SNP markers for pigmentation genes and ancestry genes to identify three major racial groupings (Li, Hu, and Wei 2013). CN103146820B (granted 2 July 2014), entitled "A method and system used for inferring whether an unidentified genetic sample from an unknown individual is from Han, Tibetan and Uyghur population," uses ninety-four SNPs to distinguish between these populations (Li Caixia et al. 2014; 一种推断未知来源个体汉、藏、维群体来源的方法和系统). The patent makes use of the criminal and terrorist antisubjects stating, "The flow of people between regions has increased and with it, foreign-related, anti-terrorism and interregional cases of crime have increased, which make case investigations increasingly more difficult" (para. 4; 同时伴随国家、区域间的人员流动加大，涉外、反恐、跨区域流动作案等复杂案件不断增多，案件侦查的难度日益加大。). According to the patent, this increased mobility and complexity mean that in cases where there are no STR matches because only a small proportion of the Chinese population is in a database, an unknown person's genetic samples from a crime scene can be analyzed for further data on their origins, which may provide direction in investigations.

The dangerous flows premise is further elaborated in patent CN104212886B (granted 22 June 2016) entitled "Method and System for Applying African, European and East Asian Population Genetic Principal Component Analysis to an Individual from an Unknown Source" (一种对未知来源个体进行非、欧、东亚群体遗传主成分分析的方法和系统). This patent uses the twenty-seven-SNP panel by Wei et al. (2016a) discussed in chapter 7. Filed in July 2014 (when Xi Jinping

Table 5.1 | List of Institute of Forensic Science Patents and Applications

Patent/ application	*Granted or priority date*	*Patent title*	*Inventors*	*Sample preparation and genetic analysis equipment mentioned*
CN101956006B	Filed 27 August 2010 Granted 16 October 2013	Method for obtaining race specific loci and race inference system and application thereof	Li Caixia, Hu Lan, Wei Yiliang	Thermo Fisher 3130XL Genetic Analyzer [0073]
CN103146820B	Filed 22 February 2013 Granted 2 July 2014	Method and system used for inferring Han, Tibetan and Uyghur population as source of individual with unknown identity	Li Caixia, Wei Yiliang, Hu Lan, Ji Anquan, Jia Jing, Li Wanshui	Qiagen QIAamp Blood Kit; SEQUENOM MassARRAY [0053]
CN104212886B	Filed 25 July 2014 Granted 22 June 2016	Method and system for applying African, European and East Asian population genetic principal component analysis to an individual from an unknown source	Li Caixia, Wei Yiliang, Ye Jian	Qiagen QIAamp Blood Kit; ABI3130 or ABI3500 Genetic Analyzer; GeneMapper ID-X software [0016]
CN106480198B	Filed 1 November 2016 Granted 17 September 2019	Method and system for individual identification of unknown samples	Zhao Lei, Li Caixia, Feng Lei, Wang Wei	Qiagen QIAamp Blood Kit; ABI3130 or ABI3500 Genetic Analyzer; GeneMapper ID-X software [0014]

continued…

Table 5.1 | ... *continued*

Patent/ application	*Granted or priority date*	*Patent title*	*Inventors*	*Sample preparation and genetic analysis equipment mentioned*
CN107400713B	Filed 18 August 2017 Granted 30 June 2020	Method and system for recognizing Tibetan population individuals on Qinghai-Tibet Plateau of China from 27 groups	Jiang Li, Li Caixia, Zhao Wenting, Liu Jing, Huang Meisha	Qiagen QIAamp Blood Kit; ABI3130 or ABI3500 Genetic Analyzer, or MassARRAY Genotyping System (US Agena Corporation) [0017]
CN108342489A	Filed 23 January 2017	Male individuals Y-SNP typing method and system	Jiang Li, Li Caixia, Zhao Wenting, Liu Jing, Zhao Lei	Qiagen QIAamp Blood Kit; Eppendorf Mastercycler (PCR), ABI3130 or ABI3500 Genetic Analyzer, GeneMapper ID-X software [0017]
CN105861654A	Filed 5 April 2016	Method and system for analyzing ten group sources for individuals of unknown origin	Li Caixia, Jiang Li, Sun Qifan, Zhao Wenting, Liu Jing	ABI3130 or ABI3500 Genetic Analyzer, GeneMapper ID-X software [0018]
CN107419017B	Filed 25 July 2017 Granted 8 September 2020	Method and system for inferring five intercontinental ethnic group sources of unknown source individuals	Liu Jing, Li Caixia, Zhao Wenting, Jiang Li, Hao Weiqi	Qiagen QIAamp Blood Kit; ABI3130 Genetic Analyzer; GeneMapper ID-X software [0055]

continued...

Table 5.1 | ... *continued*

Patent/ application	*Granted or priority date*	*Patent title*	*Inventors*	*Sample preparation and genetic analysis equipment mentioned*
CN108411008B	Filed 1 June 2018 Granted 27 July 2021	Application of 72 SNPs and associated SNP primers to help in identification of ethnic groups	Li Caixia, Xu Youchun, Liu Jing	Qiagen QIAamp Blood Kit; MassArray (Agena, US) [0084]; LuxScan-D, CapitalBio, China.[0061]
CN111073888A	Filed 9 January 2020	Southeast Asia-north population inference system based on capillary electrophoresis detection	Liu Jing, Ma Mi, Li Caixia	Qiagen QIAamp Blood Kit; ABI3130xl Genetic Analyzer; GeneMapper ID v3.2; [0052, 0070, 0090]
CN112011622A	Filed 29 May 2019	A method and system for analysing the origin of unknown individuals from African, East Asian and European groups	Li Caixia, Han Junping, Zhao Lei, Jiang Li	Qiagen QIAamp Blood Kit; NanoDrop 2000c Spectrophotometer (Thermo Fisher, US). ABI3130, ABI3130XL, ABI3500 Genetic Analyzer; ID-X software or other GeneMapper software [0017, 0070, 0085]

Note: ABI is owned by Thermo Fisher. Qiagen is a Dutch company.

declared the People's War on Terror), with the inventors listed as Li Caixia, Wei Yiliang, and Ye Jian, the patent utilizes the epideictic rhetoric of the war on terror to define the danger inherent in the increasing global movements of people:

> With the globalization of the economy, the movement of people between different countries and regions has increased, and the number of complex cases involving foreign countries, counterterrorism and cross-regional mobile crimes has increased, making it increasingly difficult to investigate these cases. For example, large cities such as Beijing, Shanghai, and Guangdong often have foreign-related cases. (Li, Wei, and Ye 2016, para. 2; 伴随经济全球化，不同国家、区域间的人员流动加大，涉外、反恐、跨区域流动作案等复杂案件不断增多，案件侦查的难度日益加大。(例如，北京、上海、广东等大城市经常出现涉外案件。)

In its minor premise, the patent then asserts, "At present, the human race is mainly divided into three major races, European, African and East Asian, and also mixed races from these major races. In our country, mixed populations are mainly European and East Asian such as the Uyghur, Kazakhs and Kirgiz in Xinjiang" (para. 2; 目前国际人种主要分为欧洲、非洲、东亚三大人种及由这些人种形成的混合人种，我国境内的混合人群主要包括新疆的维吾尔族、哈萨克族、柯尔克孜族等欧洲和东亚的混合人群). The sequence of premises leads to the specific limitations of current STR technologies and the use of individual identification databases when there is no matching profile for an unknown donor (para. 3). And the commitment phase proposes as a conclusion to these premises that ancestry information based on dividing up the world between Africa, Europe, and East Asia can help in police investigations (paras. 4–5). This patent application is significant in its use of war on terror–type rhetorical appeals, particularly how the authors associate increasing population mobility with terrorism and their specific identification of the Uyghurs, Kazakhs, and Kirgiz in Xinjiang as mixed populations.

The dangerous population flows argument appears again in the 2020 patent CN107400713B (filed in 2017) that uses eighty-seven high altitude adaptation–related SNP markers to identify Tibetans and differentiate them from twenty-seven other populations (Jiang Li et al. 2020). The dangers of increased mobility are evident in its manipulation phase in the Background of the Invention section:

> With economic globalization, the flows of people between different countries and regions has increased, and complex cases such as foreign-related, anti-terrorism and interregional crimes have been increasing, which has increased the difficulty of investigating these crimes. For example, large cities such as Beijing, Shanghai, and Guangdong often have foreign-related criminal cases. At present, the human race is mainly divided into African, [Native] American, European, South Asian, and East Asian races, and also mixed races from these major races. (Jiang Li et al. 2020, para. 2; 伴随经济全球化，不同国家、区域间的人员流动加大，涉外、反恐、跨区域流动作案等复杂案件不断增多，案件侦查的难度日益加大。例如，北京、上海、广东等大城市经常出现涉外案件。目前国际人种主要分为非洲、美洲、欧洲、南亚、东亚人种及由这些人种形成的混合人种。)

This patent application's epideictic rhetorical use of the movement of ethnic groups as a threat and then its specific focus on utilizing altitude adaptation markers to distinguish Tibetans from other populations implies that Tibetans are a threat. The Tibetan specific altitude adaptation is a type of phenotyping in that the genes that express themselves in particular physical traits, in this case genes associated with the ability to live at high altitudes without suffering the effects of altitude sickness, are transformed into a means to differentiate Tibetans from other populations.

The most detailed expression of this dangerous population flows concept is in the 2017 patent application CN108342489A entitled "Male Individual Y-SNP Typing Method and System," which seeks to use a set of ten Y chromosome SNP markers to identify male individuals, including their ancestry (Jiang Li et al. 2017). This application's "Background of the Invention" is worth quoting at length as it clearly articulates the threat minority males' mobility poses to the Chinese body politic and reflects the increasing securitization of ethnic minorities during the 2010s:

> [0002] China is a multi-ethnic country with 56 ethnic groups recognized by the government. The distribution of ethnic minorities is characterized by "large mixed living and small clusters."[1] With the rapid development of the economy and the increase in population mobility, interactions between ethnic groups are increasing, and the phenomenon of different ethnic groups living together will become more common. Since the 1990s, terrorism has spread rapidly around the world, and terrorist attacks over ethnic and territorial disputes have increased year by year. In recent years, a number of vicious

cases have also occurred in China, seriously endangering social stability and the safety of people's lives and property. Large population flows and the sophistication of terrorism bombing cases have soared, which pose a huge challenge to the ability of the public security system to handle cases. (Jiang Li et al. 2017, para. 2; [0002] 中国是一个多民族国家，通过识别并经政府确认的民族共有56个。少数民族的分布呈现"大杂居、小聚居"的特点。随着经济的快速发展和人口流动性增加，各民族之间的交流日益增多，民族间的杂居现象将更为普遍。自二十世纪九十年代起，恐怖主义在全球范围内迅速蔓延，针对民族与领土争端的恐怖袭击逐年增加。近些年我国也发生了多起恶性案件，严重地危害了社会稳定和人民生命、财产安全。人口流动大、涉恐涉爆案件的复杂程度猛增，这对公安系统的办案能力提出了巨大的挑战)

Though the application does not directly mention Uyghurs or any other minorities by name, they are alluded to in the reference to recent terrorist attacks in China. The authors argue for the use of forensic genetic ancestry methods:

> In a modern society where ethnically mixed populations are increasing, identifying the origin of criminal suspects can help reduce case uncertainty, clarify the direction of case investigation, can actively push the case investigation forward, and speed up solving the case. Inferring racial or family origin is a very useful technical analysis tool that plays a very advantageous role in solving terrorist cases that endanger national security. (Jiang Li et al. 2017, para. 2; 在民族混居情况日益增加的现代社会中，犯罪嫌疑人的群体来源的判定，将能够帮助案件的定性，明确案件的侦查方向，推动案件的主动侦查，加快案件侦破速度。种族或家系来源推断作为一种非常有利的技术分析手段，对侦破危害国家安全的恐怖案件发生起到十分积极的作用。)

This patent application intensifies the use of War on Terror type rhetorical appeals, particularly how the authors associate increasing population mobility with terrorism as a threat to national security.

Li Caixia et al., patent CN112011622B for their seventy-two-plex ancestry panel, filed in May 2019, also makes use of the dangerous flows argument:

> With the globalization of the economy, the movement of people between different countries and regions has increased, and complex cases involving foreign affairs, anti-terrorism, and cross regional

> movement are increasing, and other complex cases are increasing, and case investigation is becoming increasingly difficult. At present, the international ethnic groups are mainly divided into African, East Asian and European ethnic groups. The international ethnic groups are now divided into African, East Asian and European ethnic groups and the mixed ethnic groups formed by these ethnic groups. (Li Caixia et al. 2021 , para. 0002; 伴随经济全球化，不同国家、区域间的人员流动加大，涉外、反恐、跨区域流动作案等复杂案件不断增多，案件侦查的难度日益加大。目前国际人种主要分为非、东亚、欧洲三大人种及由这些人种形成的混合人种。)[2]

CN112011622B uses the same Uyghur (CUX), Kazakh (CKX), Tajik (CTX), and Kirgiz (CZX) subjects as Wei et al. (2016a, 32; Li Caixia et al. 2021, paras. 0040–2, 0104, 0108). It also uses 164 DNA extract samples representing two populations from Bruce Budowle of the University of North Texas (Li Caixia et al. 2021, paras. 0037, 0065). This use of these dangerous flows arguments involves an overall security emphasis on the inevitability of increasing mobility under globalization, but also suggests that these flows are threats due to the intermixing of ethnic groups. This tone is consistent with China's adaptation of US post-9/11 War on Terror rhetoric in representing the Uyghurs and other minorities as threats to social stability and national security, which is used to organize and justify their massive repression in Xinjiang.

Patent application CN105861654A, "A Method and System Using 10 Origin Groups to Analyze the Origins of an Unknown Individual" (一种对未知来源个体进行十个群体来源分析的方法和系统), has the Institute of Forensic Science in Beijing as the assignee (Li Caixia et al. 2016a). Developed by a team at the institute led by Li Caixia, the panel of seventy-four AISNPs is supposed to be able to differentiate between ten major world populations including North Asia, East Asia, Southeast Asia, and South Asia. This application is based on research published in a 2016 paper entitled "A Panel of 74 AISNPs: Improved Ancestry Inference within Eastern Asia," which was written by Li Caixia, Kang Longli, Andrew J. Pakstis, Kenneth Kidd, and other scientists and appeared in *Forensic Science Genetics International* (Li Caixia et al. 2016b). It is based on a joint research project that was mostly done with Kidd Lab cell lines (again including the Karitiana and Surui) but included a few hundred provided by the Chinese researchers (Li Caixia et al. 2016b, 102–3). The patent application makes no mention of Uyghur or other Xinjiang minorities directly; however, a number of other minorities were included. The scientists use Kidd's DNA extract from the Surui and the

Ticuna of Western Brazil as "test subjects" and their data is included in the patent application. The patent first situates ancestry research in broad terms, as medical, evolution, and forensic, then moves to the specific forensic purpose of not having an STR match on any database: "Biogeographical ancestry analysis receives extensive attention because of its contributions to bio-pharmaceutical research, personalized medicine, human history research and forensic investigations. In forensic applications, when there is no database match to a suspect or when there is no specific suspect, ancestry inference can sometimes provide critical clues for an investigation" (Li Caixia et al. 2016a, para. 0002; 生物地理祖先分析因其在生物药学研究、个性化医疗、人类历史研究和法医调查等方面的贡献受到广泛关注。在法医应用方面，当没有数据库或者没有匹配到嫌疑人或是没有特定嫌疑人时，祖先推断有时能够为调查提供关键线索。). Later, the scientists claim that this system's ability to distinguish ten major regional populations will help in antiterrorism:

> The method and system of the present invention can distinguish the populations of Sub-Saharan Africa, North Africa, Southwest Asia, Europe, South Asia, North Asia, East Asia, Southeast Asia, the Pacific Ocean and the Americas, in particular, to distinguish between North Asia, East Asia and Southeast Asia. In foreign-related, anti-terrorism cases, especially cases in which biological samples from different ethnic geographical origin populations are found for which there are no STR matches, determining the group identity of the source adds qualitative information that can clarify the direction of investigation, and so perform a critical function in a case. (Li Caixia et al. 2016a, para. 0060; 1、本发明方法和系统可以进行撒哈拉以南非洲，北非，西南亚，欧洲，南亚，北亚，东亚，东南亚，太平洋和美洲人群，尤其是实现对北亚，东亚，东南亚人群的区分，在涉外、反恐案件中，尤其对于在上述案件中提取到的不同种族地域来源人群的生物样本，在STRs分型比对不中的情况下，确定涉案人员的群体身份来源，使案件快速定性和明确侦查方向，能发挥重要作用。)

Patent CN107419017B (granted 8 September 2020), "Method and System for Inferring the Origins of an Unknown Individual from Five-Continental Ethnic Groups" (对未知来源个体进行五大洲际族群来源推断的方法和系统), was invented by Liu Jing, Li Caixia, Zhao Wenting, Jiang Li, and Hao Weiqi (Liu Jing, Li Caixia et al. 2020). This application's manipulation phase begins with "Ancestry informative markers (AIMS), which detect large differences in population distribution, can be used to infer the geographic origin of the population of DNA donors at a crime scene" (para. 2; 检测人群间分布差异大的DNA多态性位点即祖先信息位点 (Ancestry informative markers, AIMs)可以推断犯罪现场DNA供者的族群地域来源。). This is identical to the first sentence of the 2019 article associated with this patent application entitled "Exploring the ancestry differentiation and inference capacity of the 28-plex AISNPs," published in the *International Journal of Legal Medicine* (Hao et al. 2019). In the article, Li Caixia and her colleagues write in the "Acknowledgments: Special thanks are given to Professor Kenneth K. Kidd of Yale University who supplied cell line DNA samples." He provided DNA extract for fifteen populations totalling 395 individuals that was processed from cultured cell lines grown at Kidd Lab (977, 981). In the patent, Kidd is also credited as the source of DNA extract for 358 individuals representing thirteen populations (Liu Jing, Li Caixia et al. 2020, para. 46).[3] Figure 5.1 shows how Kidd Lab DNA extract samples like the Karitiana (#11 on the list) were integrated along with Uyghur (#31 on the list) and other Indigenous samples from Li Caixia's lab at the Institute of Forensic Science.

The synecdoche of region as hyperonym to categorize populations as hyponyms is evident in the screen capture of the table from patent CN107419017B. The region categories are Africa, America, East Asia, Europe, Oceania, Central Asia, South Central Asian, and South Asia, under which the populations are categorized, with in turn, the sample size (N) indicating the number of donors that represent the population. This set of categories involves a synecdochic taxonomic hierarchy of region, population, and donors (Panofsky and Bliss 2017, 68).[4]

In the patent, which was granted in 2020, some of these samples are used as test subjects to demonstrate the accuracy of the invention. In a table are the results of the 140 people, seven populations with twenty samples each, used to test the panel (Liu Jing, Li Caixia et al. 2020, paras. 80, 82). Yoruba (YOR), Karitiana (KAR), Nasioi (NAS), and European Americans (EAM) from Kidd Lab along with the Han Chinese in Guangxi (GXH), Han Chinese in Henan province (HNH), and Uyghur (UY) from Li Caixia's lab were all classified into six categories: Africa, America, East Asia, Europe, Oceania, and mixed populations (Central Asia, Central South Asia, South Asia) with the authors correctly assigned 20/20 for six of the seven populations with the Uyghurs at 17/20 (Liu Jing, Li Caixia et al. 2020, paras 46, 80). Figure 5.2, a principal components analysis (PCA) graph, includes the Uyghurs as their own category.

In the PCA graph, the focus on Uyghurs is evident at the categorical level (a form of synecdoche) with the "Uyghur" designation a particular

[0041] 实施例1、本发明五大洲际人群基因分型数据库的构建及应用

[0042] 一、材料和方法

[0043] 1、样本信息

[0044] 本发明选取公共数据库千人基因组(1000genomes)里的20个人群2000个样本及人类基因组多样性计划(HGDP-CEPH)中2个人群92个样本，检测样本包括16个人群的712份样本，共38个人群2804个个体作为验证样本。详细信息见表1。

[0045] 表1人群样本信息表

[0046]

	国家地区	人群	缩写	样本数 (N)	来源
1	非洲(AFR)	Esan in Nigeria	ESN	99	1000 Genomes
2		Luhya in Webuye	LWK	99	1000 Genomes
3		Mandingka tribe in Gambia	MAG	113	1000 Genomes
4		Mende in Sierra Leone	MSL	85	1000 Genomes
5		Yoruba in Ibadan,Nigeria	YRI	108	1000 Genomes
6		Mbuti Pygmies	MBU	38	Kidd Lab
7		Masai	MAS	20	Kidd Lab
8		Sandawe	SND	10	Kidd Lab
9		*Yoruba*	*YOR*	*37(20)*	Kidd Lab
10	美洲(AMR)	America	AMR	64	HGDP-CEPH
11		*Karitiana*	*KAR*	*54(20)*	Kidd Lab
12		Pima, Mexico	PMX	5	Kidd Lab
13		Guihiba speakers	GHB	12	Kidd Lab
14	东亚(EAS)	Chinese Dai in Xishuangbanna	CDX	93	1000 Genomes
15		Han Chinese in Beijing	CHB	103	1000 Genomes
16		Southern Han Chinese	CHS	105	1000 Genomes
17		Japanese in Tokyo	JPT	104	1000 Genomes
18		Kinh in Chi Minh City	KHV	99	1000 Genomes
19		*Han Chinese in Henan*	*HNH*	*63(20)*	Caixia lab
20		*Han Chinese in Guangxi*	*GXH*	*54(20)*	Caixia lab
21		Hakka	HKA	16	Kidd Lab
22	欧洲(EUR)	Utah residents with European ancestry	CEU	99	1000 Genomes
23		Finnish in Finland	FIN	99	1000 Genomes
24		British in England and Scotland	GBR	91	1000 Genomes
25		Iberian population in Spain	IBS	107	1000 Genomes
26		Toscani in Italy	TSI	107	1000 Genomes
27		*EuroAmer (Not Perf.)*	*EAM*	*91(20)*	Kidd Lab
28	大洋洲(OCE)	Oceania	OCE	28	HGDP-CEPH
29		*Nasioi Melanesians*	*NAS*	*22(20)*	Kidd Lab
30		Papua-New Guinean	PNG	22	Kidd Lab
31	*中亚 (CA)*	*Uyghurs*	*UY*	*237(20)*	Caixia lab
32	中南亚(SCA)	Kachari from Assam	KCH	17	Kidd Lab
33		Thoti	THT	14	Kidd Lab
34	南亚(SA)	Bengali in Bangladesh	BEB	86	1000 Genomes
35		Gujarati Indian in Houstan	GIH	103	1000 Genomes
36		Indian Telugu in the UK	ITU	102	1000 Genomes
37		Punjabi in Lahore,Pakistan	PJL	96	1000 Genomes
38		Sri Lankan Tamli in the UK	STU	102	1000 Genomes

Figure 5.1 | Screen capture of the list of population samples and their sources from Chinese Patent CN107419017B invented by Li Caixia and her colleagues. Note the number of samples from Kidd Lab (e.g., #11, Karitiana) and those from Li Caixia's lab (e.g., #31, Uyghurs).

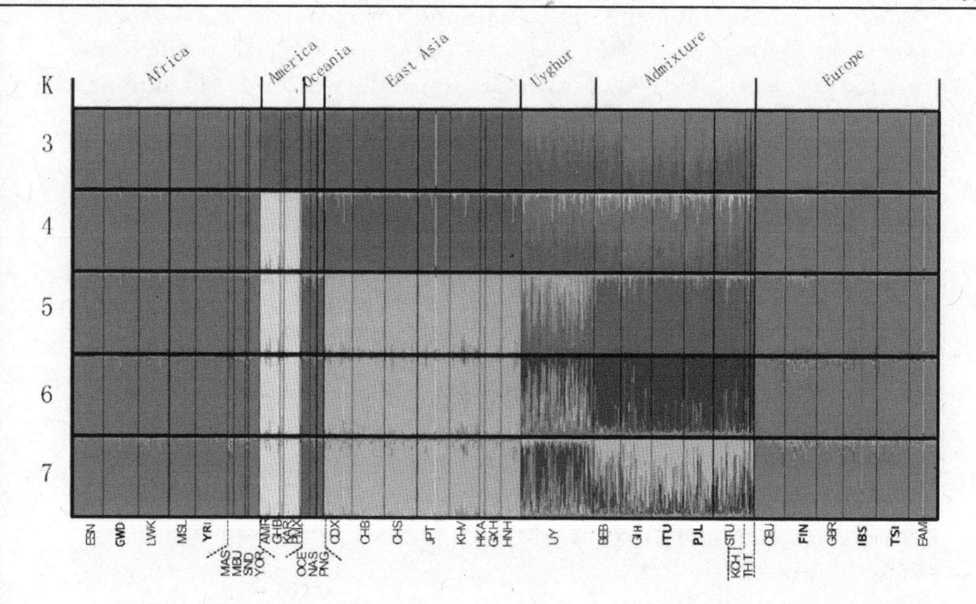

Figure 5.2 | Patent CN107419017B principal components analysis graph.

people or population whereas the rest are categorized at the continental level, Africa, America, Oceania, East Asia, and Europe, and one as "admixture" that combines these continental categories. The graph from the patent includes the Uyghur as their own category in the top bar. The position on the bar illustrates their boundary role as admixed populations between Europe and East Asia. The testing makes use of the Kidd Lab supplied DNA extract with, for example, the Karitiana (KAR) under the America category and Nasioi (NAS) under the Oceania category. A central theme in the patents is the idea of securing the flows of people within China and those coming from abroad. Being able to differentiate between the different peoples within the PRC is a central concern in the patents and applications, particularly differentiating Uyghur and Tibetans from Han Chinese. A patent involves an appeal by the inventors or their representatives to the state for recognition of a special set of rights that exclude others from using the invention. It is an exercise of state legal sovereignty. In these patents and patent applications, the

Institute of Forensic Science, part of the Ministry of Public Security, has applied for and received special legal rights to exclude others from using its inventions that differentiate Uyghurs and Tibetans from Han Chinese. These patents and applications seek to commodify a racial sorting technology developed under settler colonialism and based on research done, in part, through cooperation with foreign scientists and institutions and using US-made genetic analysis systems.

CONCLUSIONS

Article 5 of the Patent Law of the People's Republic of China states that "Patent rights shall not be granted for invention-creations that violate the law or social ethics, or harm public interests. Patent rights shall not be granted for inventions that are accomplished by relying on genetic resources which are obtained or used in violation of the provisions of laws and administrative regulations" (Chinese National Intellectual Property Organization 2020). The patents and patent applications by the Institute of Forensic Science involve the enactment of law by Chinese state institutions and attendant definitions of what constitutes harm to society. The PRC's Patent Law specifically mentions not granting patents using genetic resources that violate the law.

In Xinjiang, Chinese settler colonialism involves an important distinction from typical concepts of securitization (e.g., Copenhagen School) that narrowly focus on the designation of particular racialized groups as biopolitical threats to the population. What is occurring in Xinjiang reflects an underlying sovereignty struggle that is the root of the conflict. The patents emphasize the threats hidden in increasing global mobility and population movements, and the promise of the inventions involves using AISNP markers to identify unknown destructive terrorist and criminal bodies (those without an STR database profiles) who are threats to state sovereignty, Communist Party rule ("social stability"), and law, which will help biopolitically secure these productive flows (Bigo 2008, 36; Epstein 2007). These patents issued by the National Intellectual Property Administration to the Institute of Forensic Science and the patent applications demonstrate how categories of race operationalize and normalize differentiation of Han Chinese, Tibetans, and Uyghurs. These manhunting technologies within the Chinese security apparatus rely upon the organizing narrative schema of providing intelligence about terrorists and criminals. These organizing narratives helped orchestrate and coordinate these research networks and others in research and development of forensic genetic technologies by involved Chinese security agencies' in cooperation with Western scientists and agencies. The organizing narratives' epideictic rhetoric reproduces the moral and ideological hierarchies of the political economy of research and development in policing and national security. The use of racial categories in these technologies is intended to differentiate the populations that make up these flows. Tibetans and Uyghurs have increasingly been subjected to racially differentiating forced identification systems and security checks based on biometrics justified by the construction of them as an existential threat to the Chinese body politic. These patents/applications and journal articles claim a particular moral purpose in strengthening the manhunting capacities of security agencies to cope with the anonymity of the figures of the criminal and the terrorist. The cultural genocide that the PRC's security forces are inflicting against the Uyghurs and other Turkic peoples is a type of settler colonial imposition of violence that seeks to forcibly remake the peoples of Xinjiang into pacified subjects acceptable to Chinese Communist Party rule (Byler 2022; Sean Roberts 2020).

Western biotechnology firms have been significant enablers of this process. The table of patents at the beginning of the chapter (table 5.1) shows nine of the eleven patent filings specifically mention Thermo Fisher's ABI 3130 or 3500 capillary electrophoresis genetic analyzers. It shows how the 3130 and the 3500 and the upgraded 3130XL and 3500XL are specifically mentioned either as research instruments used by the scientists or as examples of systems upon which these ancestry tests can be conducted. As well, Thermo-Fisher PCR units and GeneMapper software are mentioned several times. Hence, these Thermo Fisher products have an important enabling role in the genetic research, with its equipment and testing chemicals central to the processing of the samples. This enabling also includes the steps in the laboratory processes that use these. For example, in patent CN107419017B (para. 55) the inventors use a Thermo Fisher produced testing kit called the SNaPshot™ Multiplex Kit to prepare the samples for testing; then use a Thermo Fisher 3130-XL sequencer to process the samples and Thermo Fisher GeneMapper software to analyze the data. Processing the samples is an important step before genetic analysis, and the Institute of Forensic Science researchers mention use of the Dutch company Qiagen's QIAamp DNA blood sample purification kits in nine of the applications and patents (see table 5.1).

The post 9/11 development by Kidd Lab of the fifty-five AISNP panel and subsequent international efforts to expand its testing and validation and other ancestry differentiation research cooperation with the

PRC's Ministry of Public Security and PRC funding agencies involves the convergence of US forensic genetic assemblages and sequencer makers with those in China. These assemblages' use of the organizing narratives of forensic genetics, with their shared quests for security against the antisubjects of the terrorist, the criminal, and the insurgent, are readily translatable and allow for the coordination by researchers, funders, companies, and security agencies from the US and China. Internally within Chinese security discourses, the same manhunt organizing narratives that shape patent applications and scientific papers, with its epideictic rhetoric of using forensic genetics to reveal the terrorist and the criminal hidden among population flows, was readily translated into patent office organizing narratives. This identity dominance is an extension of PRC political control at the genomic level over the Uyghurs and the goal of securing and governing increasing international flows of business and labour through China, as was done with the Belt and Road Initiative (Schwarck 2018).

The racial sorting technologies analyzed in this chapter mainly used ancestry estimation markers. However, an exception is the Tibetan altitude adaptation markers patent application that involves using physical trait–related genetic markers to estimate ancestry. In the remainder of the book, we will shift to phenotyping research to find genetic markers to estimate visible physical traits.

6

The VisiGen Consortium

Along with the ancestry inference marker research on the Uyghurs that intensified after the 2014 crackdown began, in 2014 the Ministry of Public Security also began to organize research on external appearance–related genetic phenotypes of Uyghurs. This would eventually expand to include large-scale projects involving tens of thousands of Western research subjects along with several thousand Han Chinese and over seven hundred Uyghurs in papers published between December 2017 to November 2019. However, to understand the origins of these large-scale projects we need to go back to the late 2000s and trace out two major assemblages. The first assemblage is the Visible Genetic Traits Consortium, which co-opted large-scale biomedical research projects into post-9/11 security state forensic genetic development in the European Union, including several prominent long-term biomedical research projects such as the Rotterdam Study that began in 1990, the well-known Twins UK project founded in 1992, and the Queensland Institute of Medical Research in Australia. The second major assemblage is based around the Chinese Academy of Sciences–Max Planck Society Partner Institute of Computational Biology in which evolutionary phenotype research begun in the late 2000s was co-opted into Chinese Ministry of Public Security research cooperation beginning in 2014. This pattern is similar to how Kidd Lab's work on genetic diversity and human evolution was co-opted into the development of ancestry informative marker panels after 9/11 in the US and China.

The 2000s saw a very rapid advance in the development of DNA arrays that can test for hundreds of thousands of SNP markers among the thousands of research subjects drawn from various large biomedical projects. In this way, these biopolitical optimizing biomedical projects (make-live) and research on evolution have been co-opted into the post 9/11 security apparatuses guiding the violence of legal and sovereign manhunts (make-die).

THE VISIGEN CONSORTIUM

A little used Google website created on 12 October 2010, introduces the VisiGen Consortium.[1] VisiGen stands for visible genetic traits. The website states, "The Visigen Consortium is an academic consortium dedicated to uncover the genetic foundations of visible traits. Visible traits are properties of human beings that can readily be seen in social interaction, such as hair or eye color, statue, facial morphology, etc." (VisiGen Consortium 2010a). The list of the original members includes several prominent genetic researchers: Manfred Kayser of Erasmus University and the Rotterdam Study; Tim Spector of the well-known Twins-UK project; Nick Martin of the QIMR in Brisbane, Australia; and Andres Ruiz-Linares of University College London and the CANDELA project, which involved Latin American subjects. This rather humble webpage outlines the beginning of a set of research networks, which from 2016 to 2019, would cooperate with Chinese genetic researchers in mass forensic genetic phenotyping projects involving tens of thousands of participants, including several hundred Uyghurs from the Xinjiang Uyghur Study. The VisiGen Consortium is an example of ostensibly large-scale health research projects being co-opted into post-9/11 security apparatuses. These major health-related projects like the Twins UK, Rotterdam Study, and those of the QIMR have publicly stated goals that emphasize the biopolitical goals of improving the health of particular national populations. However, they have been co-opted into forensic genetics as violence resources with the post-9/11 sovereign and legal goals of identifying the internal and external dangerous Other that threatens these privileged populations. In this way, the projects are coherently related in terms of their apparently contradictory goals: to biopolitically make-live in core regions and to secure these core regions by making-die in internal and external peripheries.

An early example of the post-9/11 security usage of Indigenous peoples and Uyghurs and other minorities by members of the VisiGen network is US patent application 20110312534A1 entitled "Method for Prediction of Human Iris Color" (Kayser, Liu, and Hofman 2011). The main claims of the invention are the ability to predict iris colour based on a set of six SNPs. Kenneth Kidd's submission of many cell lines to the Diversity Project including those of the Karitiana, Surui, Ticuna, and

Nasioi made them available to European and other researchers, including the VisiGen Consortium. Kayser, Liu, and Hofman (2011) followed a well-established pattern of utilizing research on Indigenous peoples' cell lines in their 2011 patent application. Before we discuss the patent, let us first consider the Diversity Project and forensic genetics.

COMMERCIALIZATION OF THE HUMAN GENOME DIVERSITY PROJECT

In the late 1990s, the Diversity Project was effectively shelved because of the extensive and effective lobbying of Indigenous activists and social justice organizations that caused major funding agencies to withdraw support (Barker 2004; Harry 1995, 2009). In 2005, one of the Diversity Project's major proponents, L.L. Cavalli-Sforza, writing in the aftermath, claimed that critics' objections of commercialization were unfounded: "they focused especially on the fear that indigenous people might be exploited by the use of their DNA for commercial purposes ('bio-piracy'). However, since its initiation, the Diversity Project has avoided commercial interests, and when the project was finally ready to be launched, it was made clear that DNA samples would be provided only to non-profit-making laboratories" (333). Cavalli-Sforza's statements were already disingenuous in 2005 since research based on the Stanford-Yale collection (which he helped organize along with Kenneth Kidd), which made up a significant proportion of the Diversity Project cell lines, had already been incorporated into patents filed by his own colleagues at Stanford University (Munsterhjelm 2014). Cavalli-Sforza's argument relies on narrowly defining commercialization as patenting of genetic sequences: "The HGDP has always opposed the patenting of DNA, to allow the study of genetic variation for fundamental research purposes" (333). He is referring to how the NIH filed patents on gene sequences from a number of Indigenous peoples. The Rural Action Foundation International's 1993 revelation of these NIH patent applications became an important part of the critique of the Diversity Project (Rural Action Foundation International 1993; Mead 2007, 34–5).

Using this narrow definition ignores the ways Indigenous peoples cell lines, data, and samples have been used as productive sources of genetic difference or as regional population representatives. In 1995, a team of researchers, including Alec Jeffreys (the inventor of DNA fingerprinting), filed a US patent application that was granted in 2001, with the major pharmaceutical company Zeneca as assignee (which after various corporate mergers is now part of AstraZeneca). US patent 6235468 claimed that usages include forensic identification processes (Baird, Royle, and Jeffreys 2001, column 15, lines 53–5).[2] The patent states, "the Karitiana DNA samples were a gift from K. Kidd and F. Black" (column 16, lines 6–7) and the "Karitiana are an inbred tribe from South America" (column 9, lines 31–4). Indeed, Cavalli-Sforza's own colleagues at Stanford University used many of these cell lines and data from the Stanford-Yale collection in a series of patent applications beginning in 1995. The assignee of the patents is Stanford University. The first patent application was US5795976 "Detection of Nucleic Acid Heteroduplex Molecules by Denaturing High-Performance Liquid Chromatography and Methods for Comparative Sequencing," which was granted in 1998 (Oefner and Underhill 1998). The application utilized data in the form of two chromatograms (graphs made by chromatography) from Karitiana, Surui, and Indigenous peoples from Colombia as examples to demonstrate the technology. Another patent granted in September 2002, US 6453244, uses research on Karitiana and other Indigenous peoples' cell lines from the Diversity Project (Ofner 2002). The 2002 patent is based on research in a 1997 paper published in *Genome Research* entitled "Detection of Numerous Y Chromosome Biallelic Polymorphisms by Denaturing High-Performance Liquid Chromatography." The article describes a novel method for detecting Y chromosome polymorphisms using 718 samples, including Karitiana, Surui, Ticuna, and Uyghurs (Underhill et al. 1997, 1003). A 2005 patent states its potential use in forensic genetic individual identification of males using Y-chromosomes: "In one embodiment, the invention provides a method for determining the ethnic origin of a male, comprising obtaining a nucleic acid sample from the male and identifying at least two polymorphic markers in the nucleic acid sample indicative of the ethnic origin of the male, using at least one primer pair from TABLE 1" (Oefner and Underhill 2005, column 4, lines 25–30). Oefner and Underhill (2005) advocated the use of the racial differentiation of crime victims: "The polymorphic sites and methods of the present invention are also useful in categorizing victims of violent crimes into ethnic and geographical groups. When a large number of victims need to be identified at a crime site, categorizing recovered victims by ethnicity can decrease the overall time for victim identification by reducing the number of comparison samples (samples from members of the victims family) to those of similar geographical origin" (column 15, lines 6–13). The mass casualty event is likely an implicit reference to the difficulties of the 9/11 identification efforts. There is no direct patent claim over Indigenous peoples' genetics, but rather the involved scientists used Indigenous peoples' cell lines, including Mbuti, Atayal, Ami,

Karitiana, and Surui, in the research process (column 15, lines 35–48). In particular, these Indigenous peoples were used to represent particular geographic regions in ancestry identification.

The proponents of the Diversity Project once proposed to sample over five hundred populations (Barker 2004, 574–5). However, today its cell lines from 1,063 individuals representing fifty-two populations are housed in a biobank (a cryogenic storage facility) in a Paris suburb (Fondation Jean Dausset n.d.a., n.d.b). Some of the only new samples that were collected for the Diversity Project were from the Chinese Human Genome Diversity Project (Chu et al. 1998, 11763; Chu 2001, 96; Fondation Jean Dausset n.d.b.). Chu (2001, 96) describes the Chinese efforts as underfunded, but nonetheless the Chinese research contributed samples from a number of Indigenous peoples and minorities, including ten Uyghurs, which were immortalized into cell lines and sent to the Diversity Project biobank (Fondation Jean Dausset n.d.b.). In contrast, the remainder of the cell lines representing populations outside China largely came from already existing collections, particularly the Stanford-Yale collection of Cavalli-Sforza and Kidd, which had been collected by various contributors in the 1980s and then immortalized into cell lines. The Diversity Project cell lines have been extensively used in forensic genetic research, with some of this research being commercialized.

THE IRIS COLOUR PREDICTION PATENT APPLICATION

Diversity Project cell lines were also used in the patent application entitled "Method for Prediction of Human Iris Colour," which sought to predict blue or brown eye colour in Europeans and adjacent populations. Erasmus University Medical Center's Manfred Kayser, Liu Fan, and Albert Hofman are listed as coinventors in the US Patent and Trademark Office application documents dated 4 March 2011. Erasmus University Medical Center is the assignee. The justification for a US patent is typically stated under the heading "Background of the Invention." The inventors argue the idea of using forensic genetics to estimate appearance in manhunts heralds a new era:

> Predicting externally visible characteristics (EVCs) using informative molecular markers, such as those from DNA, has started to become a rapidly developing area in forensic genetics. With knowledge gleaned from this type of data, it could be viewed as a biological witness tool in suitable forensic cases, leading to a new era of DNA

> intelligence' (sometimes referred to as Forensic DNA Phenotyping); an era in which the externally visible traits of an individual may be defined solely from a biological sample left at a crime scene or from a dismembered part of a missing person. (Kayser, Liu, and Hofman 2011, 1, para. 2)

The use of terms like "rapidly developing" and "[new] era" involve a type of epideictic rhetoric that emphasize the potential of these technologies as instruments of manhunting. The phrase "biological witness tool" is metonymy of genetics FOR identity, in which the "biological sample" stands FOR the victim or the perpetrator. The patent application states, "Furthermore, in disaster victim identification or other cases of missing person identification, DNA-based EVC [external visible characteristic] prediction would be useful whenever conventional STR profiles obtained do not match any putatively related individual" (20, para. 221). Again, this fits with the problems encountered by 9/11 identification efforts and other mass casualty disasters as discussed earlier. Europe is the centre of the geographic and population scope of the patent application with it stating: "Most human populations around the world have non-variable dark brown iris color while blue, green, gray and light brown colors are additionally found in people of European descent, and people originating from Europe-neighbouring regions. Thus, the DNA-based prediction of iris color may be useful in identifying persons of European and neighbouring descent, or persons residing in an area which is populated by persons of European descent" (1, para. 2).

This patent document makes express use of the word "race" as a taxonomic category to distinguish individuals from different populations. For example:

> The human from whom the nucleic acid sample is obtained can be of any race. As such, the human can be of any group of people classified together on the basis of common history, nationality, or geographic distribution. For example, the Subject can be of African, Asian, such as West Asian, Australasian, European, Middle Eastern, North American or South American descent. In certain embodiments the human is Asian, Hispanic, African, or Caucasian. (Kayser, Liu, and Hofman 2011, 3, para. 61)

There is a flexibility in racial taxonomies between geographically defined and more established categories like Caucasian. The taxonomy is also historically contextualized:

> Often the race of the human subject may not be known. The term "of European descent" means an individual who is a descendant of an individual who was born in a European country or territory in the 11th through 20th centuries, typically in the 15th through 18th centuries. Typically, at least 10%, at least 15%, at least 20%, at least 25%, 30%, 40%, 50%, 60%, 70%, 75%, 80%, 90% or 95% and up to 100% of the genetic material of a person of European descent is derived from ancestors who were born in a European country/territory or European countries/territories. The term "of West Asian descent" or "of Middle Eastern descent" can be understood accordingly. (3, para. 61)

The patent's SNPs are from research involving 6168 subjects from the Rotterdam Study (4–7). Indigenous peoples of the Diversity Project play a role in representing the populations of various parts of the world: "We also obtained the H952 subset of the HGDP-CEPH samples representing 952 individuals from 51 worldwide populations" on which to test the six selected iris colour SNPs (21, para. 225). This iris colour patent application includes a series of six maps of the world with pie charts of gene allele (variant) distribution floating over the home territories of various peoples, including the Karitiana, Surui, Nasioi, and Uyghurs, described as follows: "FIG. 4 displays the genotypes of 934 individuals from 51 HGDP-CEPH populations for each of the 6 SNPs included in the multiplex prediction test" (23, para. 231). This patent application utilizes a combination of existing racial categories like Caucasian in conjunction with the concepts of biogeographic ancestry that are more geographically specific because these categorizations are historically and geographically defined; we need to know, as Kahn (2015) asks, when are they from? The use of the Rotterdam Study subjects raises ethical questions because forensic phenotyping serves a fundamentally different purpose than biomedical research (Toom et al. 2016).

THE KEATING ET AL. PAPER

The integration of commercial endeavours and forensic genetic research, including security agencies, continued with a major cooperative paper published in 2012 on the Identitas forensic genetic testing chip. Founded in 2009, Identitas was one of the early commercial ventures that tried to exploit advances in phenotyping and ancestry by providing testing and analysis services to security agencies. The Identitas corporate webpage listed Manfred Kayser and Tim Spector as members of the company's scientific advisory board, and the VisiGen Consortium plays a prominent role in the 2012 paper.

The use of racial categories and Indigenous peoples as representatives of particular regions continued in this major early paper of the VisiGen Consortium, which included Kayser, Spector, and Martin. The paper also brought together a number of other significant figures including Kenneth Kidd, Bruce Budowle, scientists from FBI Labs in Quantico, Virginia, New York City's Chief Medical Examiner, and the Ontario provincial government's Centre of Forensic Science in Toronto.

The open access Identitas paper is entitled "First All-in-One Diagnostic Tool for DNA Intelligence," which it defines as "genome-wide inference of biogeographic ancestry, appearance, relatedness, and sex with the Identitas v1 Forensic Chip." This 2012 paper by "an international, industry-academic collaboration," claims that the new Identitas forensic chip will improve the efficiency of investigations (Keating et al. 2013, 561). The article's abstract begins, "When a forensic DNA sample cannot be associated directly with a previously genotyped reference sample by standard short tandem repeat profiling, the investigation required for identifying perpetrators, victims, or missing persons can be both costly and time consuming" (559). Having defined the quest, the authors then state how they are helping in this quest through a new technology: "Here, we describe the outcome of a collaborative study using the Identitas Version 1 (v1) Forensic Chip, the first commercially available all-in-one tool dedicated to the concept of developing intelligence leads based on DNA" (559).

A number of theorists have shown how surveillance biometrics function through various technologies to produce a "data double" of individuals across a range of measurements and data (Kruger 2013, 244, 247; Moffette and Walters 2018, 102–4). The Identitas chip involves a range of estimates of identity at both individual and population levels. This scope is summarized as: "The chip allows parallel interrogation of 201,173 genome-wide autosomal, X-chromosomal, Y-chromosomal, and mitochondrial single nucleotide polymorphisms for inference of biogeographic ancestry, appearance, relatedness, and sex" (Keating et al. 2013, 560). This expansion from individual identification involves the creation of a matrix of identity estimates at the level of the individual, such as gender estimates and appearance, but also at the level of the population, including ancestry and relatedness (Kruger 2013, 241–2). The use of the term "interrogation" may seem metaphorical, but it is actually metonymic because the 201,173 markers effectively represent a suspect in a metonymy of Genetic Markers FOR Various Traits of the

Suspect, a type of part FOR whole metonymy. In this way, the authors are answering a quest done in the name of the security and safety of the populace: "Our results demonstrate that the Identitas VI Forensic Chip holds great promise for a wide range of applications including criminal investigations, missing person investigations, and for national security purposes" (Keating et al. 2013, 560).

And it is here that we find the inclusion of Indigenous peoples as test targets: "The first assessment of the chip's performance was carried out on 3,196 blinded DNA samples of varying quantities and qualities, covering a wide range of origin and eye/hair coloration as well as variation in relatedness and sex" (Keating et al. 2013, 560).[3] This technology used 3,196 DNA samples that were extracted from cell lines or samples including those of the Karitiana and Surui from Kidd Lab as part of a total of 299 Indigenous people plus numerous people from minority groups sampled decades ago (562). VisiGen researchers Tim Spector and Nick Martin respectively provided a large number of samples from Twins UK and the QIMR in Australia (562). Significantly, as well, the SNP data, at Kenneth Kidd's request, has been included in the Kidd Lab ALFRED genetic database: "Genotype files obtained from the Visigen group contained data for 197,353 SNPs on 29 populations (including limited numbers from 27 Kidd Lab populations and two populations from others in the consortium). We have calculated and uploaded the allele frequency data for all these markers into ALFRED" (Kidd 2015c, 28). In this assemblage, Indigenous peoples and other populations functioned as test subjects for validating these commercial technologies, while also producing further publicly available SNP data.

The role of the FBI is also prominent in the paper, with a disclaimer stating, "SNP genotyping was supported in part by the FBI Laboratory Division. Names of commercial manufacturers are provided for identification only and inclusion does not imply endorsement of the manufacturer or its products or services by the FBI" (Keating et al. 2013, 569). The war on terror associations are invoked as the disclaimer continues: "This manuscript was filed under the number 12–18 at the Counterterrorism and Forensic Science Research Unit, Federal Bureau of Investigation Laboratory Division" (569). The FBI Quantico Lab's webpage describes its mission against the antisubjects of the criminal and the terrorist: "Whether it's examining DNA or fingerprints left at a crime scene or linking exploded bomb fragments to terrorists, the men and women of the FBI Laboratory are dedicated to using the rigors of science to solve cases and prevent acts of crime and terror" (Federal Bureau of Investigation 2021).

The Identitas forensic chip is based on "well-established Illumina Infinium technology" (Keating 2013, 561). And this technology is distinct from the earlier section in which Thermo Fisher capillary electrophoresis sequencing products, like the 3130XL, were significant. In these research networks, we will find that Illumina sequencing systems dominate high-resolution research that involves testing for hundreds of thousands of markers at once using arrays, a technology discussed in more detail later.

The Identitas project shows how involved scientists used Indigenous peoples' data and cell lines in development and testing processes. While some biomedical-related biotechnology development has been criticized for not providing any substantial benefits for Indigenous donors and their communities, these forensic genetic technologies will likely have a very direct impact on settler colonial governance of Indigenous peoples. The emphasis on isolated populations in genetic researchers' often demeaning Othering of the Karitiana and other Indigenous peoples and minority groups as consanguineous raises the potential for stigmatization in the use of these technologies since they produce findings on interrelatedness. The traces of this racially imbued spatiotemporality of the isolated Other is evident in the sample test output for the Identitas forensic phenotyping technology that provides a measurement of interrelatedness and whether the individual being tested comes from a "closed or isolated population" (Identitas 2016a). An example Identitas test report on their corporate website states the following about a sample: "Coefficient of Inbreeding = 0.0000" continuing that "The individual's parents do not share origins in a closed or isolated population" (Identitas 2016b). In effect, the technology provides a ready tool for stigmatization. The very wording of "Coefficient of Inbreeding" readily activates and invokes a derogatory and stigmatizing set of connotations and associations popularized during the eugenics period of the early 1900s and long wielded against racialized groups. Critically, there is a potential for pathologization of populations with higher levels of consanguinity. In this way, it enacts a particular hierarchy of mobility based on a norm of genetic heterogeneity in dominant large populations and flows of trade in which isolated populations are considered as abnormal due to their supposed lack of mobility or not marrying outside the group. This repeats long-established eugenics era concepts of isolated populations as aberrations that defy the norms of modernity. One can easily imagine scenarios in which these sorts of "coefficients" are used to demonize suspects and by extension their families and communities as genetically aberrant, particularly since these values have been pervasive in the assemblages in and through which these technologies have been researched, developed, and are now being commercialized.

CANDELA

While not included in the Keating et al. (2013) paper, what was to become an important addition to the VisiGen Consortium began in March 2010 when the Leverhulme Trust awarded Andres Ruiz-Linares a £122,000 grant for a project entitled "Network for the Study of the Evolution of Latin American Populations" (Leverhulme Trust 2010; CANDELA 2011, 2). This grant helped found the Consortium for the Analysis of the Diversity and Evolution of Latin America (CANDELA) that brought together Latin American scientists "with the aim of establishing a research resource for conducting research on genetic and social aspects of human physical appearance variation across Latin America" (2). To carry this out, "The consortium will recruit at least 1,500 research volunteers from each of five countries: Mexico, Colombia, Peru, Chile and Brazil ... Samples will be genotyped with 40 markers allowing the estimation of Native American, European and African ancestry of each individual" (2).

Ruiz-Linares and Balding then built on this initial effort with a further UK Biotechnology and Biological Sciences Research Council grant for £847,076 that ran from 2012 to 2015. There is only a single sentence on biomedical potential in the application summary, which instead emphasizes the market potential of this research on forensics. In the "Impact Summary," the authors state that the industry is still focused only on individual identification technologies of "DNA fingerprinting," which means that "Developing products for forensic DNA phenotyping would open an entirely new market sector" (UK Research and Innovation n.d.). Likely reflecting the British government's emphasis on commercial outcomes of research, the grant abstract continues, "A key element in our strategy to maximise the forensic impact of our research proposal is our collaboration within the VISIGEN consortium ... VISIGEN aims to foster research into the genetics of visible traits and promote its industrial exploitation. The consortium has initiated negotiations licence agreement with Identitas, a company interested in the development of a DNA chip (and accompanying algorithms) for forensic phenotyping in world populations. We anticipate that this license agreement will be signed while this proposal is being implemented" (UK Research and Innovation n.d.).

This entry likely alludes to the "industry-academic cooperation" mentioned in Keating et al. (2013). The abstract also anticipates potential for criticisms of its use of racial categories, stating, "There is also great public sensitivity towards this kind of research, as it can easily be perceived as reinforcing racial stereotypes and racism" (UK Research and Innovation n.d.). To counter this potential problem, the researchers will engage in social research supported by "two social anthropologists with extensive expertise [on] the social science implications of contemporary genomics research, Prof. Peter Wade (U. Manchester) and Dr. Sahra Gibbon (UCL)" (UK Research and Innovation n.d.).

The UK grant abstract's strong emphasis on forensic genetic phenotyping and its commercial potential stands in contrast to the way the CANDELA website presents the project. The CANDELA website, hosted by University College London, describes the project as concerned with how race and identity are constructed:

> We are an international, multidisciplinary consortium involving academic researchers studying the biological diversity of Latin Americans and its social context. We are currently focusing on individuals from five countries: Mexico, Colombia, Peru, Chile and Brazil. In these individuals we are performing a characterization of their physical appearance, examining their genetic make-up and social background, and evaluating their perception and attitudes regarding themselves and others. With this research we aim to probe a broad range of questions relevant to anthropological, biological and medical research. We also aim to explore the complex relationship between social and biological factors impinging on ideas about ethnic identity and race, and reflectively examine the motivations of biological research in Latin American populations. (CANDELA 2020)

This epideictic rhetoric for the University College London's CANDELA project webpage is vague about its applications with no direct mention of forensic genetic applications. This discrepancy involves an ethical and political disjuncture between the project's publicly stated goals and its proposed commercial use in forensic genetics (Toom et al. 2016).

CONCLUSION

The convergence of the CANDELA project with the VisiGen Consortium in developing forensic genetic phenotyping technologies took another important step when they began collaboration with the CAS in 2014 (UK Research and Innovation n.d.). In the next chapter, I will consider the early efforts of the Chinese Academy of Sciences-Max Planck Society Partner Institute of Computational Biology in Shanghai to develop forensic genetic phenotyping technologies before turning to their subsequent cooperation with the VisiGen Consortium in chapter 8.

7

Phenotyping Uyghurs in Shanghai

Institutionally and politically, Shanghai became a significant node in the transnational assemblages that converged with Chinese security agencies involved in developing manhunting technologies that target the Uyghurs. With heavy investments by national and Shanghai municipal governments, the city has emerged as a major site of biotechnology development in the PRC with high profile institutions including Fudan University and the Shanghai Institute of Biological Sciences. Fudan University is part of what is called the C9 League of elite research-intensive universities (Tsinghua University in Beijing is another), sometimes called China's Ivy League. One of the star institutes of these development efforts was the Chinese Academy of Sciences–Max Planck Society Partner Institute of Computational Biology in Shanghai (hereafter the Partner Institute) that ran from 2005 to 2020. Shanghai has become an important node in biotechnology by being able to mediate access to Chinese populations within transnational research assemblages. It is a graduated sovereignty arrangement in which PRC scientists (members of the Chinese Academy of Sciences) are able to organize productive international assemblages and travel due to their superior sets of rights (Ong 2006, 101; 2013, 78).

The Partner Institute represented a partnership between two of the largest and most influential scientific institutions in the world. For example, in the 2019 *Nature Index* that ranks institutions based on the number of papers published in eighty-two high-ranking journals, the CAS ranked number one while the Max Planck Society ranked number three (Nature Index 2020). The CAS has oversight of over 120 institutions employing some sixty thousand scientists and a 32.23 billion yuan budget (US$4.6 billion) in 2018 (Nature Index 2019). For example, in 2017, the Max Planck Society in Germany had eighty-four institutes and over twenty-three thousand staff, and received funding of US$1.9

billion, mainly from the German federal and state governments (Nature Index 2019). Organizational links between the Max Planck Society and the CAS began in 1974, as part of the general opening of China following the Cultural Revolution (Max Planck Gesellschaft 2014). The Partner Institute began in 2005, with Jin Li, who did advanced research with Lucas L. Cavalli-Sforza at Stanford and worked at various universities in the US, as one of the cofounders (Munsterhjelm 2014, 116–18, 172). From its opening in 2005 through to 2012, the German government provided one-third of the funding and the CAS provided two-thirds (Max Planck Gesellschaft 2014). The institute's governance structure was organized along Max Planck Society lines and there were extensive exchanges between the Max Planck institutes in Europe and the Partner Institute. In these research assemblages, the Partner Institute involved a graduated sovereignty zone in which there was a delegation of Chinese sovereignty made by the PRC government to the Max Planck Society in which the society governed the institution in conjunction with the Chinese Academy of Sciences. Through these extensive institutional networks, Chinese scientists and Western scientists made the Partner Institute into a major centre of research and production in the global forensic genetic assemblages.

The Partner Institute's symbolic significance is captured in a photograph on its webpage on the Max Planck Society website. Under the heading "Meet the President!" the photo caption proclaims, "On Wednesday, December 5, 2012, the new Chinese president Xi Jinping met with 20 foreign experts, one of them Philip Khaitovich, director of the Partner Institute. It was actually the first meeting of the president with foreigners in his new position" (Max Planck Gesellschaft 2020). According to a 2017 article on Khaitovich (who is Russian) in the Chinese Communist Party's official newspaper, the *China Daily*, "The 43-year-old, who has pushed boundaries to discover why humans are smarter than other species and how the aging process happens, attended meetings for foreign experts with Xi three times – first at the Great Hall of the People in Beijing, then in Shanghai, and finally in Moscow" (Zhou 2017). He is quoted in the article stating, "Xi is a sophisticated leader who really looks far into the future." Khaitovich was not directly involved in any Uyghur-related forensic genetic research, but press coverage of him as the director of the Partner Institute provides some insight into the important ideological and political standing of the institute.

CO-OPTION TO TARGETING UYGHURS

The Partner Institute produced a series of five English language reports that provide a useful institutional history of the development of Uyghur and to lesser extent Tibetan and other Indigenous and minority-focused genetic research. The first report for 2005–07 (the document properties creation date is 10 September 2007) makes no mention of facial genetics research, Uyghurs, Tibetans, or Xinjiang (Chinese Academy of Sciences Max Planck Society 2007). However, its next report, entitled *Research Report 2005–2009* (the document creation date is 30 December 2009), shows how the Partner Institute began facial genetic phenotyping research in 2008–09, which is a similar timeframe to the VisiGen Consortium and CANDELA project (Chinese Academy of Sciences Max Planck Society 2009). Partner Institute researchers' initial research on Uyghurs focused on evolutionary genetics that classed them as a mixed European and East Asian population. There are descriptions of several projects involving Uyghur ancestry and how this might be tied to potential health-related genetics. As such, this emphasis on potential health applications is typical of 2000s genetic research discourses that touted the wonders of this rapidly developing science (National Human Genome Research Institute 2000; cf. 2000 draft of the Human Genome Sequence completion announcement by US president Bill Clinton and British prime minister Tony Blair). In the report, under the heading "Genome-Association Study of the Human Facial Morphological Variation," facial genetics are stated in evolutionary terms: "The human face is well known for its morphological diversity between individuals and among populations. It is the most commonly used proxy for self/mutual recognition, as well as ethnicity identification. Facial morphology has also been the essential subject in traditional Anthropology. To understand the Genetics of facial polymorphism and the evolution of specific morphological traits are fundamental questions in human evolution" (73).

The face stands through a metonymic chain for individual identity and ethnic identity. Planning for what would eventually become known as the Xinjiang Uyghur Study, the report states, "We have so far sampled around 1200 individuals in Shanghai and around 1000 individuals in Taizhou, Jiangshu province, and will proceed to collect a sample of similar size in the ethnic group Uygur [*sic*] in Xinjiang province. Full facial geometry will be aligned and characteristic features intra/inter-population are to be identified and prioritized for GWAS study" (Chinese Academy of Sciences Max Planck Society 2009, 74). GWAS stands for genome-wide association study, a method that attempts to map relationships between physical traits with genes and associated SNPs at particular locations (loci) on the genome.

Three years later, the 2010–12 report discussed the progress made in these studies on Uyghurs. Citing the 2008 paper by Xu et al., which argued that Uyghurs are 50 per cent European and 50 per cent East Asian, the report states:

> This makes Uyghur a good population to carry out admixture mapping. Since our face data can be analyzed in a fully quantitative way, we can synthesize features that best describe the inter-population divergence. We can either extract the PC [principal component] axis which best segregates the Han and European whole faces, or do this for a certain feature, such as nose (Figure 6). These PC axes will then be used for the Uyghur individuals to measure their phenotypic closeness to Han Chinese or Europeans. Groups of segregating features can be defined and the DNA pooling based GWAS can be used to pin down the associated loci. (Chinese Academy of Sciences Max Planck Society 2012, 59)

In this way, we can see how the Uyghurs are positioned as admixed populations between large regional groupings of Han Chinese and Europeans. The principal component (PC) analysis seeks to identify sets of markers associated with the loci that affect particular facial features. Principal component analysis here is a type of statistical analysis used to identify the set of markers responsible for the greatest amount of variance in the facial feature (or entire face) than the set that is responsible for the second largest amount of variance and so on. The report's brief section on facial phenotyping of Uyghurs then concludes:

> Furthermore, we plan to do genome-wide genotyping on 300 Uyghur individuals. Using every 3D face with its corresponding genome-wide genotype data, we can try to find preliminary associations for many different features and then validate them in independent samples. The high-density phenotype/genotype correspondence also allows more complex quantitative genetic models to be tested. One intriguing question is to ask whether a set of "face markers" can be defined and used to predict an arbitrary individual's face. (Chinese Academy of Sciences Max Planck Society 2012, 59)

This brief summary in the report is the first reference to facial prediction phenotyping with regard to the Xinjiang Uyghur Study.

The 2012–14 Partner Institute report also mentions that research cooperation began in 2012 with Xinjiang Medical University and Xinjiang University on a project entitled "Population Admixture in Xinjiang" (Chinese Academy of Sciences Max Planck Society 2014, 216). The sampling of the Xinjiang Uyghur Study occurred in 2013–14 and, depending on the paper, involved somewhere between seven hundred and a thousand Uyghur students from Xinjiang Medical University in Urumqi. This group of subjects has formed the basis of subsequent research up to the present day. The researchers framed their research that sought to differentiate Uyghurs from Han Chinese as evolutionary and anthropological problems, but as we shall see, its "ethnicity identification" and human evolution goals would be readily translated and so co-opted into the Chinese settler colonial state's manhunting and biometric surveillance technology development targeting the Uyghurs. If we understand the concept of the People as encompassing and genetics as a metonymic conduit linking the living, the dead, and the unborn, then evolutionary research and ancestry research are vital.

During the early 2010s, Tang Kun and his team made progress in developing more efficient systems for analyzing faces by automating the process of measuring distances between different "landmarks" on the face.[1] The Partner Institute *Research Report 2009–2012*, dated August 2012, briefly outlines Tang's team progress in developing techniques for analyzing three-dimensional facial images using advanced computational methods that "make full use of the high throughput data, one has to register the thousands of 3D face images into one coordinate system," thereby allowing comparison between Han Chinese, Tibetans, Uyghurs, and Europeans (Chinese Academy of Sciences Max Planck Society 2012, 57–8). Furthermore, they state that they are engaged in research on GWASs of genetic markers involved in facial features by utilizing these four groups (58).

Guo, Mei, and Tang, in their 2013 paper, explain this new technique to map human faces. They first photographed subjects using a sophisticated imaging system called the 3dMDface, made by 3dMD LLC, a UK/US company. Utilizing a set of cameras in conjunction with software processing, the 3dMDface system produces a high-resolution three-dimensional image, including a detailed facial map called a mesh with x, y, and z coordinates, which covers 180° of the subject's face from ear to ear.

Guo, Mei, and Tang then applied a technique for the mathematical modelling of three-dimensional surfaces called thin plates splines (see figure 7.1). In mass media, thin plates splines are used in computer graphics programs that morph one photo of a face into another. In facial research, this technique aligns facial images so they can then be measured and compared. Their alignment process utilizes the tip of the nose as a central landmark. A further sixteen landmarks, including the inside and outside corners of each eye and the corners of the mouth, are then automatically labelled (see 1b in figure 7.1). Based on this set of seventeen landmarks, the face is mapped using a mesh of triangle shaped vertices that measures over thirty-two thousand (x, y, z) three-dimensional coordinates (see 1c in figure 7.1).[2] The authors described the process as follows:

> The 3D face data set. Three-dimensional facial images were acquired from individuals of age 18 to 28 years old, among which 316 (114 males and 202 females) were Uyghurs from Urumqi, China and 684 (363 males and 321 females) were Han Chinese from Taizhou, Jiangsu Province, China. Another training set, which did not overlap with the first 1000 sample faces, consisted of 80 Han Chinese, 40 males and 40 females from Taizhou, Jiangsu Province, China. The participants were asked to pose an approximately neutral facial expression, and the 3D pictures were taken by the 3dMDface® system (www.3dmd.com). Each facial surface was represented by a triangulated, dense mesh consisting of ~30000 vertices, with associated texture (Figure 1). (2013, 3)

A follow-up 2014 paper entitled "Variation and Signatures of Selection on the Human Face" combined the above facial mapping techniques with analysis of available data from the Diversity Project, including Europeans, Han Chinese, and Uyghurs. This analysis explored whether genetic variation between facial features might be the result of conventional gradual evolutionary change, such as genetic drift, rather than environmental adaptation and/or sexual selection. The researchers used the same facial imaging techniques Guo, Mei, and Tang had used in 2013, but now expanded the populations from Han Chinese and Uyghurs to include Tibetans and Europeans:

> Four hundred Han Chinese (200 females and 200 males) who were 17–25 years old were sampled in Taizhou, Jiangsu Province. Three hundred and three Uyghur (200 females and 103 males) who were 17–25 years old were sampled in Kashi [Kashgar], Xinjiang. One hundred sixty-nine Tibetans (100 females and 69 males) who were 15–22 years old were sampled in Shigatse. All participants were required to have the same ancestry over three generations. Finally,

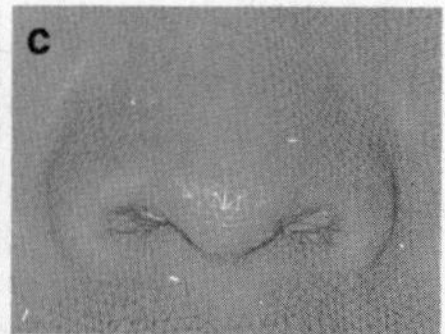

Figure 1 The surface used in our research. (a) The coordinate system used in our research (red, green and blue axes stand for x, y and z axes respectively). **(b)** An example scan with 17 landmarks marked by the colored spots. The red spots are the 6 most salient landmarks, namely the inner and outer corners of the eyes and both corners of the mouth, the blue spots indicate the other 11 landmarks used in this study. **(c)** Raw mesh details around the nose tip.

Figure 7.1 | Images and explanation of 3D facial mapping of a Uyghur male subject from Guo, Mei, and Tang (2013, 3). Note: the original article image is in colour.

> 89 individuals of self-reported European ancestry (32 females and 57 males) between 16 and 57 years old were collected in Shanghai. They were required to have complete European ancestry over the last three generations. (Guo et al. 2014, 144)

They did not conduct a genetic analysis of these participants. Rather, they utilized existing data including Chinese research on high altitude adaptation among Tibetans and the Diversity Project's research on Uyghurs, Chinese, and various European populations to identify a set of 187,290 common SNPs (145). The authors argue their study represented a significant advance, stating that "This study is, to our knowledge, the first comprehensive population differentiation analysis of the soft-tissue structures of the human face" and that their analyses "demonstrate that soft-tissue facial form does vary among the populations, and moreover provides information about population structure" (151). Tang's work, and that of Manfred Kayser, was discussed in a March 2014 *Nature* article, entitled "Mugshots built from DNA data," about Mark Shriver's paper on a simple computer model using twenty-four SNPs for twenty genes that are associated with facial shape (Reardon 2014).[3]

In addition to Tang Kun's team at Partner Institute, the Guo et al. (2014) coauthors include Mark Stoneking, an American scientist well known for his work in evolutionary genetics who is group leader at the Max Planck Institute for Evolutionary Anthropology in Leipzig, Germany. Stoneking coauthored a landmark 1987 paper that used mitochondrial DNA to estimate the last common female ancestor of all humans 140,000 to 200,000 years ago.[4] Jin Li is another influential figure. He did postdoctoral work at Stanford under L.L. Cavalli-Sforza during the 1990s. After working at a number of US universities, Jin was recruited back to China to become the codirector and cofounder of the Partner Institute from 2005 to 2009 (and is still a member). Jin's Fudan University profile begins, "Jin Li, male, was born in Shanghai in March 1963 in Shangyu, Zhejiang. Professor, Ph.D., member of the Communist Party of China" (Fudan University 2017; 金力，男，1963年3月出生于上海，浙江上虞人。教授，博士，中共党员). His CCP membership indicates his political status along with his educational credentials and status. The Fudan University profile also states, "Since May 2011, he has served as a member of the Party Committee, Standing Committee and Vice President of Fudan University" (2011年5月起任复旦大学党委委员、常委，副校长。).[5] This sentence indicates the close interrelationship between party membership and senior university administrative positions. Stoneking and Jin, along with Manfred Kayser of Erasmus University, are listed as scientific advisory board members for Tang Kun's research group in a 2014 article in the *Bulletin of the Chinese Academy of Sciences* entitled "Seeing Faces and History through Human Genome Sequences: CAS/MPG Partner Group on the Human Functional Genetic Variations" (Tang 2014, 164). The first page states, "The scientific goals of the Partner Group are to utilize large scale genomic polymorphism data to make fine inferences about human demographic history and various forms of natural selection in the human genome; and to understand the genetic basis and evolutionary mechanisms underlying the common variations in human facial morphology, both within and between populations" (161). This is a computational intensive mass data approach to try to find correlations and causal mechanisms between particular regions of SNP markers and gene expression, known as phenotypes, in facial appearance.

The Guo et al. (2014) article is an important transition as it is the first analysis of genetic data in conjunction with the facial mapping techniques from Guo, Mei, and Tang (2013). However, it still relied on use of existing genetic data from earlier research projects and the Diversity Project. The first paper using both facial measurements and genetic data from the Xinjiang Uyghur Study subjects appeared in 2016 in *Human Genetics*. Its authors include Tang Kun and Wang Sijia of the Partner Institute along with Jin Li of the Partner Institute and Fudan University and Pardis C. Sabeti of the Broad Institute at Harvard and the Massachusetts Institute of Technology. Entitled "EDARV370A associated facial characteristics in Uyghur population revealing further pleiotropic effects," the article found that a SNP variant called EDARV370A, which is prevalent in East Asian populations, was associated with multiple effects on facial characteristics, including earlobe shape and chin protrusion or reclusion (Peng et al. 2016, 102-6). The paper provides a brief account of the Xinjiang Uyghur Study recruitment process: "This study recruited 1027 Uyghur samples, who were undergraduate students from Xinjiang Medicine University, including 393 males and 634 females. The age of the samples ranged from 17 to 25 years. The study only recruited individuals whose self-reported Uyghur origins traced back to all four grandparents" (100). The authors assert they have complied with ethical standards: "The research was conducted with the official approval from the Ethics Committee of Fudan University, Shanghai, China. All the participants had provided written consents" (100). Using a Thermo Fisher SNaPshot Multiplex System, the scientists genotyped the subjects' EDARV370A SNP and ancestral origins by genotyping "24 Ancestral Informative Markers (AIMs) ... to calculate the ancestral contribution for each of the Uyghur samples" (100). This calculation was based on twenty-four ancestry inference markers selected from the International HapMap data for Chinese Han from Beijing (CHB) "as the East Asian ancestry" and Americans of European descent from Utah (CEU) "as the European ancestry" (101). They then used a software package called STRUCTURE "to estimate the ancestral proportion for each Uyghur sample based on the 24 AIMs [ancestry inference markers], using the CHB and CEU from HapMap Project Phase III as ancestral populations" (101). Metonymically, CHB stands FOR East Asian and CEU FOR European, which are well-established racial taxonomy categories. The authors argue that estimates of the "admixture time range from 800 to more than 2000 years ago" (106). As a result, "The admixture happened quite thoroughly in the population at the individual level, making most of the Uyghur individuals ~ 50:50 admixture of their East Asian and European ancestries ... This unique feature helps to reduce substructure within the Uyghur population and makes it an ideal population for association studies" due to its "high genetic diversity" compared to "a study of the same size in Han Chinese" (106). This paper and those above utilize an organizing narrative schema in which the scientists are investigating evolutionary adaptation including positive selection. Scientists construct the Uyghurs as a population based on the metonymy of genetics.

The narrative schema quest for evidence of positive selection also organizes the 2016 paper by Wu et al., which used genome-wide association scans to find variants of EDAR alleles that affected "hair straightness in Han Chinese and Uyghur." It includes a number of the same coauthors including Tang Kun, Wang Sijia, Jin Li, and Pardis Sabeti. The authors describe how "The Uyghur samples were collected at Xinjiang Medical University in 2013–2014. In total, 709 individuals (including 276 males and 433 females, with an age range of 17–25) were enrolled. The research was conducted with the official approval from the Ethics Committee of the Shanghai Institutes for Biological Sciences, Shanghai, China. All participants had provided written consent. In both studies, we collected blood samples, from which DNA was extracted" (1280). Like Peng et al. (2016), this paper also utilized the Xinjiang Uyghur Study participants but instead analyzed the 709 individuals' samples using Illumina array technologies: "All samples were genotyped using the Illumina HumanOmniZhongHua-8 chips, which interrogates 894,517 SNPs" (Wu et al. 2016, 1280). There is the familiar DNA intelligence metonymy of interrogation. As well, the Illumina's chips name is worth considering: Human, then Omni, which means *all* (originally from Latin), and Zhonghua is a Pinyin transliteration of the Mandarin Chinese word for China, so these Illumina array chips are designed specifically for the Chinese market, supposedly reflecting Chinese population characteristics.[6] According to the Illumina corporate website product description, "The Infinium OmniZhongHua-8 BeadChip delivers exceptional coverage of common, intermediate, and rare variation found within Chinese populations for genome-wide association studies (GWAS)" (Illumina 2019).

Preparing the SNP data for analysis involves a series of steps. The SNP data was first subjected to quality control, which included removing individuals with too much missing data or those who "failed the X-chromosome sex concordance check." Racial categories were at work in removing those whose "ethnic information [was] incompatible with their genetic information" and those that "failed the Hardy Weinberg deviation test" (Wu et al. 2016, 1280). "After applying these filters, we obtained a dataset of 2899 samples with 776,213 SNPs for the Han Chinese, and 709 samples with 810,648 SNPs for the Uyghurs" (1280). Eight hundred thousand SNPs might seem like a large number, but it is but a fraction of the estimated five million SNPs that are spread out across the three billion base pairs of an individual's genome. However, rapid advances in computing and analytical approaches in the 2000s enabled the development of a technique called imputation that makes use of existing data sets to estimate and so fill in SNPs that were not directly tested for by the researchers. Imputation takes advantage of the fact that groups of genes and their associated SNPs tend to be inherited together as a group called a haplotype. Therefore, the researchers used a program called IMPUTE2 that compared the eight hundred thousand or so SNPs from their test subjects with publicly available SNP data sets to estimate the nearby SNPs that were not tested for. The researchers used the 1000 Genomes Phase 3 haplotype data for "2504 individuals across the world for 81,706,044 variant positions" to impute the missing SNPs (1280). These imputed SNPs were then also subject to a selection process and, "Finally, for the Uyghur sample, a total of 6,414,304 imputed SNPs passed quality control and were combined with 810,648 genotyped SNPs for further analyses. For the Han Chinese sample, a total of 6,343,243 imputed SNPs passed quality control and were combined with 776,213 genotyped SNPs for association analysis" (1280). This imputation increased the genome-wide coverage of the SNPs by over eight times (from 810,648 SNPs to 7,153,891 SNPs), which increases the statistical power to detect associations between SNPs and particular physical traits (phenotypes). As well, imputing is much cheaper in terms of resources then actually testing for the missing SNPs, effectively leveraging the available SNP data obtained from Illumina OmniZhongHua-8 chip testing.[7] The researchers' findings did not support positive selection of EDARV370A for hair straightness among East Asians, but rather they hypothesize that "hair straightness might be a byproduct of strong selection on EDAR in East Asians for other traits" (1285).

The preceding papers involving the Xinjiang Uyghur Study are all framed in terms of researching issues of evolutionary adaptation including positive selection for visible appearance traits. The first paper to overtly utilize a manhunt forensic genetic narrative involving the Xinjiang Uyghur Study is Qiao et al. (2018), coauthored in part by Tang Kun, Wang Sijia, and Jin Li. It was received by the editors on 17 May 2018, a timeline that fits with the intensified crackdown on Uyghurs and the growing interest by Chinese researchers in the Uyghurs as a threat. Entitled "Genome-Wide Variants of Eurasian Facial Shape Differentiation and a Prospective Model of DNA Based Face Prediction," the article was published in the *Journal of Genetics and Genomics*, which is distributed by Elsevier B.V., a major academic publisher. Formerly a Chinese language journal, today it publishes only in English. Its editorial board is largely made up of academics at PRC universities, including several members of the Chinese Academy of Sciences, along with a number of US and European academics. The authors begin with a general statement of the importance of the face, which plays a "pivotal role in daily life" (419). They then move to ancestry, stating, "It has been long noted that faces bear characterized features that may surrogate one's ancestry, even in highly admixed populations," which involves the metonymic chain of facial appearance FOR genetics FOR ancestry (419). They then move to their specific project, citing the earlier paper by Guo et al. (2014) on "strong morphological divergence" between Europeans and Han Chinese in various facial features including "nose, brow ridges, cheeks and jaw" (Qiao et al. 2018, 419). These differences between Europeans and Han Chinese leads them to their quest: "Which genetic variants contribute to the substantial morphological differences among continental populations?" (419). In response, they do a brief literature review, citing various studies on genetic variation and facial morphology. They then specify their quest as identifying loci that affect "divergent facial morphological features between Europeans and Han Chinese" (420). This was carried out by using a GWAS on the "polarized face phenotypes along the European-Han dimensions, and Uyghur was used as the study cohort to dissect the genotype-phenotype association" (420). The authors tout the Uyghur as having an "ancient admixture between East-Asian and European ancestries" of about 50 per cent each, which makes them an "ideal group to study the genetic variance of divergent facial features across Eurasia" (420).

The article uses similar 3D facial mapping and array and imputation techniques to those in Wu et al. (2016). The phenotype discovery groups were composed of eighty-six "Europeans living in Shanghai" (EUR) and 929 Han Chinese from Taizhou (HAN-TZ) (Qiao et al. 2018, 420). Fifteen facial landmarks and six facial features were "extracted from" high-density 3D facial image data and a total of twenty-two facial features were identified as diverging between European and Han Chinese subjects (419–20). These twenty-two features included, for example, the distance from the nasion (the depressed point where the bridge of the nose joins the skull) to the pronasale (tip of the nose) to the subnasale (base of the nose above the lip) and the prominence of the brow ridge. According to the authors, "A candidate phenotype was chosen and termed as an ancestry-divergent phenotype if there existed a strong phenotypic divergence between EUR and HAN-TZ, and showed a wide distribution in UIG-D" (420). The 694 Uyghurs in the discovery group (UIG-D) and 171 Uyghurs in the replication group (UIG-R) were genotyped using Illumina Omni ZhongHua-8 array for 894,956 SNPs and Thermo Fisher's Affymetrix Genome Wide Human SNP Array 6.0 for 934,968 SNPs, respectively.[8] A GWAS was done of this imputed panel "for the 22 ancestry-divergent phenotypes," and six SNPs "revealed signals of genome-wide significance" (420). For example, these SNPs included rs118078182 (related to the gene COL23A1 on Chromosome 5), which is associated with the aforementioned nasion-pronasale-subnasale distances (420). The authors include computer images of faces to illustrate the effects of an A allele versus a G allele at rs118078182 on Chromosome 5 in which the A allele image has a less prominent nose compared to the G allele image (424).

Another major aspect of the paper was the authors' creation of a 3D facial prediction model for Uyghur faces using a panel of 277 SNPs, which was tested on the Uyghur replication group (UIG-R). To make the process of assessing the similarity between the real and predicted faces independent of human judgment, the authors "constructed a robust shape similarity statistic" called the shape space angle to quantitatively measure the similarity in 3D shapes (Qiao et al. 2018, 423). The model constructed "realistic 3D faces significantly closer to the actual face than random expectation" (425). To assess the usefulness of this model, they conducted a "forensic scenarios simulation" that tested the accuracy of facial prediction by randomly selecting eight images and then assigning one of them to be the "suspect" (425). The authors state the prediction model improved the odds of selecting the suspect in a "moderate yet significant" way for Uyghur males but not for Uyghur females (425). With reference to the prediction model, the authors state that it "is highly simplified and explorative. Much work is needed to improve its performance to be formally tested in real forensic scenarios" (427). As well, the paper includes several short animations of the effects of these six SNPs on facial appearance, along with a short animation demonstrating the difference between predicted and actual faces. This paper demonstrates the co-option of the Xinjiang Uyghur Study into forensic genetic phenotyping of Uyghurs.

From the inception of the Xinjiang Uyghur Study in 2009, the Partner Institute in the course of a decade became a significant centre of research on forensic genetic phenotyping in part through the use of Uyghur subjects in comparison to Han Chinese. The Xinjiang Uyghur Study began couched in evolutionary organizing narratives. However, after 2014, it was co-opted into forensic genetic phenotyping as part of the Chinese Ministry of Public Security targeting of the Uyghurs. This co-option demonstrates how genetics research on evolution and ancestry is readily integration into security apparatus based on the concept of genetics as conduits between generations that metonymically connect the dead, the living, and the unborn. The networks through which the

Uyghur genome became focuses of forensic genetic research increased in scope through cooperation with the VisiGen Consortium that began in 2014–16. The Xinjiang Uyghur Study subjects would now also be integrated into large GWASs along with cohorts in Europe, Australia, the US, and Latin America.

8

Securing Europe and China

This chapter considers how the two long-term assemblages of the Xinjiang Uyghur Study and the VisiGen Consortium (including the CANDELA project), which respectively date to the late 2000s, cooperated in a series of four studies that resulted in the publication of four papers between December 2017 and November 2019. These four papers illustrate the failure of ethics regimes to critically deal with forensic genetic assemblages as racialized sorting technology in the larger politico-economic context of increasingly genocidal forced assimilation in Xinjiang. Rather, the authors of these four papers constructed the seven hundred–plus subjects of the Xinjiang Uyghur Study as willing participants in these research projects and asserted the research was ethical, doing no harm to Uyghurs as a vulnerable population; claims that would later be challenged, as we will see in chapter 11. This research was supported by three of the largest security apparatuses in the world: China, the European Union, and the US.

BACKGROUND CONTEXT

This research cooperation occurred during the escalating repression of the Uyghurs and other Turkic peoples in Xinjiang that intensified with PRC president Xi Jinping's declaration of a People's War against terrorism and extremism in Xinjiang in May 2014. This PRC campaign of forced assimilation against the Uyghurs was further intensified when, in February 2017, the newly appointed Communist Party Secretary of Xinjiang Chen Quanguo urged officials to "round up everyone who should be rounded up." This marked the beginning of mass incarceration (Ramzy and Buckley 2019). The intensified crackdown on Uyghurs consists of two prongs: (1) intelligence-led policing that utilized a broad array of biometric surveillance technologies and (2) the use of reeducation camps to indoctrinate those identified as carriers of subversive ideologies (Greitens, Lee, and Yazici 2019, 45). These two strategies are part of an overall manhunt organizing narrative to identify, track, and capture those deemed as threats to Chinese national security.

In his discussion of biopower as power over life, Foucault argued that "killing or the imperative to kill is acceptable only if it results not in a victory over political adversaries, but in the elimination of the biological threat to and the improvement of the species or race" (2003, 256; Sean Roberts 2018b, 235). This life-and-death struggle to ensure the survival and well-being of the People is apparent in the common PRC epideictic rhetoric of a fight to the death with the three evil forces of separatism, extremism, and terrorism (Tobin 2020). For example, a 21 April 2017 news report entitled "For the Sake of the Motherland, Declare Total War on the 'Three Forces!'"(为了祖国利益向"三股势力"全面宣战！) about a mass oath swearing ceremony by Xinjiang security forces in Urumqi in April 2017 described the following resolute fight to the death:

> Zhu Changjie, vice chairman of the autonomous region, spoke loud and clear: the region's political and legal police should resolutely implement the spirit of General Secretary Xi Jinping's important speech, resolutely implement the general goal, vow to be the party's and the people's loyal guards, vow to fight the "three forces" to the end, guns are loaded, swords unsheathed, strike heavy blows, strike relentlessly, and fight for honour, righteousness and peace. We will not stop until we have won complete victory! (Sun Jian 2017; 自治区副主席朱昌杰的讲话掷地有声：全区政法干警要坚决贯彻习近平总书记重要讲话精神，坚决落实总目标，誓做党和人民的忠诚卫士，誓同"三股势力"斗争到底，枪上膛、刀出鞘，出重拳、下狠手，打出威风、打出正气、打出安宁，不获全胜决不收兵！)

This sort of sovereign call for a resolute fight to the death by a communist regime to ensure fulfillment of its vanguardist vision of the future was discussed by Mbembe (2003, 20):

> The flowering of a truly general will … presuppose[s] a view of human plurality as the chief obstacle to the eventual realization of a predetermined telos of history. In other words, the subject of Marxian modernity is, fundamentally, a subject who is intent on proving his or her sovereignty through the staging of a fight to the death. Just as with Hegel [i.e., master/slave dialectic], the narrative

> of mastery and emancipation here is clearly linked to a narrative of truth and death. Terror and killing become the means of realizing the already known telos of history.

The Chinese government's use of mass terror against the Uyghurs and other Indigenous peoples in Xinjiang is justified through this sort of telos, in support of President Xi Jinping's vision of "the Road to National Rejuvenation" and the "Chinese Dream." For example in a 1 December 2017 speech, Xi asserted the importance of this struggle for the Chinese nation: "The Chinese nation has a long history and a splendid civilization, yet it was ravaged by turmoil and upheaval of blood and fire since modern time began. But we Chinese never yielded to fate. We rose up and fought our way ahead with perseverance and, after protracted struggles, we have embarked on the broad road to national rejuvenation" (Xi 2018, 27). The pacification of Xinjiang has been a primary goal of the Chinese government under Xi Jinping, making it secure for Chinese settlement and investment. Something he makes clear in a September 2020 speech in which "Xi pointed out that the original aspiration and mission of the Party is to seek happiness for the Chinese people, including people of all ethnic groups in Xinjiang, and the rejuvenation of the Chinese nation, including various ethnic groups in Xinjiang" (Xinhua 2020).

Xinjiang had long suffered the structural violence and environmental racism of mass land expropriation, nuclear testing, mining, and industrial development typical of settler colonized regions, processes of repression further escalating since the 2009 Urumqi riots (Alexis-Martin 2019). However, in 2017, Chinese state repression shifted to an intensified necropolitics of forced assimilation and social death through state terror (Clarke 2021; Sean Roberts 2020; Kam and Clarke 2021). State measures have including mass incarceration of over one million Uyghur and other Turkic Indigenous adults in reeducation camps and placement of their children in state residential schools, coerced Han Chinese–Uyghur intermarriage, and destruction of traditional neighbourhoods as well as mosques, graveyards, and other sacred sites (Byler 2018; Clarke 2021, 9–16; Kam and Clarke 2021; Sean Roberts 2020, 228–30, 234–5). Other measures include close personal surveillance through the pairing of hundreds of thousands of Han Chinese civil servants with Uyghur families to monitor the latter, mandatory spyware on cell phones, hundreds of thousands of cameras with facial recognition spyware, checkpoints every few hundred metres in urban areas, geofencing to restrict Uyghurs' movements, and so on (Byler 2018; Clarke 2021; Sean Roberts 2020, 3, 224–7).

The Chinese government has repeatedly stated that the Uyghurs must be reformed to fit within the biopolitical norms of Han Chinese colonization. Since 2019, the China Global Television Network (CGTN) has posted extensive news reports and films that defend the Chinese government run camps. For example, a June 2021 news report on a news conference in Urumqi, in which Uyghur graduates of these camps lauded their stays, states, "Vocational education and training center graduates in northwest China's Xinjiang Uygur Autonomous Region have refuted accusations of 'genocide' in the region, saying the centers have saved them from religious extremism and helped them live better lives" (Huang Yue 2021). The racializing dehumanization of these subjugated peoples is central to the process in which the PRC asserts that the Uyghurs are a threat not only to the country but also to themselves. It is then a necropolitical logic of elimination, and the people are in a broad and sustained project to destabilize, severely curtail, or even eliminate Uyghur social community bonds and institutions that are not under strict settler state control and oversight, a form of pacification (Neocleous 2013, 8–9; Sean Roberts 2018b, 249–51; 2020, xiv–xv, 17–18; Tobin 2020, 302–3). The state security agencies' capital-imposing violence seeks to completely pacify the Xinjiang region in support of Chinese capitalist accumulation, including the Belt and Road Initiative and related expansion of settler colonization, in which a growing economy is considered a national security priority.

Under the long established racial hierarchies of Han Chinese settler colonialism (and Western imperialism), the largely Muslim peoples of Central Asia are considered suspect and predisposed to separatism, radicalism, and violence. Hence, the capacity to differentiate these populations involves the racial distinction between those who are considered normal and those who are considered abnormal and so require higher levels of surveillance and control (Kam and Clarke 2021; Sean Roberts 2018b). Chinese news reports and public and internal Chinese government documents, including the internal speeches of Xi Jinping, make frequent use of disease, pollution, and addiction metaphors affecting the body politic (Greitens, Lee, and Yazici 2019, 42–3; Leibold 2020, 57). These metaphors posit that ideologies from foreign countries are like viruses that "infect" the Uyghurs and other Turkic peoples, addictive drugs cause madness or spread like cancer, creating social instability and undermining their identification with the Chinese body politic. The following example is from the Xi Jinping regime's 2019 document *Vocational Education and Training in Xinjiang*:

> Terrorism and extremism are the common enemies of humanity, and the fight against terrorism and extremism is the shared responsibility of the international community. It is a fundamental task of any responsible government, acting on basic principles, to remove the malignant tumor of terrorism and extremism that threatens people's lives and security, to safeguard people's dignity and value, to protect their rights to life, health and development, and to ensure they enjoy a peaceful and harmonious social environment …
>
> It is hard for some people who have been convicted of terrorist or extremist crimes to abandon extremist views, as their minds have been poisoned to the extent of losing reason and the ability to think sensibly about their lives and the law. (State Council Information Office 2019b)

This pathologization of the Uyghurs and their culture securitizes them as important biopolitical threats to the Chinese people (Alexis-Martin 2019; Kam and Clarke 2021; Sean Roberts 2018b). Since 9/11, the PRC has translated US Global War on Terror discourses into its settler colonial project of constructing the Uyghurs and other largely Muslim Turkic peoples in Xinjiang as susceptible to the "three evils" of separatism, Islamic extremism, and terrorism and hence existential threats to China (Anand 2019; Smith Finley 2019; Sean Roberts 2018a; Rodríguez-Merino 2019). Internal speeches by Xi indicate he views ethnic separatism as a serious threat to the integrity of the Chinese state (Ramzy and Buckley 2019). Towards this end, Xi stated, "'We must be as harsh as them,' he added, 'and show absolutely no mercy'" towards any manifestations of such separatism (as quoted in Ramzy and Buckley 2019). The campaign of forced assimilation of the Uyghurs and other peoples of Xinjiang involves the intersection of biopolitics and necropolitics as social engineering (Kam and Clarke 2021). In particular, the designation of the Uyghurs as a threat to the Chinese state marks them as deviant and abnormal and so in need of reeducation and civilization (633).[1] This settler rationale is similar to Frantz Fanon's analysis of settler colonial discourses in *The Wretched of the Earth*:

> The colonial world is a Manichean world. It is not enough for the settler to delimit physically, that is to say with the help of the army and the police force, the place of the native. As if to show the totalitarian character of colonial exploitation the settler paints the native as a sort of quintessence of evil. Native society is not simply

> described as a society lacking in values. It is not enough for the colonist to affirm that those values have disappeared from, or still better never existed in, the colonial world. The native is declared insensible to ethics; he represents not only the absence of values, but also the negation of values. He is, let us dare to admit, the enemy of values, and in this sense he is the absolute evil. He is the corrosive element, destroying all that comes near him; he is the deforming element, disfiguring all that has to do with beauty or morality; he is the depository of maleficent powers, the unconscious and irretrievable instrument of blind forces. (1963, 41)

THE VISIGEN/XINJIANG UYGHUR STUDY PAPERS

By late 2017, there was already media coverage along with human rights reports of mass internment in Xinjiang that continued to grow in 2018 and 2019. However, between December 2017 and November 2019, four papers coauthored by members of the VisiGen Consortium, US researchers, and Chinese scientists involving over seven hundred Uyghurs of the Xinjiang Uyghur Study were nonetheless published. These articles were funded under two major European Union projects, the European Forensic Genetic (EuroForGen) Network of Excellence project and Visible Attributes Through Genomics (VISAGE) Consortium. Both projects have involved a number of significant social sciences and ethics researchers who analyze the technology in relation to the judicial and policing legislation of the European Union related to contemporary human rights and legal rights. For example, there were a number of EuroForGen project-related research meetings: "Each seminar will critically examine aspects of the potential and actual contributions of forensic genetics to the production of security and justice in the UK and other contemporary European societies" (EuroForGen 2016). Rephrased in critical terms, the seminars considered how forensic genetics has and could be involved in the production of social order through policing and judicial systems (Neocleous 2008, 2014). The EuroForGen "network includes 16 partners from 9 countries including leading groups in European forensic genetic research. It aims to create a close integration of existing collaborations as well as to establish new interactions in this highly specialized field of security. Therefore, all key players such as scientists, stakeholders and end-users (e.g., police institutions and the justice system), educational centres and scientific societies will be integrated into these activities" (EuroForGen 2019a). The VISAGE Consortium built in part on the earlier EuroForGen project and sought to develop and integrate a forensic genetic testing system using several technologies, including ancestry and phenotype, to "overcome the general limitation of current forensic DNA analysis by broadening forensic DNA evidence towards constructing composite sketches of unknown perpetrators from as many biological traces and sources and as fast as possible within current legal frameworks and ethical guidelines" (VISAGE n.d.). This composite creates what Kruger (2013) terms a *forensic-surveillance matrix* composed of individual, ancestry (population), and visible appearance phenotype estimates.

These security-related projects involved the VisiGen Consortium's co-opted biomedical projects. This is evident on the EuroForGen webpage for the Erasmus University Department of Forensic Molecular Biology headed by Manfred Kayser: "Project relevant infrastructures include the samples from Rotterdam Study, i.e., thousands of Dutch Europeans fully genotyped at genome-wide scale using microarrays and with various EVCs collected including hair structure phenotypes" (EuroForGen 2019b). The use of the term *infrastructure* is not metaphorical because the genomic and physical traits data from the Rotterdam Study subjects become actants in the production of knowledge, a similar role to Kidd Lab's conception of their collection of cell lines as resources. The webpage continues that the Department of Forensic Molecular Biology "has established close collaborations with multiple large cohorts, such as the Brisbane Twin Nevus Study from Australia, TwinsUK from United Kingdom, and Erasmus Rucphen Family study from the Netherlands. Together more than 17 thousand fully informative subjects are available for this project," rendering these subjects as violence work research and development resources (EuroForGen 2019b).

EUROPEAN UNION AND PRC COOPERATION

The period between 2014 and 2017 was one of intensifying repression by the Chinese security apparatus against the Uyghurs. Also during this period, there were a series of research exchanges and hiring of scientists between the VisiGen Consortium, the Partner Institute, and BIG that set the stage for these four research studies. In the outcomes section of the UK Research and Innovation website entry for the CANDELA Consortium grant studying over seven thousand Latin American subjects is a page on research collaboration that lists exchanges with the CAS beginning in 2014. These include "data sharing and joint data analyses" and a visiting professorship by Ruiz-Linares at the Partner Institute in Shanghai during the fall of 2014 in which he gave a talk entitled "Admixture in Latin America: Geographic Structure,

Phenotypic Diversity and Self-Perception of Ancestry Based on 7,342 Individuals – The CANDELA Study" (UK Research and Innovation n.d.; Chinese Academy of Sciences Max Planck Society 2017, 283). As noted earlier, Manfred Kayser of Erasmus University was also listed as a member of the scientific advisory panel for Tang Kun's Partner Institute projects in the 2014 CAS article on facial phenotyping (Tang 2014). According to the 2014–17 Partner Institute report, direct cooperation with the VisiGen Consortium began in 2016 (Chinese Academy of Sciences Max Planck Society 2017, 105, 276). In November 2016, Manfred Kayser gave a seminar talk entitled "Genetics and DNA Prediction of Human Appearance" at the Partner Institute in Shanghai (291). In 2017, Ruiz-Linares took a position at Fudan University in Shanghai, which is closely associated with the Partner Institute (he also maintained his University College London affiliation). Liu Fan studied with Kayser during his PhD at Erasmus and then cooperated in the early research of the VisiGen Consortium, including his joint patent application with Kayser (discussed in chapter 6). In June 2015, he was appointed under the PRC's National Thousand Young Talents Award to a full professorship at BIG while maintaining his assistant professorship at Erasmus University (Beijing Institute of Genomics 2016, 97).[2] In this way, a set of professional and personal relationships and exchanges underpinned the institutional connections that were the basis of these four studies.

During 2017–2019, despite the extensive media coverage of intensified repression in Xinjiang, including mass incarceration, the VisiGen Consortium (including the CANDELA Consortium) expanded forensic genetic phenotyping research cooperation with the network of researchers from the Partner Institute, BIG, and other CAS institutions. This cooperation was not merely a continuation of the relationships established during the early 2010s, but also involved significant changes in both research methodologies and the greatly increased number of research subjects, totalling nearly 29,000 in one paper by Liu et al. (2018). Methodologically, this coordination between these various major projects increased the numbers of research subjects available for GWASs and so increased the power of associations made between visible physical traits and genetic loci. For example, small associations between traits and genetic loci, which might not be statistically significant in a study using a smaller number of research subjects, can become significant with a larger set. Such smaller associations are important because facial and external appearance typically involve multiple genes (multigenetic), so being able to assess the potential contribution of many genes, some of which in turn also affect multiple traits, is an important part of predicting the appearance of suspects.

In this chapter, I will first analyze these four papers, published from December 2017 to November 2019, which made use of common racial taxonomies and incorporated the Xinjiang Uyghur Study subjects (see table 8.1). Then I will consider the ethical and political issues involved with this usage, particularly the disjuncture between biomedical studies and forensic genetics. The following narrative schema analysis of the four papers is not exhaustive due to the density of technical detail. Rather, I will briefly highlight the aspects that are significant to this book's focus on human hunting, particularly those involving the Uyghurs.

THE FOUR VISIGEN/XINJIANG UYGHUR STUDY PAPERS: LIU ET AL. (2018) FIND NEW LOCI

Paper #1

Coauthored by thirty-four scientists, the article entitled "Meta-analysis of Genome-Wide Association Studies Identifies 8 Novel Loci Involved in Shape Variation of Human Head Hair" was published online on 6 December 2017 in *Human Molecular Genetics*, an influential journal owned by Oxford University Press (Liu Fan et al. 2018). Its manipulation phase epideictic rhetoric uses a manhunt organizing narrative schema: "Unveiling the genetic basis of hair shape variation is relevant for understanding the molecular basis of human appearance, is potentially useful in cosmetics, and is expected to contribute towards finding unknown perpetrators of crime from DNA evidence in the emerging field of Forensic DNA Phenotyping" (560). This opening sentence asserts that this knowledge about hair shape will help advance forensic manhunting capacities. In their commitment phase, they specify the scope of their quest, noting that GWASs "have previously identified eight genes involved in human variation of head hair shape in different continental groups," and outline these various genes. However, they note that these eight genes "only explain a small proportion of the hair shape variation and trait heritability" (560). Therefore, the scientists commit to their quest, stating they will use genome-wide association scans to look for co-relations with SNP variants associated with particular genes: "Aiming to further improve the genetic understanding of shape variation in human head hair, and to find additional DNA predictors for future applications such as in forensics and anthropology, we performed a series of GWASs, replication studies, and prediction studies in a total of 28,964 subjects from 9 cohorts that include Europeans, East Asians, Latin Americans, and admixed individuals from around the world" (560–1).

Table 8.1 | Summary of the four VisiGen/Xinjiang Uyghur study papers

Lead author	Published online	Xinjiang Uyghur Study	Taizhou Longitudinal Study	Twins UK	Rotterdam Study	QIMR	CANDELA	Other contributions	Total Subjects (n)
Liu Fan et al. 2018	6 December 2017	709	2,899	3,347	2,809	10,607	6,238	Indiana University= 743	28,964
Pośpiech et al. 2018	29 August 2018	707	2,899			6,068		Indiana University=981 EUROFORGEN=1,434	12,089
Wu Sijie et al. 2018	24 September 2018	721	2,961		4,411		2,301		10,394
Xiong Ziyi et al. 2019	26 November 2019	709		1,020	3,193	1,101	5,958	ALSPAC= 3,707 (UK) 3DFacial Norms= 2,195 (US)	17,883

Each of the various projects like the QIMR and Twins UK has its own separate, brief description of its enrolment, ethics review claims, and genotyping procedures. For example, under the subheading "Chinese Taizhou longitudinal study and Xinjiang Uyghur study," the article states:

> The Xinjiang Uyghur (UYG) samples were collected at Xinjiang Medical University in 2013–2014. In total, 709 individuals (including 276 males and 433 females, with an age range of 17–25) were enrolled. The research was conducted with the official approval from the Ethics Committee of the Shanghai Institutes for Biological Sciences, Shanghai, China. All participants had provided written consent. In both Taizhou Longitudinal (TZL) and UYG, hair curliness was rated on a three-point scale (straight, wavy, and curly) by investigators. All samples were genotyped using the Illumina HumanOmniZhongHua-8 chips, which interrogates 894,517 SNPs. (572)

There is the familiar language of criminal intelligence in which the scientists are investigators using the Illumina array system to interrogate SNPs.[3]

The GWASs involved using a set of European and Australian subjects to find potential SNP markers from the QIMR, Twins UK, and Rotterdam Study (totalling 16,763 subjects), together these formed what the scientists termed the *META: Discovery*. Each cohort's GWAS data was processed using specialized software (e.g., PLINK) and the statistical technique of linear regression to look for associations between hair shape and SNPs. They identified twelve SNPs associated with particular loci/gene regions (twelve genes in total) and then tested to see if they could replicate these associations with data from subjects of each of the remaining cohorts from Holland, Poland, and the US, along with the CANDELA project that covers Latin America. They then combined the discovery and replication groups to form a "META: Non-Asian" group of 25,356 subjects. The final steps were using the eight novel SNPs in "a replication analysis in two additional East Asian cohorts," the Taizhou Han Chinese and the Xinjiang Uyghur Study (Liu Fan et al. 2018, 562). Only one of the eight new SNP markers was significant among the Taizhou Chinese, while for the Uyghur subjects, "None of the 8 novel SNPs were significant in UYG (East Asian-European admixed) despite its previously estimated 50% European genomic admixture" (562–3). Among the study's implications is: "Our data suggest that the genetic architecture of hair morphology in East Asians is substantially different than the rest of the world," as only one of the markers was mildly significant for Uyghurs and another

Table 8.2 | Liu et al. 2018 subjects and racial taxonomy

Five European cohorts	Project name	Racial taxonomic classification
N=10607	Queensland Institute of Medical Research Study (QIMR)	"North-Western European ancestry from Australia"
N=2809	Rotterdam Study (RS)	"North-Western Europeans from the Netherlands"
N=3347	TwinsUK Study	"North-Western Europeans from the UK"
N=977	Erasmus Rucphen Family Study (ERF)	"North-Western Europeans from the Netherlands" considered an isolated population
N=635	Poland (POL)	"East-Central Europeans from Poland"
Two admixed cohorts		
N=6238	CANDELA Study	"Latin Americans of estimated 48% European, 46% Native American and 6% African ancestry"
N=743	North Americans from the US	"US various origins including Europe, America, Middle East, and Asia"
Two East Asian cohorts		
N=709	Xinjiang Uyghur Study (UYG)	"Uyghurs of estimated 50% East Asian and 50% European ancestry"
N=2899	Taizhou Longitudinal Study (TZL)	Han Chinese
Total N=28,964		

for Han Chinese (569). This sort of null finding is also valid scientific data in that it eliminates potential SNPs from association with particular physical traits.

In the performance phase, the authors claim completion of their quest:

> In conclusion, with the 8 novel loci identified here and the 8 previously known loci confirmed here, we have substantially improved the human genetic knowledge of head hair shape variation in Europeans and beyond. We have increased the accuracy of predicting hair shape phenotypes from DNA genotypes over a previous model, which is relevant for forensics and cosmetics. Moreover, with newly reported hair shape genes and DNA variants we provide targets for future functional studies to further unveil the molecular basis of this externally visible trait expressing variation in people from around the world. (Liu Fan et al. 2018, 570)

The final sentence uses the metonymy of unveil as revealing something that was anonymous or hidden, particularly the face and hair as physical identifiers of a person, which fits with the two major applications of cosmetics and forensics. Metonymically, the SNP represents the gene (SNP FOR gene), and in terms of function, the SNP allele stands for how the gene phenotypically expresses itself as hair shape (allele FOR gene expression), and then genes' collective expression FOR phenotype, in this case hair shape for a particular region. The authors make use of continental racial categories including European, East Asian, Native American, and African, as outlined in table 8.2.

There is a very long list of government and private health and science funding agencies from the Netherlands, China, Australia, the UK, Russia, and Poland (Liu Fan et al. 2018, 573). The list of funding institutions includes the European Union to the EUROFORGEN (European Forensic Genetic) Network of Excellence, Manfred Kayser, and others, and the US NIJ (Grant 2014-DN-BX-K031) and the US Department of Defense (DURIP-66843LSRIP-2015) to Susan Walsh of Indiana University, as well as two major private UK philanthropic funders, the Wellcome Trust and Leverhulme Trust (Liu Fan et al. 2018, 573).

Paper #2

The 2018 paper by Liu et al. was followed up by a 2018 paper by Pośpiech et al. in *Forensic Science International: Genetics* entitled "Towards broadening forensic DNA phenotyping beyond pigmentation:

Improving the prediction of head hair shape from DNA." It is coauthored by thirty-seven scientists, including several from the Liu et al. paper: Liu Fan, Kayser, Walsh, Jin, Martin, and others associated with the VisiGen Consortium. It also included a number of other significant figures in the field of forensic genetics, including Niels Morling (University of Copenhagen), Peter Schneider (University of Cologne), Walther Parson (Medical University of Innsbruck), and others associated with the European Union–funded EUROFORGEN project. The authors used recent benchtop NGS forensic genetic systems, the Thermo Fisher Ion Torrent Personal Genome Machine, the Illumina MiSeq FGx (discussed in chapter 3), and MassARRAY array sequencing systems to test the eight SNPs and related data from Liu et al. (2018), plus a number of other SNPs identified in the literature.

The paper follows a manhunt organizing narrative schema. The manipulation phase uses the following epideictic rhetorical claim about phenotyping: "Predictive DNA analysis of externally visible characteristics (EVCs), also referred to as Forensic DNA Phenotyping (FDP), is a fast growing area in forensic genetics," which creates an epideictic rhetorical sense of quantity of the increasing tempo and significance of the field. They then define the utility of phenotyping to the police manhunt, explaining that forensic DNA phenotyping "uses DNA evidence to characterize unknown donors of crime scene traces who cannot be identified with standard DNA profiling, to allow focussed investigation aiming to find them" (Pośpiech et al. 2018, 242). DNA is intelligence to guide the investigation to find the unknown suspects, but the authors point out that it is still limited in its potential applications – "currently restricted to the three pigmentation traits, eye, hair and skin colour" – thereby identifying the gap in knowledge they will seek to address (242).

In the commitment phase, the authors note that in "recent years" there had been phenotyping studies covering a number of traits including height, male hair loss, and face morphology. They note how Liu et al. (2018) conducted a GWAS involving 28,964 subjects and found twelve genetic loci, and they state they will test these and other promising SNPs in predicting human hair shape. In the competence phase, the scientists used the Uyghur data and the Taizhou Han Chinese and QIMR SNP data from Liu et al. (2018) to model potential SNPs. They selected ninety SNPs to test on their META-Discovery group of subjects from the Rotterdam Study, QIMR, and Twins UK. Of these ninety, sixty were found to have a nominally significant association. These sixty SNPs were then refined using "6068 Europeans from QIMR, 2899 Chinese from

TZL (Chinese Taizhou Longitudinal) and 707 Xinjiang Uyghurs known to be of 50% European and 50% East Asian admixed ancestry from our previous study" (Pośpiech et al. 2018, 245) to develop two prediction models: one for straight and non-straight hair with thirty-two SNP markers for twenty-six genetic locations and another model with the categories of curly, wavy, and straight that used thirty-three SNP markers for twenty-nine genetic locations. These models were then tested on a Replication set of 2415 samples that were "collected by 10 participants of the EUROFORGEN-NoE consortium and one additional partner from the USA" (Pośpiech et al. 2018, 243; Susan Walsh at Indiana University is the US partner). These samples were tested using Thermo Fisher or Illumina next-generation sequencing benchtop systems or the MassARRAY system. The authors conclude that the thirty-two SNPs for twenty-six genetic loci they identified and tested did add some improvement in the prediction of straight versus non-straight hair, but these were only able to "explain 12.1% of hair shape variation in the current model building set of 6068 samples [Europeans from QIMR] and … the prediction outcome was correct in 46% of all straight hair classifications and 78% of all non-straight hair classifications" (250). Due to these limitations, they conclude with the epideictic rhetoric that more research is required: "Thus, despite our ability to provide an improved model and DNA marker set for hair shape prediction, providing the next step towards broadening Forensic DNA Phenotyping beyond pigmentation traits, this study also demonstrates that the search for more hair shape associated DNA variants and the investigation of their predictive value in independent samples needs to continue" (250). In the sanction phase, the paper was peer reviewed and published.

These two papers involve significant assemblages in terms of the large numbers of scientists, research subjects, and institutions operating across a total of seventeen countries. The paper by Liu et al. (2018) found a number of hair shape–related SNP markers, then Pośpiech et al. (2018) tested these markers and additional ones in a practical experiment by using widely available benchtop NGS forensic testing systems, including the Illumina FGx and Thermo Fisher Ion Torrent PGM – a large-scale research study followed by an experiment of its applicability in the field. The laboratories that did the testing on the 1,434 European subjects were part of the EUROFORGEN network of sixteen institutions from nine countries. This network was built in part through a European Union grant of €6,613,680 (out of an overall budget of €8,187,320.61) that ran from 2012 to 2016 under the European

Commission Seventh Framework Programme, "FP7 Security – Specific Programme 'Cooperation': Security." The following epideictic configured call defines the programme's quest: "European security is a precondition for prosperity and freedom. The need for a comprehensive security strategy encompassing both civil and defence security measures must be addressed." Security is "a technique of power" fundamental to the formation of capitalist social order (Neocleous 2008, 4–6). The EUROFORGEN project webpage submits to this quest for security on the European Commission CORDIS website stating the objective of the project:

> The EUROFORGEN-NoE proposal aims to develop a network of excellence for the creation of a European Virtual Centre of Forensic Genetic Research. Forensic genetics is a highly innovative field of applied science with a strong impact on the security of citizens. However, the genetic methods to identify offenders as well as the creation of national DNA databases have caused concerns to the possible violation of privacy rights. Furthermore, studies to assess the societal dimension of security following the implementation of even more intrusive methods such as the genetic prediction of externally visible characteristics are highly relevant for their public acceptance. (European Commission 2017)

This project seeks to study the transition from individual identification to population/group oriented predictive phenotyping through an integrated approach encompassing scientific research, applied practice, and ethical/legal frameworks within current European Union member states' legal systems. These two papers were a small part of their overall activities, but the inclusion of the Uyghurs as research subjects/objects of exchange nonetheless contributes incrementally to this claimed goal of improving European security.

The US security apparatus was also served by this research. Susan Walsh of Indiana University did her PhD at Erasmus University and coauthored some of the early VisiGen papers (see, for example, Keating et al. 2013). Both of the above papers credit her funding from the US NIJ for a US$1,123,404 grant entitled "Improving the prediction of human quantitative pigmentation traits such as eye, hair and skin color using a worldwide representation panel of US, and European individuals" and the US Department of Defense for a US$146,450 equipment grant, "Improving knowledge on the genetic basis of human physical appearance for human identification through next generation sequencing technologies" (Walsh 2019a, 4). In her July 2019 NIJ funding report, Walsh briefly describes the research collaboration on hair structure and the papers by Liu et al. (2018) and Pośpiech et al. (2018) that she coauthored (2019b, 6–7).

In China, Liu Fan's BIG research group's cooperation with the VisiGen Consortium was translated into public security discourses. The *2018 Beijing Institute of Genomics Annual Report* has a chronology of significant events with several under the section on the Public Security Collaborative Innovation Center, which is jointly run by BIG and the Chinese Ministry of Public Security (I will discuss this problematic relationship in detail in the next chapter).[4] Under the entry for 28 November, the paper states, "We identified 8 novel loci involved in human hair shape variation (Hum Mol Genet)," which is a reference to the Liu et al. (2018) paper discussed earlier (Beijing Institute of Genomics 2019, 89). The entry also mentions findings on hair colour variation and age-related DNA methylation among Han Chinese and concludes, "These results improved our understanding of the genetic basis of human externally visible traits and contributed to the development of forensic DNA phenotyping" (89).

Paper #3

The first two papers in this section are strongly linked through the reuse of data from theoretical research by Liu et al. (2018) and its experimental application in Pośpiech et al. (2018). The next two papers are not directly related to each other, but in making use of functional genetic experiments to test whether changes in particular SNP allele markers affect gene expression, they both involve another significant development. The involved scientists manipulate human cell lines using two technologies, luciferase assays and CRISPR genetic editing, to assess whether particular SNPs affect the expression of particular genes.

The first of these papers, by Wu et al. (2018), was published in *Public Library of Science Genetics* and entitled "Genome-wide association studies and CRISPR/Cas9-mediated gene editing identify regulatory variants influencing eyebrow thickness in humans." It is coauthored by Wu Sijie, Liu Fan, Jin Li, and Tang Kun, along with other BIG and Partner Institute colleagues, European scientists, including Manfred Kayser, and Latin American scientists, such as Andres Ruiz-Linares. Rather than the manhunt narrative schema, the paper's manipulation and commitment phases utilize a schema of scientific discovery in which the scientists are seeking to understand whether the shape of the eyebrow is subject to positive selection in different populations (2).

The paper also puts the materials and methods section after the results discussions. Instead of European populations as the main target, there is a different geographically defined racial logic in the competence phase's subnarrative sequence that begins with GWAS scans of 2961 Han Chinese of the Taizhou Longitudinal Study as a discovery cohort to identify SNPs. These SNPs were then tested in replication cohorts of the CANDELA study, the Rotterdam Study, and the Xinjiang Uyghur Study. In the enrolment subnarratives, the "721 Uyghurs (including 282 males and 439 females, with an age range of 17–25) were enrolled at Xinjiang Medical University, Urumqi, Xinjiang Province, China, as part of the Xinjiang Uyghur Study (UYG) in 2013–2014" and are described as "an admixed East Asian-European population" (Wu et al. 2018, 12). The paper claims that all research subjects gave informed consent. This 2013–14 enrollment date would become important in the controversies of late 2021, as we will see in chapter 11.

In the phenotyping, "Eyebrow thickness (i.e., density) was rated by eye on a three-point scale (TZL, UYG, RS: scarce, normal, and dense; CANDELA: low, medium, and high), following an established standard and based on photographic imagery" (Wu et al. 2018, 12). In the genotyping and imputation competence subnarrative, "For TZL and UYG, blood samples were collected, and DNA was extracted. All samples were genotyped using the Illumina HumanOmniZhongHua-8 chip, which interrogates 894,517 SNPs" (13).[5] Utilizing a series of software programs to analyze the data, including one called DeepSEA that uses a deep learning algorithm, four SNPs were eventually selected. All four SNPs are in noncoding regions, i.e., they were not in genes. There are only 20,000 to 25,000 genes, so while most SNPs are not directly involved in protein production, a high percentage are in the regions involved in transcription, which regulates the expression of genes (11).[6]

In a set of functional genetic experiments, the researchers used two different technologies to directly manipulate human cells to test the relationship between alleles in gene expression. The first of these, called a luciferase assay, involves isolating an SNP marker associated with the expression (transcription) of a gene (SOX2) on the third chromosome (Wu et al. 2018, 15). The second technology used in the functional genetic experiments was the recently developed gene editing technology called CRISPR. The authors chose four markers, including one on chromosome 3 (rs1345417) and one on chromosome 5 (rs12651896), that had high correlations to the expression of two different genes (SOX2 and

FOXD1) to be tested on human cells using CRISPR. CRISPR is a powerful and controversial new gene editing technology that uses immune proteins from bacteria to edit human or other species' cells.

In another subnarrative, the scientists then sought evidence of positive selection utilizing various statistical approaches to analyze four genes called SOX2, FOXD1, FOXL2, and EDAR. However, they conclude, "Our results thus provide no evidence that eyebrow thickness is under strong positive selection in human populations" (Wu et al. 2018, 9). The authors then discuss the heterogeneity of alleles among different populations, including a FOXL2 variant "associated with eyebrow thickness in Latin Americans" that they found no evidence for "in East Asians or Europeans" (11). In the concluding paragraphs, the authors consider their findings in relation to various articles from the literature. In the article's performance phase, the authors claim completion of their quest stating: "In conclusion, we identified three novel genetic variants near SOX2, FOXD1 and EDAR that influence eyebrow thickness. Furthermore, we found evidence for population heterogeneity in the genetics of eyebrow thickness. Finally, our results suggest that eyebrow thickness may not be subject to strong positive selection" (12).

The ability to now routinely insert targeted sequences into human cells to assess their role in gene expression is a significant development in forensic genetic phenotyping research. This ability to directly test on human cells is a significant change in research practice, which previously depended on correlations suggested in other studies. It illustrates the importance of the fungibility of cells to biotechnology development, in some ways similar to earlier immortalization processes using Epstein-Barr virus that produced the Karitiana and Surui cell lines.

"Largest set … thus far"

The general culture of complacency and indifference in forensic genetic research is exemplified in the November 2019 article "Novel Genetic Loci Affecting Facial Shape Variation in Humans," published in *eLife*, an online journal supported by the Howard Hughes Medical Institute, the Wellcome Trust, and the Max Planck Society (see figure 8.1). The paper is coauthored by forty-nine scientists, including VisiGen's Kayser, Spector, Martin, and Ruiz-Linares in conjunction with Tang Kun of the Partner Institute and Liu Fan of BIG and Erasmus University. This paper would prove to be the last of this assemblage involving Uyghurs before ethical and political scrutiny of forensic genetic research involving Uyghurs reached a critical stage in late 2019, as we will see in chapter 11.

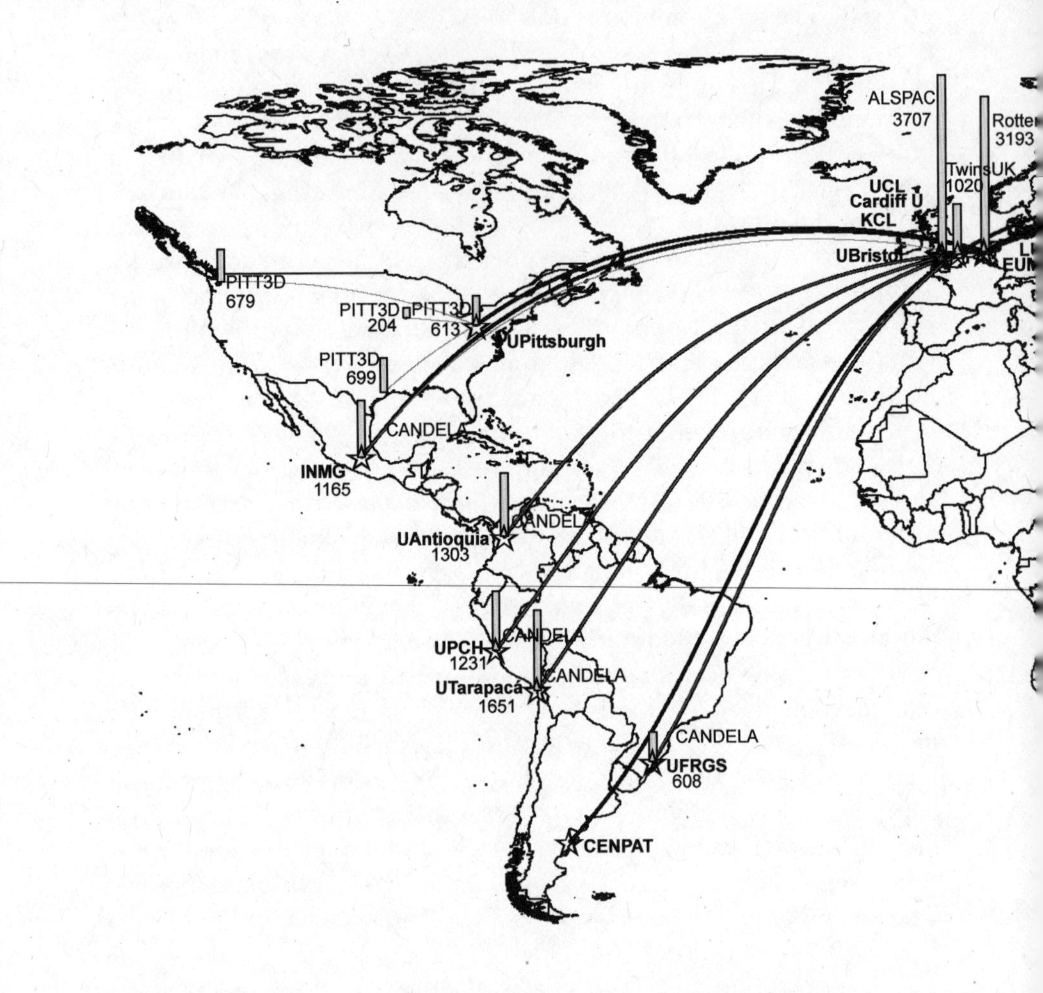

Figure 8.1 | Map of the assemblage of institutions and number of research subjects from each study involved in Xiong et al. (2019). The involved scientific institutions are marked by stars.

The paper's manipulation phase describe the human face as a "multi-dimensional set of correlated, mostly symmetric, complex phenotypes with high heritability," noting further that there is facial similarity between twins and to a lesser extent other relatives and "stable facial features within and differences between major human populations, and

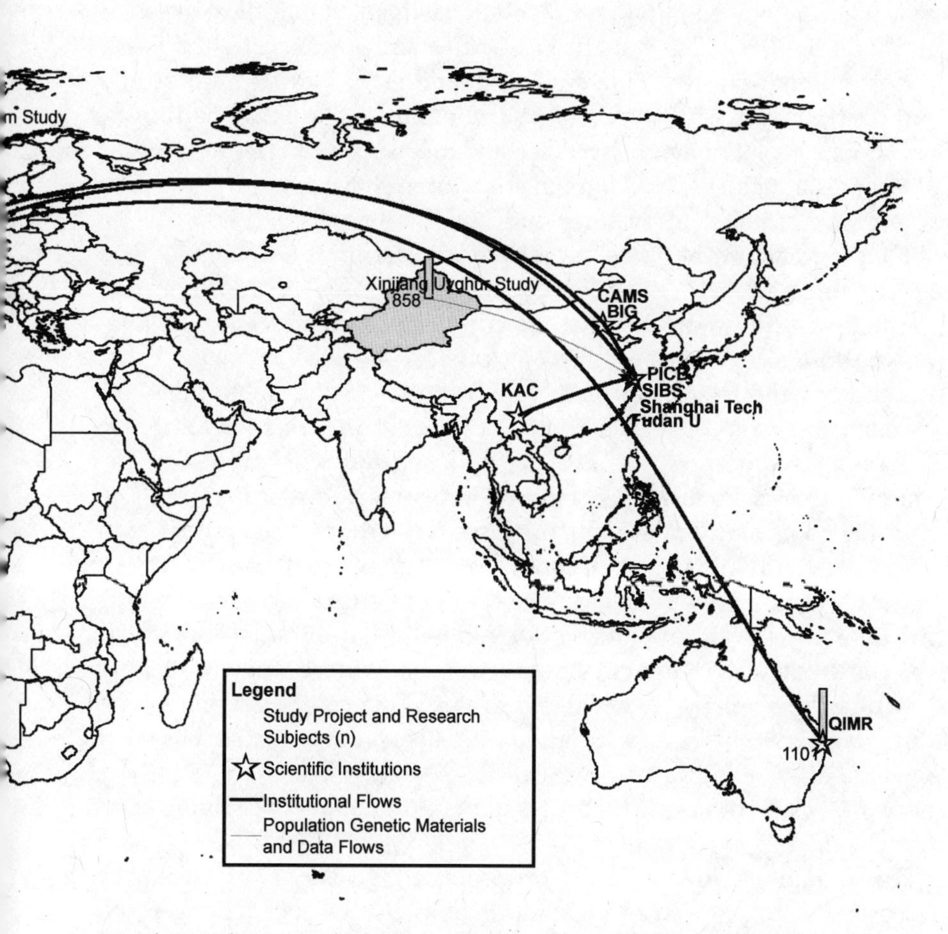

Map by Mark Munsterhjelm 2022. Based on Xiong et al. 2019.
The Pittsburgh 3D Study numbers of participants are approximate

the enormous diversity among unrelated persons almost at the level of human individualization" (Xiong et al. 2019, 2). This statement is followed by epideictic rhetoric calling for more research: "Understanding the genetic basis of human facial variation has important implications for several disciplines of fundamental and applied sciences, including human genetics, developmental biology, evolutionary biology, medical genetics, and forensics" (2). This sentence associates and so translates the paper and its findings into the different disciplines of the various authors and embeds the paper within their respective projects' organizing narratives, which span evolutionary, biomedical, and forensic genetic research, in this sense translating the project into an attractive passage point towards their respective research goals (Cooren and Taylor 2000, 185).

For the commitment phase, the scientists accept the quest, stating, "Led by the International Visible Trait Genetics (VisiGen) Consortium and together with its study partners, the current study represents a collaborative effort to identify novel genetic variants involved in human facial variation in the largest set of multi-ethnic samples available thus far" (Xiong et al. 2019, 3). The authors use the epideictic rhetoric technique of quantity, "the largest set … thus far," a superlative form to show the importance of the research assemblage (O'Gorman 2005, 28–32).

In the competence phase, for the discovery groups, the authors used 10,115 "North-Western Europeans" from the Rotterdam Study (3,193), TwinsUK (1,020), Avon Longitudinal Study of Parents and Children near Bristol in the UK (3,707) and 2195 people of "European ancestry from the United States" in the Pittsburgh 3D Facial Norms study to identify genetic markers related to facial appearance. They mapped seventy-eight distances between thirteen facial landmarks including the subnasale at the base of the nose (just above the lip), the sides of the nose, tip of the nose, the inside and outside corners of the eyes and mouth (Xiong et al. 2019, 3–4).

Their analysis found "24 face-associated genetic loci" and they selected the top SNP marker associated with each loci (Xiong et al. 2019, 5). These twenty-four SNP genetic markers were then tested on replication groups (cohorts) of 7,917 individuals. These included the CANDELA project (5,958) from Brazil, Chile, Mexico, Colombia, and Peru, who are described as "Latin Americans with ancestry admixture estimated at 48% European, 46% Native American and 6% African"; the Xinjiang Uyghur Study (858), who are described as "Uyghurs from China with ancestry admixture estimated at 50% East Asian and 50% European," and subjects of "North-Western European ancestry from Australia" in the QIMR Study (1,101) (Xiong et al. 2019, 6). All of the cohorts were genotyped using various Illumina systems and their data was imputed using the 1000 Genomes reference set. For example: "Xinjiang Uyghur (UYG) study. The UYG samples were collected at Xinjiang Medical University in 2013–2014. In total, 858 individuals (including 333 males and 525 females, with an age range of 17–25) were enrolled. The research was conducted with the official approval from the Ethics Committee of the Shanghai Institutes for Biological Sciences, Shanghai, China. All participants had provided written consent. All samples were genotyped using the Illumina HumanOmniZhongHua-8 chips, which interrogates 894,517 SNPs" (Xiong et al. 2019, 17). After removing samples with missing data, "ethnic information incompatible with their genetic information," and so on, they had a "dataset of 709 samples with 810,648 SNPs for the Uyghurs" (17).[7]

The GWASs for face phenotypes were conducted on the SNP data and the seventy-eight 3D facial distance measurements (which serve as phenotypes) using linear regression software like PLINK and Rare Metal Worker to look for relationships between SNP locations and particular facial distances (Xiong et al. 2019, 18–20). An overall meta-analysis was carried out that combined the GWAS "results for samples of European origin" i.e., the Rotterdam Study, Twins UK, QIMR, and Pittsburgh 3D Norms (20). The replication study found ten SNPs that were significant, six of which were new and four that had been described in other studies.

The authors assert that the use of a large meta-analysis was a successful approach because the studies of the individual groups testing for the SNP marker variants were underpowered; that is, none of the groups were big enough to show significant associations between markers and facial distance measurements. However, when the data from all the separate groups was combined and analyzed, the greatly increased number of subjects showed the effects of the SNPs (Xiong et al. 2019, 7).

The fungibility of human cells was important to the use of five SNPs for three different loci that were selected for luciferase assays. These assays involved the use of custom-made genetic fragments (which you can order online) that were inserted into cells (using a process called transfection) along with firefly and renilla (sea pansy) bioluminescent enzymes (luciferase) to see if there was an effect on gene expression (Xiong et al. 2019, 22). Two different cell lines were used; the first were "derived from human induced pluripotent stem cells" from the skin of a fifty-seven-year-old man established in the 2010s and the second were derived from melanoma skin cancer cells from a thirty-one-year-old Caucasian man that was established in the mid-1970s (Peebles, Trisch, and Papageorge 1978; Xiong et al. 2019, 22). The authors identified alleles of SNPs in a region (4q28.1 INTU) on Chromosome 4 that were associated with "reduced length of some nose features," while in a region on Chromosome 2 (2q36.1), some SNP alleles were associated with "increased length of some nasion-related features" (Xiong et al. 2019, 12). Nasion refers to the bridge of the nose.

In future studies, they plan to use further "direct functional testing of the associated DNA variants in suitable cell lines or mouse models." However, they note that genetic markers identified in their study only "explained small proportions of the facial phenotype variance with up to 4.6%" (Xiong et al. 2019, 15). Which means that there is still much to be "discovered in future studies" and that this information "may provide the prerequisite for practical applications of predicting human facial information from genomic data such as in forensics or anthropology. On the other hand, the increasing capacity of revealing personal information from genomic data may also have far-reaching ethical, societal and legal implications, which shall be broadly discussed by the various stakeholders alongside the genomic and technological progress made here and to be made in future studies" (15). This brief mention of the potential implications of the research for individual privacy fits with genetic research and the EU's general emphasis upon individual rights, but they never question the implications for the Uyghurs as a vulnerable people under increasingly genocidal repression.

The paper was positively sanctioned. The *eLife* open access website (funders include the Wellcome Trust and the Max Planck Society) that published it also included the peer review letters, which deal with various technical issues such as statistical thresholds (eLife 2021). None of the peer review opinions mention anything about the ethics of using the Uyghurs as research subjects nor of the political situation in Xinjiang.

ADVANCING ASSEMBLAGE GOALS

This research study is significant in how it translated phenotyping research of the face into a diverse range of the coauthors' respective research projects. The organizational complexity of the paper is evident in the forty different funding entries for the various coauthors (Xiong et al. 2019, 24–6). These include European Union Horizon 2020 counterterrorism funding, Shanghai municipal government, the CAS and other Chinese national government agencies, Netherlands government agencies, Medical Research Council of the UK, and Ruiz-Linares and Balding's UK Research and Innovation grant for the CANDELA project from 2013 (discussed in chapter 6). The paper is listed on several of the authors' respective project websites. The institutional actants' spatial temporality might be interpreted under some simple globalization theses that the state is abrogating its sovereign functions to the scientists. In some regards, this is true, in the sense that the scientists' networks exceed their respective national state boundaries or those of the European Union. However, state institutions perform sovereign decisions over funding and research ethics approvals that are covered by various pieces of national or European Union legislation, depending on the jurisdiction. Furthermore, laboratory processing was done in the country where the research subjects were sampled. This graduated sovereignty performance sees the laboratories under national and/or European Union regulatory frameworks and ethics authorizations and funding grants. The informed consent and ethics review board claims as sovereign decisions that enacted various national legislation, regulations, and international bioethics norms made by the various projects were accepted by the other members of the assemblage, a type of external recognition.

Within forensic genetic research assemblages, scientists are productive neoliberal subjects able to readily organize and articulate production relations that span multiple forms of time-space based on their authority within transnational science (Ong 2013, 78). Scientists constitute themselves as scientific researcher subjectivities by constituting Indigenous peoples as their objects of research (Munsterhjelm 2014; Reardon and TallBear 2012). By virtue of their capacities to enact leading edge technologies, there is a gap between regulatory regimes catching up, so in the absence of law, scientists exercise prerogative powers done in the name of the common good (Arnold 2007; Munsterhjelm 2014). As well, genetic researchers are foundational mediators and policy advisors and implementers of these technologies as expert witnesses, researchers, public speakers, and so on, hence the sovereign decisions over who has what rights and obligations are dispersed throughout the assemblage. In the Xiong et al. (2019) paper, the materials and methods section describes how each of the projects has been approved by local institutional review boards, at the university or institute, for example, which would place them under a national legal and ethical framework. However, this ethical claim is nonetheless reliant on it being recognized as legitimate by the other actors in the international assemblage.

EU DELIVERABLES: VISAGE ADVANCING SECURITY RESEARCH

In these four papers' assemblages, the researchers, through the funding process, become entrepreneurial under the competitive grant system, and like Kidd's funding reports earlier, we can see how researchers from the US and Europe translated this cooperation into their respective security-related research projects. These papers' research studies involve assemblages that traverse and articulate together multiple security apparatuses of the European Union, the United States, and China. It is this capacity to translate and thereby span and articulate such disparate apparatuses that makes these research assemblages very effective at gathering necessary resources and authorizations to carry out their respective research projects.

For the VisiGen Consortium, these papers represent a series of efforts to expand the scope of forensic genetic phenotyping beyond eye, hair, and skin colour. They are initial research efforts, which will nonetheless inform police practice through their direct inclusion in forensic genetic systems or through subsequent appropriation from the literature by security agencies scientists. In this way, there is the dispersed enactment of the police manhunt as the sovereign decision over the friend/enemy distinction. Erasmus University is a major production node in these networks in terms of research and training of personnel. The number one entry on the Xiong et al. (2019) funding list was the European Union grant entitled "Visible Attributes through Genomics: Broadened Forensic Use of DNA for Constructing Composite Sketches from Traces" (VISAGE).

The VISAGE project seeks to improve police and security agency capacity to engage in the sovereign decision over the friend/enemy distinction, which interacts with the biopolitical protection of favoured populations in the larger organizing narratives of the European Union's Horizon 2020 research funding program. The VISAGE project is funded under a Horizon 2020 program, "3.7. Secure societies – Protecting freedom and security of Europe and its citizens," which defines the parameters of its quest in its call for submissions: "This Challenge is about undertaking the research and innovation activities needed to protect our citizens, society and economy as well as our infrastructures and services, our prosperity, political stability and wellbeing" (European Commission 2022; n.d.). A subcategory within this program is "3.7.1 – Fight crime, illegal trafficking and terrorism, including understanding and tackling terrorist ideas and beliefs," which VISAGE is embedded under (European Commission 2022).

A central goal of the VISAGE project is a "Working Package" of genetic tests, including "enhanced age prediction," "enhanced ancestry prediction," and "enhanced appearance prediction" (VISAGE 2019b). A VISAGE funding report explains how findings from Liu et al. (2018) and Pośpiech et al. (2018), funded under earlier EU grants for the EUROFORGEN, were integrated into the Work Package:

> For head hair shape, EMC [Erasmus Medical Center] together with other partners from within and outside VISAGE previously published a marker search study [Liu Fan et al. 2018] and more recently published a predictive DNA marker set [Pośpiech et al. 2018] based on work done as part of the EUROFORGEN_NoE markers and prediction model for head hair shape were made fully available for use in VISAGE, as previously described in the Periodic Report. (VISAGE 2019a, 5)

The report identifies how the two studies were integrated into the VISAGE Work Package 4 (WP4), something that is restated in another report (D4.1): "WP4 adopted an already published model for hair structure prediction that was developed outside VISAGE (Pośpiech and others, 2018)" (VISAGE 2019b, 34).

The Deliverable D2.2 report has a three-page summary of what would become the Xiong et al. (2019) paper. However, the facial shape SNP markers from Xiong et al. were not sufficiently detailed so they would not be included in the prediction model, though the paper was summarized and included as a deliverable for the project (VISAGE 2019a, 33–4; VISAGE 2019b, 14–17). This paper was translated in the report as a milestone (MS) or goal in the project, asserting that: "The facial shape associated SNPs identified in this work were delivered for facial shape prediction analysis in WP4 (see D4.1), thereby fulfilling MS7 in time. Based on the outcomes of the prediction analysis in WP4 (see D4.1) it was decided not to include facial shape in the VISAGE tool developments in WP3 & 4" (VISAGE 2019a, 17). The D4.1 report states that the predictive value was too low: "The extremely small amount of facial shape variance they explain demonstrate that many more facial shape associated SNPs need to be identified in larger GWASs with more statistical power in the future, and tested for their predictive value, which may eventually bring facial shape prediction from DNA to a similar level as described for the other appearance traits above. Therefore, we decided not to use facial shape in the VISAGE tool development in WP3 and WP4" (VISAGE 2019b, 33–4). In this way, the research on the Uyghurs done in conjunction with the VisiGen Consortium was translated into European Union security systems for forensic genetic phenotyping. The authors also assert the need for larger scale studies in the future to find more SNPs associated with facial shape.

BEIJING INSTITUTE OF GENOMICS ANNUAL REPORT MENTIONS

In the 2019 *Beijing Institute of Genomics Annual Report*, once again Liu Fan's VisiGen cooperation is cited in conjunction with the Ministry of Public Security's Institute of Forensic Science:

> LIU Fan's group, with multi-international cooperation and the physical evidence identification center of the Ministry of Public Security, has made greatly important progress in the genetic basis of human morphological characteristics and the DNA portrait of forensic science, including the identification of several genes that affect human facial morphology (ELIFE, 2019; Hum Genet, 2019), uncovering genetic heterogeneity of height and optimize the prediction model for tall stature, which lays a foundation for forensic genetics research on adult height in East-Asian (Int J Legal Med, 2019) (Beijing Institute of Genomics 2020, 35)

"ELIFE, 2019" refers to Xiong et al. (2019) while the other two papers, in *Human Genetics* and *The International Journal of Legal Medicine*, both involve 715 Uyghurs in Tumxuk in southern Xinjiang who were sampled by the Ministry of Public Security and Bingtuan Security Bureau and would become the focus of considerable controversy and eventual article retractions, as we will see in chapter 11. Both Liu et al. (2018) and Xiong et al. (2019), which tested the seven hundred–plus subjects of the Xinjiang Uyghur Study, were translated into the forensic genetic research goals of the Public Security Collaborative Innovation Center, a collaboration between BIG and the Ministry of Public Security. Therefore, the VisiGen Consortium and its partners' cooperation with PRC-based researchers of the Partner Institute and BIG using Uyghurs was translated into advancing the respective security apparatuses of the European Union, the US, and PRC.

CONTRADICTION BETWEEN THE ETHICS OF BIOMEDICAL AND FORENSIC GENETIC RESEARCH

There is a fundamental ethical conflict when health-oriented genetic samples or databases containing data originally taken for health purposes is used for forensic purposes or otherwise adapted into police forensic investigations. In 2016, in a sharp rebuke of an article by Manfred Kayser advocating the widespread adoption of forensic genetic phenotyping in police investigations, a group of concerned academics, including Amade M'charek, Troy Duster, Helena Machado, and Erin Murphy, involved in critical research on forensic genetics wrote about a number of problems with Kayser's advocacy (Toom et al. 2016). These problems include routine phenotyping and ancestry research leading to police targeting "suspect populations" and the use of biomedical information as violating ethics of confidentiality and trust. Another issue was how biomedical studies were being used for forensic genetic research:

> Research and validation of markers for FDP [forensic DNA phenotyping] technologies has sometimes utilized biological material collected for biomedical purposes. A major challenge here pertains to the ethical practice of research around the donation of biological material: did donors who volunteered DNA for biomedical research also consent to its use in the development of FDP technologies for forensic uses? Research and development of biomedical applications in a healthcare or disease context and of forensic applications in a criminal justice context have very different aims and are subject to different ethical regimes of law, science, and practice. (e2)

Biomedical subjects' participation and donation of samples is an altruistic gift of life being made to another. Such altruism involves an epideictically-defined ideal that persuades the donor to contribute to the common good. One of the key aspects of informed consent is the ability of the donor/participant subject to determine the purposes to which their samples and data will be used.

The various health research projects publicly define their respective missions and project goals in biomedical terms of advancing human health, not forensic genetic research and development. The Rotterdam Study is a longitudinal health study with nearly fifteen thousand people from the Ommoord district of Rotterdam in the Netherlands: "The study targets cardiovascular, endocrine, hepatic, neurological, ophthalmic, psychiatric, dermatological, otolaryngological, locomotor, and respiratory diseases." (Ikram et al. 2018, 807). Similarly, a Twins UK webpage states its purpose as: "TwinsUK aims to investigate the genetic and environmental basis of a range of complex diseases and conditions" (Twins UK 2020). As well, according to the QIMR webpage, "The Genetic Epidemiology Laboratory seeks to identify the particular genes involved in complex disease aetiology" using longitudinal studies that follow thousands of research subjects for years and even decades (Queensland Medical Research Institute 2020). The three papers involving the QIMR research subjects (Liu Fan et al. 2018; Pośpiech et al. 2018; Xiong et al. 2019) state they provided informed consent. However, this informed consent was actually phrased in terms of a general blanket approval, "which includes sharing of genome data with collaborators and for unspecified research projects, subject to the new proposed research having ethical approval" (QIMR, email to author, 30 September 2021). None of these research projects' epideictic rhetoric mentions forensic genetic research directly. The QIMR webpage alludes to forensic genetics; for example, the listing of international research cooperation includes Manfred Kayser and Tim Spector and papers such as the facial shape paper by Xiong et al. (2019) that involves the Xinjiang Uyghur Study subjects (Queensland Medical Research Institute 2020).[8] The CANDELA project website's epideictic rhetoric, as discussed earlier, is similarly misleading in omitting any direct mention of forensic genetics; though Ruiz-Linares and Balding's funding application emphasized the market potential of forensic genetic phenotyping and ancestry research and development (UK Research and Innovation n.d.). In short, there is a radical disjuncture between the main stated public goals of these various projects and their usage in forensic genetics phenotype research involving Uyghurs, which contributes to the development of expertise and technology that targets Uyghurs. This research cooperation was another instance of the co-optation of biomedical health research projects into forensic genetic research and development as violence resources.

The above scientific cooperation that combines numerous research studies together to increase the potential power of associations between visible external traits and SNPs from particular genetic loci has multiple ethical and political problems. First, ethically, this joining together involves a hierarchy of research subjects between those subjects from Europe and Latin America who have autonomy and the right of withdrawal versus Uyghurs (and also likely the Taizhou Han Chinese) who do not have such rights in practice. In these assemblages, the scientists assert that the Partner Institute enroled the Uyghurs as research subjects capable of exercising free and full informed consent in 2013–14, implicitly asserting they are equal in their legal and ethical recognition to that exercised by other such subjects, such as the Australian, British, or Dutch citizens in the QIMR, Twins UK, or Rotterdam Study, respectively. Second, the research violates the ethical standard of nonmaleficence because Uyghurs are disproportionately burdened by the development and implementation of racial differentiating genetic technologies that will reinforce the biometric surveillance and manhunting regime of Chinese apartheid and cultural genocide in Xinjiang. The papers assert that the involved Uyghurs have and continue to willingly and freely submit to the scientists' larger quests including the development of forensic genetic phenotyping technologies that can be used against Uyghurs in Xinjiang. Implied here is the claim that Uyghur subjects could withdraw their consent and data, something that is not tenable under the current repression in Xinjiang. This assertion contradicts the inequality of potential negative risks and impacts of this research upon the Uyghurs. Third, the tens of thousands of subjects from several health research studies are being used not for health research purposes but as violence resources contributing to the development of racialized manhunting technologies to be used against Uyghurs. In practice, all three of these major issues intersect.

There have been multiple systemic ethical failures and violations that a critical situational ethics analysis that considered the intensification of repression in Xinjiang would have readily revealed. However, scientists, by accepting Chinese researchers' informed consent and ethics claims over the Xinjiang Uyghur Study participants and incorporating them into the research project with the tens of thousands of other donors, made these other donors into helpers in the development of PRC manhunting technologies and expertise. The failure of the scientific researchers to distinguish forensic genetic research from biomedical purposes makes the biomedical donors into agents in the development of manhunting technologies that utilize racial taxonomies. However, the VisiGen Consortium and its colleagues would finally be confronted over their use of the Xinjiang Uyghur Study in late 2021 and 2022, as we will see in chapter 11.

9

Convergence: The Partner Institute, the Beijing Institute of Genomics, and the Institute of Forensic Science

The international joint research projects discussed in the last chapter, published from late 2017 to late 2019, appear to be examples of routine academic cooperation, but the mentions of the significance of the papers by Liu et al. (2018) and Xiong et al. (2019) to the collaboration between BIG and the Chinese Ministry of Public Security are indicators of a much more problematic set of relationships within the PRC. The remainder of the book will show how scientists at the Partner Institute and BIG engaged in long-term research cooperation with the Ministry of Public Security's Institute of Forensic Science in Beijing on forensic genetic phenotyping and ancestry research that began around 2014. We will consider how the networks of Liu Fan of BIG and Erasmus University, Tang Kun of the Partner Institute, and Li Caixia of the Ministry of Public Security converged from the mid-2010s onward as part of a larger program that co-opted and integrated evolutionary and forensic genetic phenotyping research into public security–related biometric research and development mandated under the PRC's 13th Five-Year Plan (2016–2020).

Li Caixia and her colleagues at the Ministry of Public Security's Institute of Forensic Science published their first paper mentioning the possibility of conducting forensic genetic phenotyping on Uyghurs in 2015 (Sun et al. 2015). It is a broad survey of research done up to that point on inferring ancestry, physical traits, and potential genetic diseases or disorders of suspect DNA. It also tries to identify future trends in these areas. The authors cite a range of work done in China, Japan, Europe, and the US, including two papers on facial morphology coauthored by Mark Shriver and Peter Claes as well as several on hair and eye colour prediction by the Erasmus University Medical Center affiliated researchers Susan Walsh, Liu Fan, and Manfred Kayser. They discuss as well the significance of using a range of genetic markers including SNPs for "Forensic DNA workers [who] can use these to infer different regional racial populations, determine DNA populations by detecting DNA information, infer their physical characteristics, and provide direction and clues for case investigation" (233; 法医dna工作者可以藉此进行不同人群的种族地域推断， 通过检测dna信息判断dna供者所归属的种群，推断其体貌特征，为案件侦查提供指向性的线索。). Utilizing a taxonomy of well-established global skin colour–coded continental categories that are further divided into large regional categories, the authors construct Uyghurs and Kazakhs as mixtures of Europeans and East Asians:

> The large populations of humans mainly include white Europeans, black Africans, yellow East Asians, brown Oceanians, and American Indians. There are also a large number of sub-populations in different continents, such as Northern Europe, Central and Southern Europe, East Asia, Southeast Asia and the South and North of China. In intercontinental regions, a large number of mixed races have been formed due to marriage and integration of populations. For example, Xinjiang is located in the hinterland of the Eurasian continent where Europeans and East Asians meet, forming mixed racial groups such as the Uyghurs and Kazakhs. (Sun et al. 2015, 233; 人类大的种群主要包括欧洲白人、非洲黑人、东亚黄种人、大洋洲棕色人、美洲印第安人等。在各大洲的内部又存在大量各具差异的亚人群, 如北欧、中欧和南欧, 东亚、东南亚以及我国的南北方人群等。在洲际之间由于人群的婚配、融合等又形成了大量的混合人种, 例如我国新疆地处欧亚大陆腹地, 是欧洲人与东亚人交汇的地方, 形成了维吾尔族、哈萨克族等混合人种。)

Furthermore, different populations have different physical characteristics, so detecting a set of DNA markers with strong differences between populations will allow for estimating the most likely ethnicity of the donor, which "indirectly obtain the general physical characteristics of the DNA donor" (233; 进而间接获取DNA供者的大致体貌特征). This is another example of indirect phenotyping in which ancestry provides the basis to infer potential visible traits (Hopman 2020, 431–2; Koops and Schellekens 2008, 161–4).

The authors contend that there must be more attention to three major aspects: first, discovering more relevant genetic loci and their respective contribution to each facial morphology trait; second, facial morphology prediction algorithms and models; and third, analysis of results. As well, they advocate further research into how height is affected by the environment and age research such as menthylation and telomere length (Sun et al. 2015, 234). The authors conclude the article stating:

> Exploring the relationship between genes (including DNA structure and later modifications, etc.) and phenotypic characteristics, diseases, etc., to further unravel the mysteries of life has always been a hot topic of research for biologists. For forensic geneticists, it is not enough to know only these correlations, but it is also necessary to combine the various genetic markers found in these basic science fields to measure them, in order to infer and depict the phenotypic characteristics of the donor from a DNA sample. (234–5; 探索基因（包括dna结构、后期修饰等）与表型特征、疾病等的关联性, 进一步揭示生命奥秘, 一直是生物学家们的研究热点。对于法医遗传工作者来说, 仅仅知道这些关联性是远远不够的, 还需要通过综合这些基础科学领域发现的各种可能具有关联性的遗传标记, 对其进行测算, 实现由检材dna到对其供者即生物个体的表型特征推断与刻画。）

This concluding paragraph involves the translation of biopolitical medical and biological research on disease and phenotype into the sovereign human hunting violence work of forensic genetic estimation of the physical traits of the donor from their DNA. There is a very strong moral claim being made here that forensic genetics research is not just an academic exercise but rather has real-world consequences. Such epideictic rhetoric is common in the conclusions of scientific research articles because the authors are attempting to convince readers that they have made a contribution to the academic field and to society.

The 2015 paper provides a roadmap of what was to come in these research efforts. It reflected long-term planning and research cooperation to advance this integrated research agenda that would build upon existing strengths developed at the Partner Institute, the Shanghai Institute of Biological Sciences, BIG, and other CAS affiliated research institutes, universities, and various Ministry of Public Security labs.

The initial June 2016 reports by the US government–funded Radio Free Asia on implementation in Xinjiang fit with calls during the same period in Chinese press reports and articles by scientists from the Ministry of Public Security that advocated the use of forensic genetic phenotyping and ancestry as part of an integrated suite of biometric technologies for tracking, identifying, and capturing criminals. These efforts reflect the increasing translation and adaptation of biometrics in Chinese security discourses. In October 2016, the Chinese Ministry of Science and

Technology, which is a major research funding agency, in consultation with the Ministry of Public Security issued a call for technologies aimed at improving public security (Ministry of Science and Technology 2016, 1–2). This call, entitled "'Public Safety Risk Prevention and Control and Emergency Technical Equipment' 2017 Key Project Guide," integrates forensic genetic phenotyping, including medical conditions, with other biometric technologies:

> 2.2 Study of the Key Techniques of Detailed Description and Precise Identification of Criminal Suspects.
>
> Research content: research on the key technology of inferring the biometric characteristics of suspects in the case and relevant medical condition associated molecular marker tests; research on high-sensitivity testing techniques for internal and exogenous features of bodily fluid stains; research on dynamic biometrics and target object identification technology based on video; research on technologies to integrate applications such as palm print, footprint, face, iris, voiceprint, etc.; develop fast, portable, high-throughput forensic DNA detection equipment and reagent consumables; based on the integrated application of the above technologies, the development of application systems to provide detailed traits and accurate identification of criminal suspects. (Ministry of Science and Technology 2016, 4; 2.2 犯罪嫌疑人特征精细刻画与精准识别关键技术 研究内容：研究案事件嫌疑人生物特征及相关医学症候关联分子标记检验推断关键技术；研究体液斑痕中内、外源性特征成分高灵敏度检验技术；研究基于视频的动态生物特征识别与目标对象鉴定技术；研究指掌纹、足迹、人脸、虹膜、声纹等多生物特征综合应用技术；研制快速、便携、高通量法医 dna 检测设备与试剂耗材；基于以上技术的集成应用，研发犯罪嫌疑人特征精细刻画与精准识别应用系统。)

In this way, forensic genetic phenotyping is an integral part of a range of biometric technologies, including facial and gait recognition, to create a composite profile of suspects (Kruger 2013). The list of criteria includes something that is extremely controversial in the West: the use of disease traits in identifying characteristics. This is the subject of various arguments about the right not to know about potential genetic predispositions and the use of patient medical records to search for suspects (Toom et al. 2016).[1]

The Ministry of Science and Technology published an article on the 10–11 October 2017 meetings for the launch of projects by the

Ministry of Public Security Institute of Forensic Science. The article includes a section on "Research on key technologies for detailed characterization and precise identification of criminal suspects" (犯罪嫌疑人特征精细刻画与精准识别关键技术研究). The article describes how forensic genetic phenotyping will contribute to the goals of the Chinese Communist Party Central Committee's 13th Five-Year Plan, which covers the years 2016 through 2020. The project, entitled "The Key Technologies for Detailed Description and Precise Identification of Criminal Suspects" focuses on the "study of geographic [origin], age, face, height, disease, and other characteristics and precise identification techniques to comprehensively build a new technical system for finding, characterizing, and identifying suspects to achieve a 'precision strike' against criminal suspects" (Ministry of Science and Technology 2017; 犯罪嫌疑人特征精细刻画与精准识别关键技术 ("犯罪嫌疑人特征精细刻画与精准识别关键技术研究"项目，主要研究地理、年龄、面貌、身高、疾病等特征刻画和精准识别技术，全面构建查找、刻画、识别嫌疑人的新技术体系，实现对犯罪嫌疑人的"精准打击"。). The above project title and description clearly construct forensic genetic phenotyping and ancestry research within an overall suite of surveillance and manhunting technologies to create a composite of biometric measurements for identifying, tracking, and capturing criminal suspects. The above news report includes a photo that shows Ye Jian of the Institute of Forensic Science on the far right as part of a panel (Ministry of Science and Technology 2017). This major project, which has the same title as the October 2017 news release, was funded by a three-year 17.87 million renminbi (US$2.7 million) research grant (2017YFC0803500) to Ye Jian by the Ministry of Science and Technology (Sciping.com 2018). This grant is subdivided into a series of smaller grants focusing on the various technologies listed above. Of particular significance to this book's analysis is grant 2017YFC0803501 on forensic genetic identification, which appears on Liu Fan's 2017 Erasmus University curriculum vitae with the same title as grant 2017YFC0803500, in many of the papers by the Ministry of Public Security's Li Caixia et al. on ancestry SNPs, and in Liu et al. (2018) and Xiong et al. (2019) on phenotyping done with the European Union–funded VisiGen and VISAGE Consortiums (Liu 2017).[2]

The implementation of these efforts was outlined in a Zhao Xingchun et al. 2018 article entitled "Forensic Scientific Research and Development in the 13th Five-Year Plan" in the Chinese language journal *Forensic Science and Technology*, which is published by the Institute of Forensic

Science under the supervision of the Ministry of Public Security. The authors included Ye Jian, who is deputy director of the institute. The article sets out a broad overview of the state of forensic science in China, how it is becoming more professionalized with increasing use of high technology, and how it is playing an increasingly important role in fighting crime, "maintaining overall social stability, serving economic and social development and ensuring that the people live and work in peace and contentment" (维护社会大局稳定、服务经济社会发展、保障人民安居乐业中发挥着越来越重要的作用。). The 13th Five-Year Plan is a decisive period in "China's economic and social development. It is also a critical period for comprehensively deepening the reform of public security work," which involves significant opportunities and challenges (87–8). The authors then set out to define the current state of forensic science in China, outlining both its strengths and shortcomings. They cite one of the institute's successes: "for example, we have started SNP technology-based racial inference research, and we have achieved the identification of black, white, yellow, Chinese, Tibetan, and Uyghur races" (以特征刻画研究为例，当前已着手开展基于SNP技术的人种推断研究，目前已经实现黑白黄人种、汉藏维种族鉴别), citing Li Caixia et al.'s 2013 patent application for CN103146820B to differentiate Han, Uyghurs, and Tibetans (91). The authors also describe the future goals of the 13th Five-Year Plan; the following excerpt is worth quoting at length because it clearly situates forensic genetic phenotyping and ancestry research within a multifaceted larger biometric identity dominance strategy and manhunt organizing narrative schema:

> The "Study of the Key Techniques of Detailed Description and Precise Identification of Criminal Suspects" (Project 35) has been established and launched. The description of suspects is one of the important research objectives of criminal technology during the "13th Five-Year Plan" period. The research will focus on describing and extracting more information on suspects from physical evidence and trace information, so as to narrow the scope of investigation of cases. The key breakthrough directions are: inferring human phenotypic traits of biogeographic/age/face/height, biological tissue source inference, microbiological-based drug addiction inference, key technique of analyzing bodily fluid stains and describing criminal suspect appearance; as well as accurate individual identification based on second-generation sequencing technology and multi-class genetic markers, video-based dynamic biometric identification [gait analysis] and target object identification, suspect multi-biological feature

> collection device and database construction, quick multi-channel forensic DNA on-site inspections and other key technologies to identify suspects. (Zhao Xingchun et al. 2018, 90; "犯罪嫌疑人特征精细刻画与精准识别关键技术研究" (项目35) 已立项并启动。嫌疑人特征刻画是刑事技术"十三五"时期重要研究目标之一，研究将从物证、痕迹信息中，着重描述和挖掘嫌疑人更多信息，缩小案件侦查范围。重点突破方向为：生物地理/年龄/面貌/身高等人类表型特征推断、生物斑迹组织来源推断、基于微生物组学的新型毒品吸食人员推断、体液斑痕中摄入成分分析与供体特征刻画等犯罪嫌疑人特征精细刻画关键技术，以及基于二代测序技术和多类遗传标记的精准个体识别、基于视频的动态生物特征识别与目标对象鉴定、嫌疑人多生物特征综合采集设备及数据库构建、快速多通道法医DNA 现场检验等犯罪嫌疑人精准识别关键技术。)

Biogeographic ancestry and forensic genetic phenotyping for height, age, and facial appearance are defined within an overall integrated biometric identification system to create a multifaceted biometric profile of potential suspects (Kruger 2013; M'charek 2008) within an overall surveillance and manhunting system.

PUTTING THE FIVE-YEAR PLAN INTO PRACTICE

The theme of bridging scientific research with forensic practice is evident in the organizing narratives of the research cooperation between the Partner Institute and the Ministry of Public Security (MPS) Institute of Forensic Science's Evidence Identification Center from the mid-2010s onward. The 2014–17 Partner Institute report mentions that someone from Wang Sijia's dermatogenomics (skin genetics) research group gave an invited talk to the "MPS Evidence Identification Center, Beijing, China, August 2014" (Chinese Academy of Sciences Max Planck Society 2017, 106).[3] Long-term cooperation on forensic genetic phenotyping is evident in a May 2016 ad on a Chinese government website for a postdoctoral position at the Partner Institute. The ad lists Tang Kun of the Partner Institute and Li Caixia of the Institute of Forensic Science as the joint supervisors. The postdoctoral position's job description provides a summary outline of the organizing narratives of the research cooperation entitled "Research on the DNA of Physical Appearance Traits" (基于DNA的面貌特征刻画研究):

> 1. Research content: facial feature landmarks and facial difference comparison in the Chinese population; screening and evaluation of SNPs related to facial morphological features in the Chinese

> population; facial morphology prediction algorithm and model research; landmark system and predictive model prediction accuracy and recognition rate evaluation test study. Through research, we will establish a DNA characterization technology system suitable for the Chinese population, and explore the potential information from biological evidence to improve the evidentiary value of biological evidence. (Chinese Postdoctoral Science Foundation 2016; 1、研究内容：中国人群面部特征点标记、面部差异比对研究；中国人群面部形态特征相关的SNPs位点的筛选与评估研究；面部形态预测算法与模型研究；位点体系与预测模型的预测准确性和识别率评估测试研究。通过研究将建立适合我国人群的dna面貌特征刻画技术体系，深入挖掘生物物证的潜在信息，提高生物物证的证据价值。)

Metonymically, the concept of the Chinese people is implicit in their usage of the term *Chinese population*, and it is this population from which and over which the Chinese government under the Chinese Communist Party claims its authority.

The 2014–17 Partner Institute report, published in May 2017, cites the early successes of Tang Kun's Functional Human Genetic Variation research group and its plans for cooperation with the Institute of Forensic Science:[4]

> Our GWAS of common human facial variation was among the first to identify the genetic determinants of facial shape. Furthermore, a prediction model was proposed and proved to be able to predict one's face to a significant extent based purely on DNA. This marked a milestone in the forensic technology. In the long run, we are going to collaborate closely with national forensic institutes, such as the Center for Material Evidence Authentication, Ministry of Public Security, to expand the training samples, improve the prediction model and eventually to be able to apply the face prediction methods in real forensic scenes. (Chinese Academy of Sciences Max Planck Society 2017, 204)

The above Partner Institute version also constructs this relationship as a translation of the advanced research done at the Partner Institute, with the Ministry of Public Security's Institute of Forensic Science, into technologies that can be implemented in routine forensic genetic practice. These relations are important to translating leading edge science into practical forensic genetic testing that can be used in routine policing and counterterrorism efforts.

Uyghurs are a central focus in this comprehensive phenotyping research and development effort under the 13th Five-Year Plan. The first papers appear together in 2017 in an issue of the Chinese language journal *Forensic Science and Technology*, which is published by the Institute of Forensic Science under the supervision of the Ministry of Public Security. Funded in part by Ministry of Science and Technology grant 2017YFC0803501, Zhao Wenting (including Li Caixia) et al. (2017) follow the commonly used organizing narrative schema of the STR profile of the unknown sample not matching a database, which as "criminal intelligence" limits police abilities to solve a case, so "The unknown 'suspect X' is a major obstacle to uncovering the facts of a case and solving the case, often leading to unsolved cases and a backlog of cases" (260; 这些未知身份的" 嫌疑人X" 给案件的侦破和案情的还原带来了很大的阻碍, 甚至常常形成悬案、积案。). However, the authors assert that "human appearance is mainly determined by genetic information" so forensic genetic phenotyping can help by acting like a "paintbrush" allowing "criminal investigation experts to draw a facial portrait of the 'suspect X,'" thereby providing "key support for the resolution of difficult and unresolved cases," which has led to it becoming a "hot topic in the field of forensic science in recent years" (260; 人类外观特征主要是由遗传信息决定的, 能否让dna成为刑侦专家手中的画笔, 为难以找寻的" 嫌疑人X" 画一幅面部肖像, 从而为破获疑难案件和悬案积案提供关键支持, dna分子画像技术因此而成为近年来法庭科学领域的前沿研究热点。).

In the competence phase, the authors engage in a survey of the available literature on the state of forensic genetic phenotyping, looking, for example, at the work of Kayser and Liu Fan on the Irisplex and Mark Shriver's work on facial SNPs, and note how much of the early knowledge about facial genetics originated in studies about facial deformities. They mention the 3D facial mapping techniques developed by Guo, Mei, and Tang (2013). As well, they specifically discuss the research of Tang Kun of the Partner Institute on Uyghurs stating "At present, Tang Kun's laboratory through the analysis of the differences in facial features between European and Chinese Han people, combined with the results of GWAS data of Uyghurs, a typical mixed Eurasian population in China, found more than 400 SNPs loci related to Uyghur facial features and built a 3D face prediction model" (Zhao Wenting et al. 2017, 261; 目前, 唐鲲实验室通过对欧洲、中国汉族面部特征的差异分析, 结合我国典型欧亚混合人群– – 维吾尔族GWAS数据结果, 找到了400多个与维族面部特征相关的SNPs位点并建立了3D面部预测模型。). The paper concludes that while facial DNA studies have had some initial success, face shape involves multiple genes that are not fully understood, so the technology is not ready yet for deployment:

> However, as the research on genetic loci related to the face of the population continues to advance and the prediction models continue to be optimized, DNA molecular profiling technology will definitely have a major breakthrough in the future. The idea of using the biological evidence left behind at a crime scene to reveal the unknown "Suspect X" is very likely to become a reality, and become a new means for forensic DNA to fight crime. (Zhao Wenting et al. 2017, 262; 不过, 随着人群面部相关遗传位点研究的不断推进和预测模型的不断优化, dna分子画像技术在未来必定会有较大的突破。利用犯罪现场遗留的生物物证让未知的"嫌疑人X"现身的设想极有可能变成现实, 成为法医dna打击犯罪的新型手段)

This article provides an introduction to forensic genetic phenotyping, setting up the other two articles on phenotyping in the same issue.

The second article in this journal issue specifically targets Uyghurs. Coauthored by Li Caixia, Tang Kun, and Ye Jian, it is entitled "Experimentation on Human Facial Prediction by Relevant DNA SNPs" (Liu et al. 2017). This article is significant because it involves direct cooperation between the Partner Institute in Shanghai with the Institute of Forensic Science in Beijing in studying genetic-based facial appearance estimates. It expressly builds on the Partner Institute's earlier research on Uyghurs, including the aforementioned 2014 paper by Guo et al. Again the article's manipulation phase describes research on forensic genetic phenotyping as a way to provide leads to help police investigations: "Depiction of individual's EVCs (externally visible characteristics) based on DNA analysis is gradually becoming a forensic research hotspot because it can provide effective leads to police investigation when conventional DNA STR data do not match either known individuals or any criminal DNA database. Compared with the other morphologic features, human facial morphology is better conserved and less affected by environmental factors. Moreover, the combination of gene analysis and image technology has greatly promoted research on facial morphological inference" (Liu et al. 2017, 265). The authors consider the rapid advances in genetic analysis and imaging technology, particularly three-dimensional approaches.

In the competence phase, the authors first cite some of the existing literature, including Kayser's HIrisPlex for hair and eye colour and Parabon Nanolabs forensic genetic phenotyping service to police. They also summarize the 2014 *Nature* report "Mugshots Built from DNA Data" on Kayser et al.'s findings of facial SNPs, Mark Shriver's facial prediction model, and Tang Kun's lab development of facial mapping techniques, citing the paper by Guo, Mei, and Tang (2013) and their analysis of Uyghur facial features: "Through selection of 350 facial morphology-related SNPs analyzed from sequencing 24 Chinese males (18 Uygur and 6 Han), the relevant SNP phenotypes were obtained so that a model of facial morphologic prediction was built up based on such one previously-developed" (Liu et al. 2017, 264). The research team's paper utilizes a Uyghur specific panel of 277 SNPs from a Qiao et al. (2016) paper, which is a 2016 pre-print of the Qiao et al. (2018) discussed earlier (Liu et al. 2017, 266, 269). To these 277 SNPs they added a further seventy-three SNPs derived from a literature review for a total of 350 SNPs associated with facial shape. These were tested on eighteen Uyghur and six Han male subjects who gave informed consent. The methodology is similar to the papers discussed earlier on GWASs based on 3D imaging of the subjects' faces, their age, body mass index, and blood sampling. A significant difference is the sequencing equipment that the researchers used, an Illumina HiSeq X10 system, which is a networked set of ten Illumina sequencers worth about US$10 million (Zhao Wenting et al. 2017, 265).[5] This is the only instance of the use of the Illumina HiSeq X10 system by Chinese authors analyzed in this book so it may have been subcontracted out. Nonetheless, it points to these researchers being able to access leading edge American-made sequencer systems. The resulting predicted faces were compared to the actual subjects using statistical analysis and the authors state that although the predicted models were closer to the donors than to random faces, there is still a long way to go. They describe the four main problems as: a lack of research on genetic loci; insufficient genome wide data associated with facial features; a prediction model that used a small number of samples; and finally, the need for an increase in the number of factors in building models, such as skin colour. They finally state: "In conclusion, this article has made a preliminary attempt on DNA facial molecular portraits, and successfully obtained the genetic prediction face of the sample donor through the DNA typing data, and the recognition accuracy rate is better than the random accuracy rate. In the future, investment will be increased, the number of samples will be continuously increased, and genetic loci that more accurately represent facial morphological characteristics will be found, and the prediction model will be optimized and there will will surely be a major breakthrough in DNA molecular profiling technology" (Liu et al. 2017, 260, 总之，本文对 DNA 面部分子画像

进行了初步尝试，成功通过 DNA 分型数据获得了样本供者的遗传预测脸，且识别准确率优于随机准确率。未来将会加大投入，不断增加样本数量，找出更准确代表面部形态特征的遗传位点，进而优化预测模型，DNA 分子画像技术定会有较大的突破。).

Facial age estimation is the topic of the second paper coauthored by Ye Jian, Li Caixia, and Tang Kun, entitled "Research Progress on Human Facial Age Estimation and Synthesis of Age-correlated Appearance," which appeared in this issue of *Forensic Science and Technology* (Pan et al. 2017). The abstract begins, "Human facial age estimation and aging progression are being increasingly paid much attention because of their huge value in the aspects of forensic investigation, missing people searching, security control and surveillance" (270). This article provides an overview of the state-of-the-art technology in face age estimation and aging and its potential applications to criminal investigations, surveillance, and identity confirmation. It does not directly mention Uyghurs.

However, in 2018, Li Caixia, Ye Jian, and Tang Kun coauthored another paper, which appeared in the *Journal of Forensic Medicine*, entitled "Age Estimation and Age-Related Facial Reconstruction of Xinjiang Uygur Males by Three-dimensional Human Facial Images" (Pan et al. 2018). This paper did not involve genetic testing, rather the scientists analyzed the faces of 105 Uyghur men aged seventeen to fifty-seven, focusing on the ways the face could be analyzed utilizing Artec Studio software to predict aging: "Objective. To search age-correlated facial features and construct an age estimation model based on the three-dimensional (3D) facial images of Xinjiang Uygur males, and to structure individual face images of old age and young age … Methods. Pretreatment was performed to collect 105 3D facial images of Xinjiang Uygur males aged between 17–57 years by Artec Studio software. The facial images were transferred to high-density 3D dot matrix data by FaceAnalysis software, and each image could be represented with 32,251 vertexes" (363). The research team applied the facial mapping techniques developed by the Partner Institute's Guo, Mei, and Tang (2013), discussed in chapter 7, to analyze the facial features of Uyghurs. In effect, the Partner Institute's phenotyping research on Uyghurs was now being incorporated into Ministry of Public Security technology development efforts.

THE BEIJING INSTITUTE OF GENOMICS

In June 2015, Liu Fan was appointed as a professor at BIG (Beijing Institute of Genomics 2016, 97). As discussed earlier, he worked in the Visigen Consortium as a postdoctoral researcher and later as an assistant professor at Erasmus University Medical Center in the Netherlands. He was then recruited under the Thousand Talents Plan, a Chinese government effort to attract scientific experts from prestigious foreign institutions to work in China (9).[6] Liu's return placed his extensive experience, expertise, and research network in the service of these efforts to develop China's phenotyping research.

The timing of Liu Fan's recruitment and return may have been influenced by the signing of a research cooperation agreement between BIG and the Ministry of Public Security Institute of Forensic Science in Beijing. In BIG's 2015 annual report's chronology of events, "December 23. Beijing Institute of Genomics and Material Evidence Identification Center of the Ministry of Public Security formally signed an agreement for 'Forensic Genome Collaborative Innovation Center' and had an opening ceremony" (Beijing Institute of Genomics 2016, 96–7). It features a photo of Ministry of Public Security officials in police uniforms, including Ye Jian of the Institute of Forensic Science in Beijing on the left facing BIG officials, which includes the president of BIG Xue Yongbiao. In the "Message from the Director," Xue lauds the agreement: "BIG also officially signed an agreement with the Ministry of Public Security Forensic Identification Center for building a 'Forensic Genome Collaborative Innovation Center.' Extraordinary development in technology and application is achieved by co-constructing of a series of public security urgent demanded cutting-edge technologies, developing a public security science and technology guarantee system with independent intellectual property rights in China to meet the major needs of the country and industry, promote the overall development of the discipline and improve the level of forensic genomics technology" (v). The goal here is clearly to translate leading-edge science into police practice. The idea of having intellectual property rights in China is significant in reducing external costs for licensing and purchase of foreign supplies from the US and Europe and in asserting control over the technology assemblage, a dependency that would have significant future consequences after 2019. Another indicator of the closer ties between BIG and the Institute of Forensic Science is evident in the role of Ye Jian of the Ministry of Public Security's Institute of Forensic Science. Ye was not on BIG's Organizing Committee in the 2015 *Beijing Institute of Genomics Annual Report*, but was listed under the committee members in the annual reports for 2016 to 2020, the most recent available at the time of writing (Beijing Institute of Genomics 2016, 2017, 5, 2018, 11, 2019, 9, 2020, 6, 2021, 6).

The *2018 Beijing Institute of Genomics Annual Report* makes several references to group studies by its personnel of the political thought and Chinese Communist Party ideology of Xi Jinping, indicating a strong institutional demand for political and ideological conformity among its members. BIG is different from the Partner Institute in that it is wholly under Chinese governance and so its publications directly mention adherence to Xi Jinping Thought. Various public statements of allegiance to Xi Jinping are evident in the *2018 Beijing Institute of Genomics Annual Report*; in the report's "Address from the Director," Xue Yongbia proclaims that, "In 2018, all BIG staff have made new achievements under the guidance of Xi Jinping's Thoughts on Socialism with Chinese Characteristics for a New Era" (1). Such public statements of ideological dedication to Xi Jinping are becoming increasingly common in Chinese scientific discourses. Whereas the 2016 annual report mentions one study activity of Xi Jinping's ideology, in contrast the 2018 annual report mentions five such activities (95, 96, 97, 99). For example: "The Party Committee of BIG held the fourth extended learning conference of party committee center group, specializing in studying Xi Jinping's Thirty Speeches on the Socialist Thought with Chinese Characteristics for New Era. Party Secretary Wang Liping gave a detailed explanation on the Thirty Speeches' major contents and learning requirements, and delivered a thematic report on the Speech 'Adhere to the Overall Concept of National Security,' while Director Xue Yongbiao made his keynote speech on scientific research works" (97).

The sacred dead of the Chinese people were also honoured in a visit to a Chinese Civil War memorial: "More than 120 people including Party members and Youth League members of BIG went to the Beijing Xishan Unknown Heroes Memorial Square and held an educational activity themed 'Remain true to our original aspiration and keep our mission firmly in mind.' The participants have strengthened their ideals and beliefs by honoring heroes" (Beijing Institute of Genomics 2018, 96). This visit to a national memorial for the sacred dead (Anthony Smith 2000) is a public ritual of commitment to the sovereignty of the Chinese nation. This sort of public display of patriotism, with the group photograph of the attendees standing on the monument behind the Chinese flag as a totem (Marvin and Ingle 1999) symbolizing the nation, is a ritual of reverence for the sacred dead of the Chinese Civil War, typical of the recent ascension of Xi Jinping Thought, which increasingly emphasizes nationalism. Chinese Communist Party ideologies use the sacred dead of the Chinese Civil War as central actants as those who sacrificed their lives for the Chinese nation rather than a Chinese empire that it inherited from the Ching Dynasty.

The *2019 Beijing Institute of Genomics Annual Report* provides further evidence of the increasing assertion of Chinese Communist Party control over academic institutions. Under the heading "Work of the Party

Committee" is a chronology of events that begins with the following 29 January 2019 entry: "BIG held the 2018 Leading Group Democratic Life Meeting. The Party members of the leading group discussed problems and deficiencies, clearly defined the direction of improvement, and carried out criticism and self-criticism earnestly and frankly" (Beijing Institute of Genomics 2020, 90). The idea of academics engaged in public self-criticism has echoes of the persecution of intellectuals during the Cultural Revolution. The 25 April 2019 entry more specifically mentions adherence to Xi Jinping Thought, the first of five events in the report: "BIG held an extended learning meeting of party committee center group, specializing in studying the keynote speeches of General Secretary Xi Jinping on cadre work and the 'Regulations on Assessment of Party and Government Leading Cadres'" (90). This emphasis on leading cadre again seems to indicate increasing Chinese Communist Party assertion of control over academic researchers.

The English translation of the *2019 Beijing Institute of Genomics Annual Report*'s "Address from the Director" by Xue Yongbiao is different from that of the Chinese version. The English version omits the paragraph on adherence to Xi Jinping and Chinese Communist Party ideology, whereas the second paragraph in the Chinese version reads as follows:

> The Institute has always been guided by the party's political construction, resolutely implemented the spirit of General Secretary Xi Jinping's important instructions and the Party's central decision-making and deployment, implemented the work requirements of the party group and branch group of the school, and carried out the educational theme of "not forgetting the original intention and remembering the mission" and emphasizing the idea of "innovating technology, serving the country, and benefiting the people" to researchers. In 2019, the political leadership role of party committees, the battle fortress of party branches, and the role of party members as a model vanguard was exerted. The selection of "models around" and the cultural forum of "inheritors" will be carried out to strengthen the construction of innovative culture and create a clean and healthy research environment. (研究所始终以党的政治建设为统领，坚决贯彻习近平总书记重要指示批示精神和党中央决策部署，落实院党组和分院分党组各项工作要求，扎实开展"不忘初心、牢记使命"主题教育，提升广大科研人员"创新科技、服务国家、造福人民"理念。2019 年进一步发挥党委政治引领作用、党支部战斗堡垒作用及党员模范先锋作用，开展"身边的榜样"评选活动及"传承者"人文论坛，加强创新文化建设，营造风清气正的科研环境。)

In contrast, the Partner Institute does not have any mention of Party-related activities in its 2014–17 report, the most recent available, though the 2017–20 report may have some if it becomes available. The only overt mention of the Party is in the list of media stories, which includes two 9 May 2015 stories by the Xinhua News Agency about the Partner Institute director meeting Xi Jinping (Chinese Academy of Sciences Max Planck Society 2017, 294). I have done various searches of the Partner Institute website and the Internet in general, and I have not been able to find any accounts of institutional rituals of patriotism and adherence to Xi Jinping Thought in the form of study groups or speeches at the Partner Institute. This absence may be due to the Partner Institute being a joint partnership with the Max Planck Society and employing a large number of foreign scientists to improve productive connectivity to global genetic research networks, which make the Partner Institute in some ways analogous to a special economic zone (cf. Ong 2006, 102–11). Xi appears to hold the Partner Institute in high regard, given he has met the Partner Institute director on three occasions as discussed earlier. Though it may be less subject to such demonstrations of Party loyalty, the Partner Institute began cooperation with the Institute of Forensic Science in 2014, which fits with the timeframe of BIG's cooperation with the Institute of Forensic Science, formalized in their December 2015 agreement.

Strong adherence to Xi Jinping's policies, including supporting national security, means involved scientists must advance these agendas in the development of forensic genetic manhunting technologies. The facial forensic genetic phenotyping research on Uyghurs in the above research has depended in large part on the Xinjiang Uyghur Study with its sampling in 2013–14. However, a new project begun during the escalation of repression in 2017 involving 715 Uyghurs in the Bingtuan administered city of Tumxuk is significant because it is specifically conducted for the purposes of forensic genetic phenotyping. It involves the coordinated efforts of the Partner Institute, as a joint Max Planck Society and CAS institute, and BIG in conjunction with Chinese security agencies, in particular, the Ministry of Public Security's Institute of Forensic Science and the Bingtuan Public Security Bureaus in Urumqi and Tumxuk.

10

The Tumxuk Uyghurs

Tang Kun, Liu Fan, and Li Caixia's first coauthored paper in 2018 involved a new group of 715 Uyghur subjects from the Bingtuan administered city of Tumxuk in western Xinjiang's Kashgar Prefecture. This chapter explains how these study projects on the genetics of Uyghur height and facial features emerged out of growing cooperation between the Ministry of Public Security's Institute of Forensic Science and the Bingtuan, with an agreement announced in February 2017, which included the construction of a large forensic genetics lab for the Bingtuan Security Bureau in Tumxuk.

THE SITUATION IN TUMXUK

Southern Xinjiang's Tarim Basin is the traditional Uyghur homeland where they are still a majority. Controlled and administered by the Bingtuan's Third Division, Tumxuk is located in the southwest of Xinjiang, not far from the Kyrgyzstan border, 300 km east of the ancient city of Kashgar and over 3700 km by road west of Beijing. In 2015, its population was 163,101, comprised of 60,914 Han Chinese and 101,042 Uyghurs, with a small number of other minorities (Statistic Bureau of Xinjiang Uyghur Autonomous Region 2017). Given the intensity of the Chinese government's suppression of independent reporting that has accompanied its campaign of forced assimilation in Xinjiang, there is not a lot of information on conditions in Tumxuk. However, comparing a number of stories from different sources provides a sketch of the Bingtuan Third Division's apparatus of repression.

The Uyghur intellectual Ilham Tohti and other activists ran the Uighurbiz.net website (2006–2013), which was an independent source of Uyghur news, information, discussion forums, and advocacy. During

2013, Uighurbiz.net had several stories on Bingtuan security forces in Tumxuk demolishing Uyghur homes and expropriating their land (Uyghur Human Rights Project 2018, 27–9). The website also documented how some Uyghurs who participated in protests in Tumxuk against the Bingtuan's continued house demolitions were subjected to police intimidation and harassment, fired from their jobs, and arrested (29). One such story, published on 26 July 2013 and entitled "Two Uighur brothers whose house was demolished in Tumxuk were arrested" (图木舒克房屋遭强拆的维吾尔兄弟二人遭抓捕) features a photo of one of the brothers, Mehet Imin, and his wife and their three young children crouched down in front the ruins of their demolished home (Uighurbiz.net 2013; Uyghur Human Rights Project 2018, 30).[1]

Coverage of such abuses stopped abruptly near the end of 2013, when Chinese authorities shut down the website. On 15 January 2014, Tohti was arrested and charged with advocating separatism through the website. In November 2014, he was sentenced to life imprisonment by a court in Urumqi. Seven of his students who worked on the website were also arrested and sentenced to prison terms of three to eight years (Committee to Protect Journalists n.d.). Within the available space for political discussion on government policy and the experience of settler colonialism in all its forms, the website and the brief opening that the widespread adaptation of the smart phone in the early 2010s allowed was slammed shut with Xi Jinping's declaration of the People's War on Terror (Byler 2018). The government's arrest of Tohti and his trial for separatism were part of this crackdown on what limited public sphere the Uyghurs had for independent communication and political debate (Byler 2018).

The increasing repression of any independent political advocacy for Uyghurs is further evident in the 2017 arrest of Huang Yunmin, a former judge in the Bingtuan's Third Division (the Bingtuan has its own judicial system). Huang, who is a Han Chinese settler and a former Communist Party member, was sentenced to ten years for "incitement to racial hatred and ethnic discrimination" by a Bingtuan court in Tumxuk, likely in retaliation for his advocacy on behalf of local agricultural labourers and others on Bingtuan farms (Qiao 2017).

The PRC media also provides evidence of the escalating repression in Tumxuk, including implementation of a web of security checkpoints utilizing biometric surveillance systems. A 27 September 2017 *Global Times* (owned by Chinese government's *The People's Daily*) article reported that Tumxuk had received ten terrahertz body scanners for use at the security check points around the city and that this number would eventually increase to thirty-five (Zhao Yusha 2017).[2] This application of high technology body scanners is justified using the war on terror rhetoric that constructs Xinjiang's minority groups as potential threats: "Tumxuk and Kashgar are the gateways and major battlefields for Xinjiang's security work. The two cities are inhabited by many ethnic minority groups which move freely in large numbers every day, Qin Wenrong, a Xinjiang military officer, told the *Global Times*" (Zhao Yusha 2017). Politically and geographically, the officer repeats the dominant Chinese government war on terror conception of Xinjiang as a frontier region threatened by terrorist infiltration from Central Asia. The threat of terrorists, what Epstein (2007) terms *destructive bodies*, hidden among minority groups as dangerous populations and hence threats to the legitimacy of the Party's rule is evident in the attempt to link the repression in Tumxuk to China's larger security crackdown in Xinjiang ahead of events with large symbolic importance for the Party: "Qin said that the Tumxuk government has invested heavily in these scanners to enhance their counter-terrorism work ahead of the upcoming 19th National Congress of the CPC [Communist Party of China], which will convene in Beijing on October 18 [2017]" (Zhao Yusha 2017). Held every five years, the Chinese People's Congress is a grand political spectacle of the Communist Party, which representatives of minority groups attend, frequently dressed in the respective traditional clothing of their people. The main political significance of the 19th People's Congress was that it symbolically represented the consolidation of Xi Jinping's power by approving removal of term limits and thereby giving Xi a formal mandate to rule indefinitely.

Body scanners are a surveillance/capture device used to sort flows of Uyghurs as dangerous populations. However, the *Global Times* article portrays the technology as more user-friendly to Uyghurs, stating: "'We used to get into conflict with people when we examined them with handheld scanners, especially women,' Xu noted. A local Uyghur named Rozi, who had to undergo the scanner several times when passing through the station, said the new scanner is more convenient as it saves him time and protects his privacy" (Zhao Yusha 2017). The article normalizes the surveillance of Uyghur as threats, saying that it is done in a nonintrusive and culturally respectful manner that tries to minimize inconvenience. The article also cites Qin Wenrong, the Xinjiang military officer, saying "that the Tumxuk government conducted massive security checks throughout the city a few days ago and urged government officials to pay attention to suspicious signs, such as a sudden increase of strangers in homes." This article shows the normalization of the escalating repression in Tumxuk in which high technology scanners and mass security sweeps are supposed signs of the beneficent determination of the settler state against terrorism.

There has been some PRC news coverage of the reeducation camps in Tumxuk. A 4 August 2015 article published by Bingtuan Television entitled "Legal Education and Training School Officially Started" (法制教育培訓學校正式開課) describes the opening of a reeducation center run by the Bingtuan Third Division in Tumxuk (Qing 2015). In an accompanying photo, several prisoners are listening to a lecture by the instructor, who sits under a banner written in Uyghur and Mandarin that says "Legal Lecture" (法制讲堂). The article begins:

> "Religious extremism has 15 major manifestations. Among them, the first one is to advocate not obeying anyone except God, openly resisting government management, distorting and denigrating party and state policies ..." Director of the Political and Legal Office of the Jiashi General Office Kurban Jesiti is teaching a course about "15 Major Manifestations and Signs of Religious Extremism," during the legal education training class, the participants are listening carefully and recording the key points. (Qing 2015, ellipsis in original; 「宗教極端思想滲透有15種主要表現，其中，第一種表現是鼓吹除了真主以外，不服從任何人，公開抵制政府管理，歪曲、詆毀黨和國家政策 ... ...」伽師總場政法辦主任庫爾班·熱西提正在圍繞《宗教極端思想滲透15種主要表現形式及苗頭》進行授課。在法制教育培訓課上，學員們正在認真聽講並將重點內容記錄下來。)

By these criteria, any disagreement with state policies and willing display of noncompliance is considered a manifestation of extremism. That the instructor is Uyghur is typical of the way local Indigenous people are integrated into the security apparatus, particularly in lower-level positions (Byler 2022, 47). The article is an example of the earlier stages of the reeducation camps, when incarceration tended to be of shorter duration and more targeted, before the mass incarceration of 2017 began.

In 2017, the Chinese government enforced mass forced assimilation upon the Uyghurs and other Indigenous peoples. A May 2017 article by Bingtuan Satellite Television (2017) entitled "The Bingtuan De-Extremeification Legal Rule Publicity Team delivers lectures in the Third Division including key companies, primary and secondary schools, and the Rule of Law Education Transformation Center" (兵

团去极端化法治宣讲团深入三师重点连队、中小学校、法治教育转化中心宣讲) describes some Third Division mass propaganda activities during this early stage of the crackdown that marked the beginning of mass incarceration. The article states that the activities included over three thousand people in elementary schools and secondary schools along with a reeducation camp. It has a set of photos of male prisoners at a reeducation camp sitting and listening to lectures with the caption: "The Third Division of the Tumxuk City Rule of Law Education and Training Center for the 50th Regiment, the 51st Regiment, the 53rd Regiment and other regiments' rule of law education training class, teachers talk about their real experiences as teaching examples to guide the students to sharpen their vigilance to distinguish between right and wrong" (Bingtuan Satellite Television 2017; 第三师图木舒克市法治教育培训中心及五十团、五十一团、五十三团等团场法治教育培训班，讲师团老师以身边人说身边事、身边事教育身边人，引导转化学员擦亮眼睛，明辨是非。).

As part of the large push during early and mid-2017 to expand the internment camp system, bidding was opened for the construction of new camps. Adrian Zenz (2018) of the Jamestown Project documented a dramatic set of links to various bidding contracts, including one from the Bingtuan Third Division. One tender, dated May 2017, is entitled "The Third Division's 51st Regiment Legal System Education and Training Center Construction Project" (Bidcenter.com.cn 2017). The project description lists the objectives: "Project scale: Reconstruction of fences and gates, installation of video surveillance security system, renovation of dormitories and canteens and restaurants, renovation of indoor plumbing and electrical wiring, outdoor surface hardening, supporting infrastructure and equipment purchase" (Bidcenter.com.cn 2017; 工程规模：围墙大门改建、视频监控安防系统安装、宿舍楼和食堂餐厅室内房间改建、室内水暖电管线改造、室外地坪硬化、基础设施配套及设备购置。).

Satellite imagery reveals the ways the prison camp infrastructure in Tumxuk has developed over the last decade. The Australian Strategic Policy Institute's (2022) Xinjiang Data Project (see table 10.1) provides the coordinates of several of the Bingtuan prison installations. Using the Google Earth historical imagery function to view the coordinates allows you to see how these installations rapidly expanded after 2017.

This limited set of examples, while incomplete, nonetheless indicates increasing repression in Tumxuk, typical of the larger repression in Xinjiang, which intensified through 2014–17 with intensive security checks and the massive expansion of internment camps after 2017. The

Table 10.1 | Tumxuk Bingtuan Detention Centre map coordinates

Detention centre type	*Bingtuan unit*	*Map coordinates*
Third Division Tumxuk Prison. Tier 2 Reeducation Facility	23rd Company of the 51st Regiment	39.91163536, 79.0236617
Tier 2 Reeducation Facility	44th Regiment	39.85139679, 79.02327614
Tier 2 Reeducation Facility	2nd Company of the 50th Regiment	39.93598165, 79.38247743
Tier 2 Reeducation Centre	44th Regiment, Third Division	39.85370553, 79.02317332
Tier 4 Prison	6th Company of the 52nd Regiment	39.876435, 79.125725
Tier 3 Detention Centre	3rd Company of the 44th Regiment	39.794866, 79.07824

brief account above clearly indicates that the Uyghurs of Tumxuk are subjected to China's intense surveillance, mobility control measures, and reeducation camps. Therefore, any claims of informed consent and ethical research under such repressive conditions are invalid.

INSTITUTE OF FORENSIC SCIENCE AIDING THE BINGTUAN'S SOUTHWARD DEVELOPMENT

Another indicator of the increasing securitization in Tumxuk is the construction of a forensic DNA lab there. The Institute of Forensic Science's Physical Evidence Center signed a cooperation agreement with the Bingtuan Public Security Bureau on 22 February 2017 and planning for the construction of the first project in Tumxuk began (Wei 2017). According to a Bingtuan news report on this cooperation agreement, "The Material Evidence Identification Center of the Ministry of Public Security is the largest, strongest and most professional comprehensive physical evidence inspection and identification institution in the field of criminal technology in China" (公安部物证鉴定中心是我国刑事技术领域规模最大、实力最强、专业最全的综合性物证检验鉴定机构), with expertise in over twelve major areas including forensic injury and DNA (Wei 2017). The brief article concludes that the Evidence Center had been "directly involved in the field investigation, inspection and examination of almost all major serious incidents and disasters over the years in our country" (直接参与了我国不同历史时期几乎所有重特大案事件、灾害事故的现场勘查和检验鉴定工作) including the "3.14 Riots in Tibet" and the "7.5 Incident in Xinjiang" along with several earthquakes. The 14 March 2008 protests in Lhasa Tibet and the 5 July 2009 riots in Urumqi were two of the most significant uprisings against PRC settler colonial rule in recent decades.

As part of this agreement, the Evidence Center cooperated with the Bingtuan Third Division's Tumxuk Public Security Bureau with the construction of a 350 square metre forensic genetic lab (China CNTC International Tendering 2017). Public tenders for equipping the lab were published in late September 2017 and it was completed at a cost of over 4 million yuan (approximately US$570,000) according to a 5 February 2018 Bingtuan news report, including a Thermo-Fisher 3500 genetic analyzer (6; Wei and Wang 2018). The news report was quite clear about the overall ideological role of the Ministry of Public Security cooperation with the Bingtuan as part of the settler colonization of the southern Xinjiang region:

> According to reports, the Ministry of Public Security's Physical Evidence Identification center assisted in the construction of the DNA Laboratory of the Third Division Public Security Judicial Identification Center. The Ministry of Public Security following the guidelines of Xi Jinping's new era of socialism with Chinese characteristics, thoroughly implemented the spirit of the 19th Party Congress. Thoroughly implementing the central government's strategy of ruling Xinjiang, and the mandate of the Bingtuan, focusing on the main target of Xinjiang, using effective "boxing combinations" in counter-terrorism and maintaining stability, under a strategic cooperation framework agreement between the Ministry of Public Security Material Evidence Identification Center and the Bingtuan Public Security Bureau, in order to implement the Bingtuan's southward development strategy. (Wei and Wang 2018; 据介绍，此次公安部物证鉴定中心援建三师公安司法鉴定中心dna实验室，是公安部以习近平新时代中国特色社会主义思想为指引，深入贯彻落实党的十九大精神，贯彻落实党中央治疆方略和对兵团的定位要求，聚焦新

疆总目标，打好反恐维稳“组合拳”的生动实践，也是公安部物证鉴定中心与兵团公安局在战略合作框架协议下，贯彻兵团向南发展战略的具体举措。)

In these Bingtuan representations, the significance of the forensic genetic lab in the Bingtuan southward colonization strategy targeting the Uyghur homeland in southern Xinjiang's Tarim Basin is further elaborated in another February 2018 news story on the opening of the Tumxuk laboratory:

> The DNA Laboratory is the second divisional DNA laboratory of the Bingtuan Public Security System and is the first divisional DNA laboratory of the Southern Xinjiang Bingtuan Public Security Bureau. It has greatly improved the criminal science and technology level and equipment level of the division's public security, which will benefit the first division of Alar City and the 14th Division Kunyu City by providing an objective, accurate and powerful scientific basis and technical support for case investigation and litigation, which plays an active role in maintaining social stability and long-term peace and security. (Qi 2018; 该dna实验室是兵团公安系统第二家师局级dna实验室，是兵团南疆公安机关第一家师局级dna实验室。它的投入使用，大大提升了师公安刑事科学技术水平和装备水平，将惠及第一师阿拉尔市、第十四师昆玉市，为案件侦破和诉讼提供客观、准确、有力的科学依据和技术支持，为维护社会稳定和长治久安发挥积极作用。)

Maintaining social stability is a central justification for PRC's large-scale repression in Xinjiang (Sean Roberts 2020, 175). One article on the opening of the lab mentioned that Zhao Qiming, the director-general of the Ministry of Public Security's Institute of Forensic Science in Beijing, attended and features a photo of him (Qi 2018). These articles show the direct role of the Evidence Center in reinforcing the Bingtuan's institutions in Xinjiang. This announcement indicates that there were strong links established that likely supported the Evidence Center's cooperation with the Third Division on what might be called the Tumxuk Uyghur project, which we turn to now.

THE TUMXUK UYGHUR PAPERS

The first paper utilizing samples from 715 Uyghurs from Tumxuk appeared in November 2018 in the Chinese language journal *Hereditas* (Beijing), entitled "The Effect of EDARV370A on Facial and Ear

Morphologies in Uyghur Population" (Li Yi et al. 2018). Among the coauthors are Tang Kun, Liu Fan, and Li Caixia, which institutionally brings together, respectively, the Partner Institute, BIG, Erasmus University, and the Institute of Forensic Science.[3] Three other coauthors, Li Yi, Li Dan, and Xiong Ziyi are from the 2019 *eLife* paper on facial prediction by Xiong et al. (2019) and Pośpiech et al. (2018) discussed earlier. This paper is also significant because it introduces the new set of samples collected from 715 Uyghurs (688 men and 27 women) from Tumxuk during this period of intensified repression. The grant, 2017YFC0803501, co-owned by Liu Fan and Ministry of Public Security personnel (likely Li Caixia), is also credited as a funding source, as it is in the article by Xiong et al. (2019) and the VisiGen/Xinjiang Uyghur Study papers discussed earlier.[4] In this way, there is a seamless integration of personnel and organizational resources between large international projects and the more Uyghur-specific research being done in China. Another coauthor, Liu Haibo, is from the Bingtuan Public Security Bureau in Urumqi.

The article was received by the journal on 19 September 2018 and accepted on 1 November 2018, so the sampling and research were likely done during 2017–18, part of the larger escalation in the scale of genetic research targeting Uyghurs after 2014 such as we saw with the Wei et al. (2016a, 2016b) and Jiang et al. (2018a) papers on AISNP panels discussed in chapter 5. The article begins by discussing the EDARV370A variant of the EDAR gene located on Chromosome 2. They state it has had a "strong positive natural selection in East Asia" while having a very low frequency among European and African populations and that these differences help explain ectodermal differences in external appearance including hair thickness, chin, and earlobe attachment (Li Yi et al. 2018, 1025).[5] They then briefly discuss the role of EDARV370A in ectodermal diseases and mention its forensic genetic utility: "At the same time, EDARV370A has potential application value in forensic science as a molecular marker affecting multiple human appearance phenotypes. For example, using this gene as a molecular marker can infer important personal information, such as appearance from DNA samples, which can improve initiative and intelligence in criminal investigation work" (Li Yi et al. 2018, 1025; 同时,edarv370A作为影响多个人类外貌表型的分子标记,在法医学上具有潜在的应用价值,例如,利用该基因作为分子标记可实现对dna检材进行所属外貌等重要个人信息的推断,增加刑侦工作的主动性和智能性。). The authors then assert that the Uyghurs are an ideal population to study: "Because EDARV370A has its highest level of polymorphism in the mixed populations of Asia and Europe, the

Uyghur group as a typical mixed East Asia and Europe group is the ideal target for studying EDARV370A and its associated features. Therefore, this study further studied the EDARV370A candidate genes in the face and ear morphology of 715 Uyghur samples from Xinjiang" (1026; 由于 EDARV370A 在亚欧混合人群中的多态性最高，维吾尔族群体作为一个典型的东亚和欧洲混合的群体，是研究 EDARV370A 及其关联特征的理想对象，因此本研究进一步对 715 例新疆维吾尔族样本进行了面部和耳朵形态的 EDARV370A 候选基因研究。).

For the competence phase, in the materials and methods section, the authors outline the enrolment process: "The samples were collected from Tumxuk City, Xinjiang Uyghur Autonomous Region, with a total of 715 Uyghur individuals, including 688 males and 27 females. The ages ranged from 16 to 59 years old with an average age of 35.5 years old" (Li Yi et al. 2018, 1026; 样本采集自新疆维吾尔自治区图木舒克市，共计 715 例维吾尔族无关个体，其中男性 688 例，女性 27 例。年龄为 16~59 岁，平均年龄 35.5 岁。). Statistically and demographically, this sampling is very different from the Xinjiang Uyghur Study, which was disproportionately women. This emphasis on males seems to follow the stereotype of Uyghur men as potential threats (Tobin 2020, 305–6). The subjects were screened using a number of criteria: "Individuals included in the study met the following requirements: (1) parents and grandparents were Uyghurs; (2) did not receive hormone therapy; (3) did not have thyroid disease, pituitary disease or tumor; (4) did not have a drug induced growth problems, such as dwarfism, gigantism or acromegaly" (1026; (纳入研究的个体均符合以下要求：(1) 父母及祖父母均为维吾尔族；(2) 没有接受激素治疗；(3) 未患有甲状腺疾病，脑垂体疾病或肿瘤；(4) 没有因药物作用引起的生长发育问题，例如侏儒症，巨人症或肢端肥大症。). In this way, the donor criteria define Uyghurs through parentage and try to sample "normal" facial genetic variation among the Uyghurs rather than that affected by physical disorders or diseases (1026).

Information associated with the blood samples is collected during the enrolment process. They also collected data on physical appearance, using their height and weight measurements to calculate body mass index (BMI) (Li Yi et al. 2018, 1028). And here they made use of a state-of-the-art Artec Spider handheld scanner to convert their faces into 3D facial data using the mapping methods developed by Guo, Mei, and Tang (2013) discussed earlier (Li Yi et al. 2018, 1026). The authors assert that the enrolment process was conducted in an ethical and legal manner: "The study has been reviewed by the Ethics Committee of the Ministry of Public Security's Physical Evidence Identification Center, and all subjects have signed informed consent" (1026; 本研究已通过公安部

物证鉴定中心伦理委员会审查，所有受测者都签署了知情同意书). The enactment of law and legal processes of informed consent is a central claim of the Chinese government agencies.

Assessing the subjects for various facial and ear traits, the genotyping involved the following: "DNA samples were obtained from peripheral venous blood, collected in EDTA-Vacutainer tubes and stored at −20°C" (Li Yi et al. 2018, 1026; DNA 样本来源于外周静脉血，用 EDTA-Vacutainer 试管收集放置于−20°C下保存。). This sampling involves a part of the metonymic chain of the Uyghur research subjects represented by their blood samples that are transformed into a DNA sample. The samples were processed with reagents and tested using an Illumina Infinium Global Screening Array 650K model (GSA) to genotype 700,078 autosomal SNPs, including chromosome 2's EDARV370A (rs3827760) (1026–7). The statistical analysis again involves the use of PCA to assess the relative effect of each component on seven facial and ear measurements and the relationship between features and the donors' EDARV370A alleles: GG, AG, or AA. The scientists found that EDARV370A was associated with chin shape phenotype and three ear phenotypes.

In the discussion section, the scientists reiterate that the Uyghurs are a European and East Asian population that mixed about eight hundred to two thousand years ago and therefore an ideal group for conducting GWASs (Li Yi et al. 2018, 1030). For example, they restate their finding that people who are homozygous for 370A (both genomes have the same allele) tend to have a different chin shape than those who are homozygous for 370V (which is common in Europe and Africa).

The article concludes by asserting that understanding how EDARV370A affects eight facial traits identified by the authors will contribute to the development of forensic genetic phenotyping to depict human appearance, which will help guide police investigations with "more initiative and more intelligence" (Li Yi et al. 2018, 1031；更加主动、更加智能的方向发展。). Therefore, the article frames the research findings in terms of their potential practical impacts on police investigations, a type of manhunt.

BEIJING INSTITUTE OF GENOMICS DATABASE ENTRIES

The next mention of the Tumxuk Uyghurs appears as a set of entries on a database run by the Beijing Institute of Genomics Datacenter in early 2019. The data is for a subset of 612 Tumxuk Uyghur subjects (590 males and 22 females) from a project (PRJCA001171) entitled "The candidate gene analysis for human facial morphology in Eurasian population"

(Beijing Institute of Genomics Datacenter 2019b). The data provider is listed as Li Caixia of the Ministry of Public Security while the submitter is listed as Li Yi of BIG (2019a). Li Yi is also a coauthor on the 2018 *Hereditas* (Li Yi et al. 2018) and 2019 *eLife* paper (Xiong et al. 2019). The project description states, "To explore the genetic mechanisms of human facial morphology, several variants for about 700 Uyghur individuals were genotyped on Illumina Infinium Global Screening Array 650K. At the same time, various facial phenotypes were derived from all samples" (Beijing Institute of Genomics Datacenter 2019b). The project listing contains 612 entries for the Uyghur subjects (2019c). The entry for one of the subjects reads as follows (Beijing Institute of Genomics Datacenter 2019a):

> Accession: SAMC051815
> Sample name: uyg1
> Title: uyg1
> Organism: Homo sapiens
> Description: Normal Uyghur individual living in Tumxuk City in Xinjiang Uyghur Autonomous Region, China
> Biomaterial provider: Caixia Li; Beijing Engineering Research Center of Crime Scene Evidence Examination, Institute of Forensic Science

This database entry is the first of 612 and shows how the research subject is objectified as "biomaterial" provided by Li Caixia.

Li Caixia, Liu Fan, Tang Kun, Li Yi, Xiong Ziyi, and others published an article using the Tumxuk Uyghurs on 25 April 2019 entitled "EDAR, LYPLAL1, PRDM16, PAX3, DKK1, TNFSF12, CACNA2D3, and SUPT3H gene variants influence facial morphology in a Eurasian population" (Li Yi et al. 2019). The article appeared in *Human Genetics*, an influential journal published by Springer Nature, a major international academic publisher. This paper uses the same subset of 612 of the 715 Tumxuk Uyghurs as the 2018 *Hereditas* paper. The article's manipulation phase is, "In human society, the facial surface is visible and recognizable based on the facial shape variation which represents a set of highly polygenic and correlated complex traits. Understanding the genetic basis underlying facial shape traits has important implications in population genetics, developmental biology, and forensic science" (681). The article does not use a manhunt narrative schema, but rather outlines findings from a number of GWASs on "associations between DNA variants and normal facial variation," reporting "a total of 125 SNP for 103 distinct genomic loci with genome-wide significant association to a number of different facial features" (681). It notes that these studies have focused on European populations and states that "whether these findings are generalizable in Asian populations remains unclear." It then defines the paper's quest: "Here, we investigated the potential effects of the 125 facial variation-associated SNPs on facial morphology in a European–Asian admixed population" (682). In this way, it adapts a quest for knowledge about how genetics affects facial morphology in admixed populations, which relies on a taxonomy that reifies European and East Asian as dominant continental populations.

The materials and methods section begins with their ethics statement: "This study was approved by the Ethics Committee of the Institute of Forensic Science of China, and all individuals provided written informed consent" (Li Yi et al. 2019, 682). However, the remainder is different from any proceeding papers, which may indicate they are reacting to increasing global attention to the Uyghurs: "The participants were all volunteers. The consent was discussed in their native language and the signature was in their native language" (682). This mention of the subjects being volunteers and the informed consent being conducted in the Uyghur language is unique among the papers I have read, perhaps reflecting growing international criticism of their research. The remainder of the samples section states where the donors are from and the selection conditions: "We sampled a total of 612 unrelated Eurasian individuals living in Tumxuk City in Xinjiang Uyghur Autonomous Region, China," along with the conditions outlined in earlier papers about having Uyghur parents and grandparents and no hormonal disorders (682). Throughout the article, the authors try to represent the Uyghur donors as Eurasian, using this term fifteen times while using Uyghur only three times, designations that involve synecdochic metonymy (Category FOR Members of Category; a Genus FOR Species relation). The scientists screened potential donors by asking them questions about their ancestry, which shows the synecdoche of Uyghur FOR Eurasian (member of category FOR category) in the ancestry criteria: "All individuals met the following conditions: (1) their parents and grandparents were both of Uyghur origin" (682). Furthermore, the authors screened the donors as prototypical members and hence representative of the larger classification of Uyghurs in their physical appearance: "(2) they had not received hormone therapy; (3) they had no thyroid disease, pituitary disease, or tumors; and (4) they had no medical conditions affecting growth and development, such as dwarfism, gigantism, and acromegaly" (682). This selection process involves a synecdochic metonymy of Typical Member of Category FOR Category in which these criteria for selection are based on concepts of prototypical members of this category and so representing normal variation in the population (Bierwiaczonek 2020, 227; Gibbs 1999, 66). The samples section also explains how the scientists used an Artec Spider 3D scanner and software to create a 3D image of each donor (Li Yi et al. 2019, 682).

The phenotyping section states that the authors used an updated version of the system set used in the 2013 article by Guo et al. based on "x-y-z coordinates of 17 facial landmarks" (Li Yi et al. 2019, 682). The "DNA samples were genotyped on an Illumina Infinium Global Screening Array 650 K," which can test for around 650,000 markers at a time (682). The scientists then did a series of statistical analyses for co-relations between particular alleles with facial distance phenotypes based on physical measurements made using these seventeen facial landmarks. Of the 125 SNPs identified in previous GWAS studies, the researchers identified eight SNPs (listed in the paper's title) as having a significant role in face shape among the Uyghur subjects. For example, like the *Hereditas* paper findings, the EDARV370A alleles were influential, this time on eight facial landmark distances (682). The researchers then engage in a discussion of the existing literature on the various SNPs.

The funding credits include grant 2017YFC0803501 along with the National Natural Science Foundation of China, the Institute of Forensic Science, and Fan Liu's Thousand Talents Plan award. Throughout, the paper refers to the Institute of Forensic Sciences, but there is no mention of the Ministry of Public Security. As well, Li Caixia is a corresponding author and uses her Tsinghua University email address rather than her Ministry of Public Security email address (as she does in other articles as well). The article repeatedly refers to Uyghurs as Eurasians, emphasizing they are an admixed population. This designation is based on European and East Asian functioning as larger racial groups that define the Uyghurs as boundary figures between the two.

THE HEIGHT PREDICTION PAPER

In addition to facial appearance, height is another one of the targeted phenotypical traits under the 13th Five-Year Plan (Zhao Xingchun et al. 2018). There is a Beijing Institute of Genomics Datacenter database listing with a submission date of 22 January 2019 for another project, PRJCA001236, entitled "Predicting adult height from DNA variants in a European-Asian admixed population" (Beijing Institute of Genomics Datacenter 2019d). The project description states its goal: "To evaluate the predictive power of height-associated SNPs from GIANT study in 2014 in Uyghurs. Several variants for about 700 Uyghur individuals were genotyped on Illumina Infinium Global Screening Array 650K and adult height phenotype was acquired using a stadiometer from all sample." GIANT, Genetic Investigation of Anthropometric Traits, is a consortium of US and European researchers that produced a 2014 paper of nearly seven hundred height-correlated SNPs (Wood et al. 2014). Again, funding acknowledgements are made to grant 2017YFC0803501 and the Ministry of Public Security, and the data provider is Li Caixia of the Ministry of Public Security. The place of origin of the subjects is not mentioned but it appears to use the Tumxuk Uyghurs.

Li Caixia and Liu Fan were among the coauthors of a 2019 paper published in the *International Journal of Legal Medicine*, entitled "Predicting Adult Height from DNA Variants in a European-Asian Admixed Population" (Jing et al. 2019). Another coauthor, Mi Ma, has a joint appointment with the Institute of Forensic Science and Bingtuan: "Xinjiang Production and Construction Corps Seventh Division Public Security Bureau, Ürümqi, China" (1667). In addition to Ministry of Public Security and Bingtuan researchers, Liu Fan also lists his dual appointments with BIG and Erasmus University Medical Center. The paper was received 27 November 2018, accepted 5 March 2019, and published online 12 April 2019. It appears in another Springer Nature published journal entitled *International Journal of Legal Medicine*, which is an influential forensic genetic journal. They mention Uyghurs several times in the abstract. This paper adapts a conventional forensic genetic hunting narrative with its manipulation phase: "Predicting appearance phenotypes from genotypes is highly relevant in forensic genetics, particularly when standard forensic STR-profiling is uninformative. So far, the most successful examples are restricted to the prediction of human pigmentation traits, such as eye color, hair color, and skin color, since models with a limited number of SNPs can provide highly accurate prediction results" (1667–8). They then move into a brief literature review about height as a trait involving many genes, citing a 2014 paper on height among Europeans in which Liu Fan was the lead author. In their commitment phase, they claim: "Here, we evaluate the predictive power of the 697 height-associated SNPs by focusing on a European-Asian admixed population, in which genetic admixture has been estimated about 50% European and 50% Asian. We address the effect of population heterogeneity on height prediction and explore the practical potentials of height prediction in non-European populations in forensic settings" (1668). With its construction of the Uyghurs as a half-European half-Asian admixed population and the idea of applying the findings of research done on European populations to those outside of Europe, this commitment phase is very similar to that of the above 2019 paper on facial morphology.

They then briefly recount some of the literature to define their quest's significance in relation to the field. "We sampled a total of 689 unrelated Uyghur males living in Xinjiang Uyghur Autonomous Region, China" along with the same conditions of having Uyghur grandparents and no hormonal disorders as the earlier two papers (Jing et al. 2019, 1668). "The current study included 687 males (mean age = 35.57, SD = 10.15) from a European-Asian-admixed Uyghur population with a previously estimated 50:50 ancestry ratio" (1669). The 687 men are likely the Tumxuk Uyghurs, because the total number of subjects is nearly the same as the 688 in the *Hereditas* paper and the average age of 35.57 with a standard deviance of 10.15 is nearly the same as the *Hereditas* paper findings of 35.6 years old with a standard deviance of 10.17 (Li Yi et al. 2018, 1030). Removing the twenty-seven women subjects, who account for only 3.7 per cent of the 715 donors, would not change the average age and standard deviation very much. The samples were again processed using the "Illumina Global Screening Array (GSA) (Illumina, San Diego, CA, USA)" (Jing et al. 2019, 1668). The data was processed through reference to the 1000 Genomes Project's Phase 3 data to impute 6,775,821 SNPs. Using 581 height-related SNPs from the GIANT study, the prediction analysis used the set of potential SNPs to determine if they were correlated with height in Uyghurs. The authors claim to have completed their quest: "Overall, our study represents a pioneering effort in genomic profiling of adult height in a non-European population. The prediction accuracy obtained in the Uyghurs was lower than that obtained in the Europeans but fine-tuning the prediction modeling could considerably improve the prediction accuracy" (1676).

The authors then conclude with an epideictic rhetorical call to action for research into height-related markers: "We emphasize the need of large collaborative efforts in the East Asian populations to deliver a comprehensive set of Asian-specific height-associated genetic markers, which is expected to boost the accuracy in genomic profiling of human adult height in Asians as well as worldwide and finally contribute to the integrated prediction of human appearance traits in forensic applications" (Jing et al. 2019, 1677). This concluding paragraph sets out the need for a comprehensive phenotyping effort, one that fits not only transnational forensic genetic discourses but also translates into and submits to the PRC's Five-Year Plan public security technology development efforts. In these articles, scientists have successfully organized circuits of knowledge production that enroled Uyghurs and took their blood samples in Tumxuk in southern Xinjiang, processed them in labs in Beijing and Shanghai, and were peer-reviewed and published in genetic research journals by European-based Springer Nature. However, their epideictic rhetorical call for international cooperation would receive a sharp rebuke, as we will see in the next chapter.

Again, this 2019 *International Journal of Legal Medicine* height paper mentions grant 2017YFC0803501 and the Ministry of Public Security funding and ethics approval (Jing et al. 2019, 1677). The height paper states, under the heading "Compliance with ethical standards," that "The study was approved by the ethics committee of Institute of Forensic Science, Ministry of Public Security, China. All participants provided written informed consent" (1677). Furthermore, "The authors declare that they have no conflicts of interest" (1677). These claims of compliance with ethical standards and no conflict of interest are untenable. The Ministry of Public Security has been actively engaged in the violence work of repressing the Uyghurs and this journal article is part of the development of phenotype and ancestry research for the purposes of differentiating Han Chinese and Uyghur. The Tumxuk papers involve the performance of sovereign decisions in the racialized hierarchy between the various actants. The researchers are able to organize and articulate the research assemblage without concern for any Uyghur resistance. Uyghur subjects in contrast are subjected to a severe security apparatus in their everyday lives, one that will often geofence them in their home village or neighbourhood, depending on their *hukou* (home registration), with extensive travel restrictions. In contrast, the researchers have *hukou* residence permits in Shanghai and Beijing, a function of their privileged status along with other forms of state recognition. This hierarchy is also evident in the differential obligations, in which Uyghur subjects would be obliged to participate or else attract suspicion thereby bringing danger to themselves and their families.

THE DISTRIBUTED PERFORMANCE OF SOVEREIGNTY

The performance of sovereignty and the recognition of who speaks for whom is distributed across these racializing assemblages in various ways. The Chinese state asserts genetic sovereignty over Chinese people including Uyghurs and other Indigenous peoples. Therefore, all of the processing of Chinese samples was done within the boundaries of China (with the exception of the ten Uyghur cell lines in the Diversity Project collection). However, the networks of the Chinese scientists of the Partner Institute and BIG are international in scope. It is the Western scientists' unquestioning recognition of the right of Chinese scientists to represent the Uyghurs within the research assemblages spanning transnational science, research institutes, funding agencies, review boards, and various security agencies that allows the research assemblages to maintain its ethical claims. Not questioning ethical claims about the Uyghurs fits with the way the manhunt narrative reduces research to a technoscientific process apparently outside of the political sphere. That is, the norm of improving police abilities to generate investigative leads functions as the epideictic rhetoric central to the organizing narratives of these research assemblages. However, external forces would soon intervene, causing significant disruptions to these assemblages.

11

Destabilizing Research Assemblages

In the previous chapters on ancestry and phenotype research, scientists have hierarchically coproduced themselves as scientists through processing Indigenous peoples as their objects of research. This authoritarian relationship depends upon its general acceptance within the assemblages of forensic genetics and genetic research in general, including funding agencies, publishers, journals, and so on. Within these hierarchies, scientists are neoliberal self-actualizing entrepreneurial subjects able to articulate and organize international assemblages spanning transnational science, biotech firms, state security and funding institutions, and Indigenous peoples (Ong 2006). Involved scientists are privileged under contemporary neoliberal governance by exercising what sociolegal theorist Mariana Valverde (1996) terms a *despotism of the self*, but this coproductive self-actualization relies upon despotism over Indigenous peoples through the denial of their rights and self-determination by using these peoples' genetic materials as resources in research and development (Dean 2007, 120–1). However, these extended international assemblages are vulnerable to various external disruptions and pressures that shape the potential agency of actants, including research subjects, scientists, and institutions (Callon 1986, 219–21; Munsterhjelm 2014, 147–60). For example, Indigenous protests and international organizing eventually shelved the Human Genome Diversity Project, leaving it with only 1,064 cell lines representing fifty-two populations, a fragment of its proponents' original aspirations of tens of thousands of samples representing five hundred to seven hundred populations (Barker 2004; Harry 2009; Reardon and TallBear 2012). These disruptions can also occur at a more microlevel, such as the US National Health Institutes' Coriell Cell Repositories stopping its sales of the Karitiana and Surui cell lines given to them by Kidd Lab in the early 1990s, over twenty years prior. I noticed this in the spring of 2015 and sent an email to Coriell Cell Repositories' customer service department to ask why the cell lines and DNA extract were no longer listed.[1] In a 29 June 2015 response, they stated that "These lines are no longer available because they were withdrawn from the Repository upon request of the tribes." This withdrawal directly contradicts Kidd and his colleagues' claims of using the Karitiana and Surui with their informed consent. However, that said, Coriell continues to sell Nasioi and Maya cell lines and DNA extract from these cell lines, among others (such as the Ami and Atayal of Taiwan), that Kidd Lab also gave them, which shows the political selectiveness of this sort of removal (Munsterhjelm 2014, 167–70). In actor network terms, the stability of forensic genetic assemblages requires not only routine acceptance and translation by internal participants of research goals and research subjects' informed consent, but also that conflicting interpretations be omitted and/or actively resisted, in what can be termed a *sociology of estrangement and disassociation* (Callon 1986, 219–21; Cooren and Taylor 2000, 180–5; Galis and Lee 2014, 155–6).

It is at this point that some of the research in the preceding chapters enters the story directly. This disruption became evident in the press coverage of my research by the *New York Times* and National Public Radio (NPR) in the United States and its uptake in US Congressional hearings and government sanctions against the PRC as part of the growing concern over international forensic genetic cooperation with PRC public security forces during 2017–19. These articles and resulting debates indicate how scientists and biotechnology firms can be forced to account publicly for some of their actions under certain geopolitical circumstances. In this case, there are escalating tensions in the current strategic/military and political economic relations of the PRC and the US. Stated in broad terms, these include China's direct challenge to American military hegemony in Southeast Asia with China's construction of military bases on islets in the South China Sea that effectively assert Chinese territorial sovereignty in a strategic region adjacent to major shipping routes, the so-called nine-dash line, areas of which are also variously claimed by Vietnam, Taiwan, Brunei, the Philippines, and Malaysia. As well, political economic tensions have escalated over the long-running trade surplus that China has with the US. Crucially, China's political economic and extraterritorial expansion into central Asia and in general globally with the Belt and Road Initiative challenges US influence. The US's continued support of Taiwan as an independent Chinese settler colonial state is another long-term major point of contention (Munsterhjelm 2014). These intertwine with the overwhelming evidence revealing the extent of the PRC's cultural genocide apparatus in Xinjiang, which has been variously translated by a heterogeneous assemblage of Uyghur activists and diaspora organizations, human rights organizations, activist academics, mass media outlets, and Western governments. Internationally, Uyghur diaspora organizations, human rights groups, and academics have challenged the PRC's genocidal campaign against the Uyghurs and other Turkic peoples in Xinjiang. These efforts began to have a major impact in 2018 as the Chinese government, which had first denied the existence of Xinjiang's reeducation camps, was forced to acknowledge these camps when confronted with extensive satellite imagery, testimonials of former prisoners, news reports, and other evidence.

I began focusing on forensic genetic research involving Indigenous peoples in early 2014 (see, for example, Munsterhjelm 2015) and began this book project in 2016. In June 2017, I read a May 2017 Human Rights Watch report and coverage in *Nature* on Thermo Fisher's sales to Xinjiang security bureaus, which cited the research of the Belgian computational biologist and ethicist Yves Moreau (Cyranoski 2017; Human Rights Watch 2017a). I then started to more closely analyze the Chinese university and security agency connections that I had noticed in the research papers involving Kenneth Kidd's fifty-five AISNP panel (see, for example, Pakstis et al. 2015a). I wrote a short opinion piece that was published in April 2018 on a weblog affiliated with the Society for the Social Studies of Science. It discussed the use of Indigenous peoples and Uyghurs in Kenneth Kidd's AISNP research cooperation with Li Caixia of the Ministry of Public of Security along with Bruce Budowle's keynote speech in October 2017 at the Thermo Fisher sponsored forensic genetic conference in Chengdu, China (Munsterhjelm 2018). This article did not attract any media attention or other responses. To disrupt these assemblages would require a stronger assemblage in terms of both evidence and involved organizations.

In June 2018, I found a patent application filed by Institute of Forensic Science's Li Caixia and her colleagues using Uyghurs and I began to research further. In August 2018, I contacted Human Rights Watch's Maya Wang and Sophie Richardson. In September and October 2018, I found further patent filings for a total of nine (three patents and six patent applications) by Li Caixia et al. with the Institute of Forensic Science as assignee, most of which used and targeted Uyghurs, Tibetans, or other minority peoples. I assembled summaries of these into an information kit, which also included findings from my April 2018 article and an initial analysis. Wang introduced me to Wee Sui-Lee of the *New York Times* and we began cooperation in November 2018. During further research in the fall of 2018, I found the afore-discussed mass transfers of DNA extract samples by Kidd Lab to Li Caixia and her colleagues at the Ministry of Public Security.

Since 2017, Thermo Fisher had been under strong criticism for its continued sales of genetic sequencers and other equipment to Xinjiang security agencies from activist academics, Human Rights Watch, and other social justice organizations. As well, Republican members of US Congress, such as Mark Rubio (who also cited the Human Rights Watch reports) and Chris Smith (Strumpf and Khan 2018a), who conceptualized repression in Xinjiang as an issue of religious persecution, which fits with right-wing anticommunist ideologies prevalent in the US. Rubio sent a letter about Thermo Fisher's sales in Xinjiang to US Commerce Secretary Wilbur Ross, who responded in a 11 June 2018 letter, stating that "The items are low-technology products that are available from worldwide sources, including indigenous Chinese sources, and have numerous legitimate end-uses, including in education, medical research and forensics" (2018b). The US Commerce Secretary's response downplayed the significance of the Thermo Fisher equipment and refused to enact any restrictive measures against the company.

Such high profile criticism of Thermo Fisher continued and the corporation finally conceded in a 20 February 2019 press release entitled "Updated Statement on Xinjiang" (Thermo Fisher 2019).[2] The press release follows a standard public relations crisis management playbook, with an admission of wrong doing committed followed by demonstrations of reform and a claim of redemption that signifies the ultimate moral commitment of the company to serving science and Humanity (Munsterhjelm 2014, 108, 157). This public announcement fits with liberal governance's cycle of problematization, reformation, and replacement (Dean 1999, 190–1). In the press release's manipulation phase, public criticism of Thermo Fisher for its Xinjiang sales is implied and the release's text begins with the commitment phase in apparent response to external criticism: "As the world leader in serving science, we recognize the importance of considering how our products and services are used – or may be used – by our customers" (Thermo Fisher 2019). The company makes an appeal to "serving science" as a type of epideictic rhetoric of moral submission thereby reasserting the morality of its resulting fundamental fiduciary duty. The competence phase claims a knowing-how-to-do and an able-to-do modality: "We undertake fact-specific assessments and have decided to cease all sales and servicing of our human identification technology in the Xinjiang region – a decision that is consistent with Thermo Fisher's values, ethics code and policies." It does not directly mention the wrongdoing of Chinese security agencies, but rather says that sales in the Xinjiang region are inconsistent with its corporate values. The remainder of the press statement reiterates their dedication to the use of forensic genetic equipment in the sovereign friend/enemy decision-making process to protect populations (knowing-how-to-do modality): "We are proud to be a part of the many positive ways in which DNA identification has been applied, from tracking down criminals to stopping human trafficking and freeing the unjustly accused." In the announcement's performance phase, it claims moral redemption with the epideictic rhetoric of serving Humanity using its corporate branding slogan (Thermo Fisher 2019): "We are committed to continuing to deliver those benefits to our customers, consistent with *our mission to enable our customers to make the world healthier, cleaner and safer*" (Munsterhjelm 2014, emphasis added; Thermo Fisher 2020). This decision to stop selling scientific equipment to Xinjiang security agencies shows how this US corporation affected the capacity of Chinese security agencies to research, develop, and implement forensic genetic technologies in the biopolitical distinction between Han Chinese as productive bodies and Uyghurs and other Indigenous peoples as destructive bodies (Aradau and Van Munster 2007; Epstein 2007). This capacity to distinguish enacts racial caesurae that guides Chinese government sovereign decisions to biopolitically secure its settler populations against those peoples deemed as threats (Aradau and Van Munster 2007; Kam and Clarke 2021; Epstein 2007; Sean Roberts 2020). This impact demonstrates how graduated sovereignty functions, in which recognition of such decisions are distributed across the disparate time-spaces of assemblages that exceeds the borders of the PRC.

Sui-Lee Wee's article, "China Uses DNA to Track Its People, with the Help of American Expertise," was published the next day, on 21 February 2019. The article used my findings about Kenneth Kidd's cooperation with Li Caixia, the provision of DNA extract samples, the use of these DNA samples in Chinese patent applications filed by Li Caixia and her colleagues at the Institute of Forensic Science, Li Caixia et al.'s inclusion of Uyghur data in the ALFRED, Bruce Budowle's cooperation and the central role of Thermo Fisher equipment in the research process, and how China represented some 10 per cent of its annual sales of US$20 billion in 2017 (Thermo Fisher 2018a, 6; Wee 2019). The article begins by using the visual epideictic rhetoric of a photo of a man looking out a sunny window with the caption, "Tahir Imin, a 38-year-old Uighur, had his blood drawn, his face scanned and his voice recorded by the authorities in China's Xinjiang region" (Wee 2019). Rather than recount the entire article, I want to focus on how this article and a follow-up July 2019 podcast done by NPR's *Planet Money* destabilized the Institute of Forensic Science's organizing narrative schemas as a global assemblage in the production of AISNP forensic genetic research and development.[3] This destabilization includes how involved assemblage actants like Kenneth Kidd, Thermo Fisher (at least publicly, though not in practice), and Bruce Budowle reacted by dissociating themselves from the Institute of Forensic Science's research assemblage.

Manipulation Phase Criticisms

Acting on behalf of the People and Humanity, the security apparatuses of the US and People's Republic of China are both senders of a quest for biometrics to improve security against terrorism and criminals. The President's DNA Initiative by the US Department of Justice in 2004 first called for alternative markers including ancestry, phenotype, and ethnicity. As chapter 10 showed, the Chinese government has similarly called for biometric research and development of technologies, including forensic genetic ancestry and phenotyping targeting Uyghurs and other Indigenous peoples because they are viewed as potential threats (see figure 11.1).

CRITICISM C1: *Aiding a genocidal regime*. Challenging the ethical claims of forensic genetic researchers, Wee's (2019) article begins with the personal experience of a Uyghur man named Tahir Imin, now exiled in the US, who was forced to undergo a "free health checkup" that involved biometric profiling including a voiceprint, fingerprints, 3D facial scan, and blood sampling. When he asked how he could see his results, he was told to go see the police, thus demonstrating it was for security purposes, not its purported health intentions. Similarly, the NPR podcast and *All Things Considered* broadcast utilized the experience of a Uyghur graduate student in the US with the pseudonym Alim who was arrested upon his return to Xinjiang for a summer visit. He was forced to go through similar biometric profiling processes and then incarcerated in a reeducation camp for four weeks (National Public Radio 2019). These personal experiences are a form of testimony by Uyghurs that sharply criticized the epideictic rhetoric of the Chinese state's call for action. Both personal testimonies are used to represent the experiences of mass incarceration of upwards of one million people in reeducation camps. The estimates of over one million people in the camps function as epideictic rhetoric that utilizes quantity, which together with the personal testimonies, illustrate

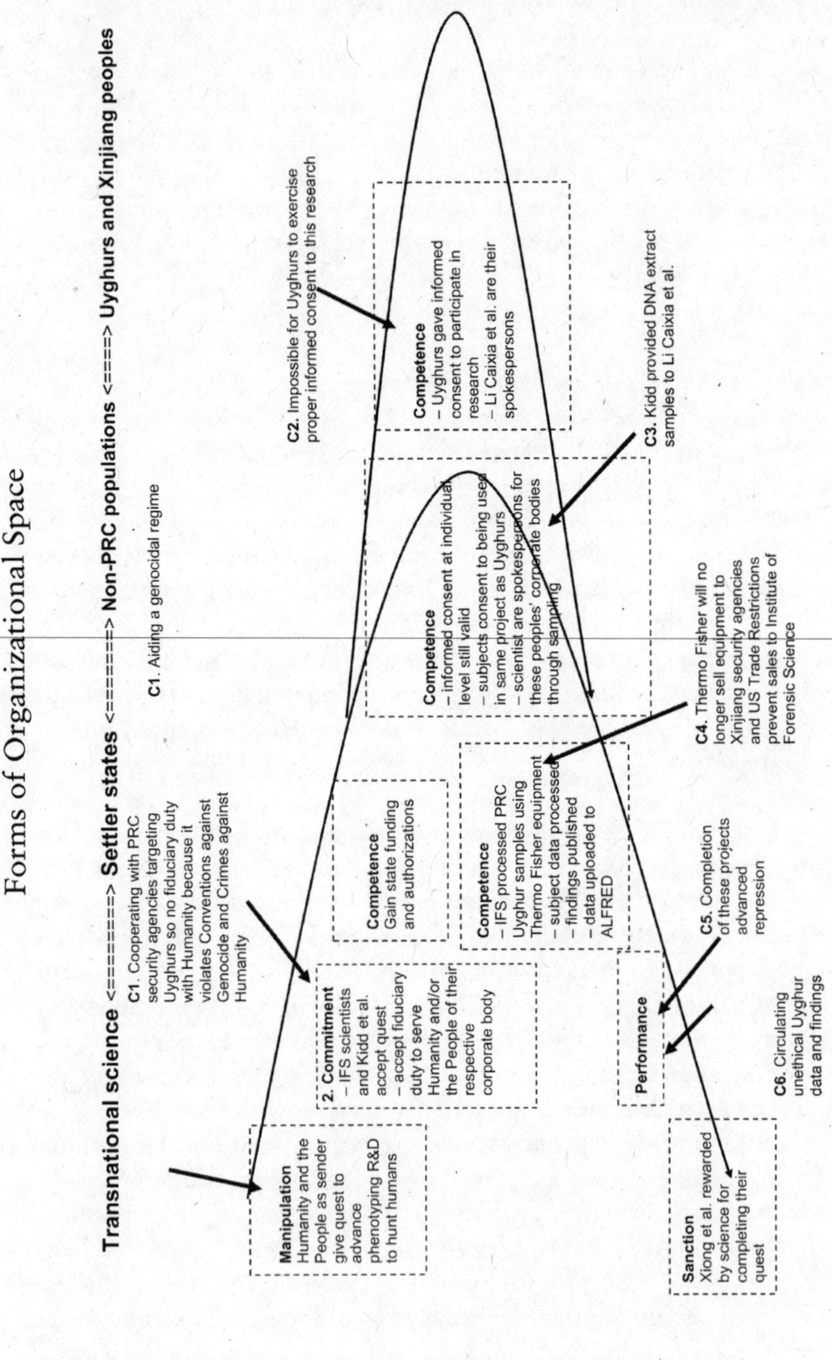

Figure 11.1 | Diagram of how external criticisms destabilized Kidd Lab and Institute of Forensic Science Cooperation organizing narratives.

the scale and severity of the genocidal crisis the Chinese government is inflicting on the Uyghurs and other Turkic peoples in Xinjiang. The *New York Times* article quotes anthropologist Darren Byler of Simon Fraser University stating that the Uyghurs had no choice but to submit to the mandatory health check-ups of the "Physicals for All" program.[4] Both articles emphasize the unjustness of the Physicals For All program as part of the racialized biometric profiling of the Uyghurs and other Turkic peoples. The NPR report included interview segments by Ailsa Chang with Sophie Richardson of Human Rights Watch in which Richardson discussed the Chinese security agency's use of biometrics such as gait recognition and centralized databases for social control (Chang and Fountain 2019a).

CI DEFENCE: *Knowledge of populations is necessary*. In response, Kidd defended the epideictic rhetoric of the state's quest to know its populations in an effort to secure them: "He said governments should have access to data about minorities, not just the dominant ethnic group, in order to have an accurate picture of the whole population" (Wee 2019). In practice, Kidd essentially asserts that state security agencies need to know the genetic traits of different populations to have the capacity to secure them.

Commitment Phase Criticisms

The commitment phase involves the move to a specific project of advancing knowledge on Uyghurs and other minority groups from the sender's general quest for security in the manipulation phase. As we saw in earlier chapters, both Kenneth Kidd and Li Caixia are receiver-subjects whose quests are embedded within their respective security apparatuses. Kidd's quest through his submission for NIJ grants for research on IISNPs and AISNPs is embedded within the US security apparatus. These efforts built on his significant earlier research on genetic diversity and his expertise in forensic genetics. Li Caixia's quest for Uyghur ancestry markers is embedded within the Chinese security apparatus with its general quest for a biometric suite of technologies of which forensic genetic ancestry is one of her specific projects. Her funding and authorizations come from the Ministry of Public Security. This submission and fiduciary duty to their respective security apparatuses is minimized in international research papers. Instead, scientists utilize the highly translatable generic quest for security on behalf of Humanity to improve police manhunting capacities and so make police investigations more effective and efficient.

CRITICISM C2: *Cooperating with PRC security agencies.* International research assemblages are an important means through which the researchers' goals and fiduciary duties to their respective security apparatuses can be achieved. This cooperation was the central target of criticism in the *New York Times* article and NPR podcast, particularly Kidd's decision to work with the Ministry of Public Security. Cooperation began in 2010 when Kidd received an offer of an all-expenses-paid trip to China by the Ministry of Public Security (Wee 2019; National Public Radio 2019). I was quoted as saying this international cooperation "legitimizes this type of genetic surveillance" (Wee 2019).

C2 DEFENCE: *International cooperation is normal.* Kidd defended his cooperation with the Ministry of Public Security, saying that he cooperates with police agencies all over the world and considered the police agencies in China were no different (Wee 2019). In general, the scientists seem very happy to submit to the quest given to them by the People or Humanity to improve manhunting technologies. However, as Kidd's comments indicate, they are less willing to accept the fiduciary responsibility to ensure their research is ethical in practice and in its potential impacts on vulnerable peoples. Kidd's blame-avoidance is quite similar to Budowle's epideictic rhetorical claims. Wee shows how Budowle tries to disassociate himself from the Ministry of Public Security: "Jeff Carlton, a university spokesman, said in a statement that Professor Budowle's role with the ministry was 'only symbolic in nature' and that he had 'done no work on its behalf.' 'Dr. Budowle and his team abhor the use of DNA technology to persecute ethnic or religious groups,' Mr. Carlton said in the statement. 'Their work focuses on criminal investigations and combating human trafficking to serve humanity'" (Wee 2019). This statement is somewhat misleading because Budowle did supply Li Caixia with DNA extract samples, as discussed in chapters 4 and 5. As well, he attended conferences, like those in Xi'an in 2015, where he was pictured beside Ye Jian of the Ministry of Public Security, and gave a keynote address in Chengdu in 2017, which was sponsored by Thermo Fisher and PRC universities and included researchers from Chinese security agencies, including Ye Jian, as discussed in chapter 5. The statement reiterates the submission to Humanity by Budowle and his team of scientist, a central epideictic rhetorical claim that is frequently emphasized in ethical controversies (Munsterhjelm 2014, 114–16).[5]

The 5 July 2019 NPR's *Planet Money* podcast and an abridged version broadcast nationally in the US on the 18 July 2019 edition of NPR's *All Things Considered* news program focused on Kidd's research cooperation as an extension of his larger AISNP panel project. Kidd frames his AISNP research and development project as part of his long-term research into human genetic variation:

> FOUNTAIN: But in 2010, there was still a big hole in the global genetic picture that he was trying to assemble – a place where he didn't have a lot of genetic samples from yet. And that place was China.
> CHANG: There's, like, more than a billion people there. I imagine there's a lot of genetic diversity that's worth mining (laughter).
> KIDD: Yes, there is.
> FOUNTAIN: What Ken says he didn't know, what he couldn't know, was how China would end up using his research.
> CHANG: I mean, did you have any concerns at the outset?
> KIDD: Not at the outset. (Chang and Fountain 2019a)

Kidd in this way claims a type of innocence that he could not foresee what would happen in Xinjiang, in part because access to these populations involved cooperation with PRC security agency researchers.

Competence Phase Criticisms

CRITICISM C3: *Informed consent violations in the enrolment subnarrative*. A central criticism was of how Kidd and other scientist's routinely and uncritically accepted Chinese scientists' claims of informed consent by the Uyghur subjects and the validity of ethics oversight by the Institute of Forensic Science.

C3 DEFENCE: *Deny fiduciary duty*. Kidd denied that he had a fiduciary duty to exercise due diligence in confirming the Ministry of Public Security's informed consent claims: "As for the consent issue, he said the burden of meeting that standard lay with the Chinese researchers, though he said reports about what Uighurs are subjected to in China raised some difficult questions. 'I would assume they had appropriate informed consent on the samples,' he said, 'though I must say what I've been hearing in the news recently about the treatment of the Uighurs raises concerns'" (Wee 2019). Kidd's claim that an institution involved in the repression of the Uyghurs must bear the burden of meeting an international research ethical standard to do no harm is an attempt to deflect blame and also deny the fiduciary duty and due diligence that he and other researchers had in ensuring that their research did no harm.

Kidd's recourse to uncritically accepting PRC sovereign decision-making processes may be the politically expedient requisite for doing research with PRC institutions, but it is not mean he and other scientists were excused from ethical and political accountability. Arguably, this denial demonstrates complacency and perhaps willful ignorance in not questioning the ethical claims of the security agency of an authoritarian government that has been engaged in a forced assimilation campaign against the Uyghurs that began in 2013–14 and escalated into mass incarceration in 2017. This enrolment criticism challenges the politically and organizationally expedient assumption of equality of recognition of rights among all enroled research subjects, which contributes to the routine failure to consider vulnerable peoples.

In his interview with NPR's Ailsa Chang, Kidd maintained that he had done nothing wrong, stating that he could not know what would happen in the future (Chang and Fountain 2019). When asked about informed consent, he stated that, according to Li Caixia, "the samples were collected with signed, informed consent." When Chang responded, "And you didn't question that?" Kidd replied, "On what basis would I question it?" (Chang and Fountain 2019a). Kidd said he accepted Li's claims because it is normal practice in genetic research and there was no way to verify whether there had actually been informed consent anyways. Such uncritical acceptance of informed consent claims from a settler colonial authoritarian state security agency is typical of complacency. Due diligence in research has long established criteria requiring caution in any research involving vulnerable populations. In particular, scientists, journals, and institutional review boards must actively consider informed consent and ethics claims rather than passively accept them (e.g., Declaration of Helsinki). The increasingly racialized nature of forensic genetic technologies since 9/11 makes them inherently risky due to their potential use against minorities and Indigenous peoples.

CRITICISM C4: *Kidd provided DNA extract samples to Li Caixia et al.* Kidd's provision of the 2266 DNA extract samples representing forty-six populations provided significant capacities (fulfilling able-to-do and knowing-how-to-do modalities) to Li Caixia et al.'s research efforts.

C4 DEFENCE: *Kidd is "not particularly happy."* Kidd defended his sharing of DNA extract samples, saying, "I had thought we were sharing samples for collaborative research" (Wee 2019). "Dr. Kidd said he was 'not particularly happy' that the ministry had cited him in its patents, saying his data shouldn't be used in ways that could allow people or institutions to potentially profit from it. If the Chinese authorities used data they got from their earlier collaborations with him, he added, there is little he can do to stop them. He said he was unaware of the [patent] filings until he was contacted by *The Times*" (Wee 2019).

Kidd in effect accuses the Ministry of Public Security of deception by not informing him of their use of the DNA extract samples in the patent applications. There is some inconsistency between Kidd supporting the use of his fifty-five AISNP panels in Illumina and Thermo Fisher forensic genetic testing systems (which includes Illumina's US patent) and his criticism of the Institute of Forensic Science researchers filing patents using data derived from joint research with him and the DNA extract samples he provided.

The particular property type assertion of fiduciary responsibility that Kidd exercised over the 2266 DNA extract samples is evident in how he asked Li Caixia to stop using the samples. Chang asked Kidd, "Have you asked scientists at the Chinese Ministry of Public Security to stop using the genetic samples you provided them?" To which he replied "Yes." Chang followed, "Did they say they would stop?" Kidd's response, "I've gotten no reply," shows the independent agency of the DNA extract that continues to be used by Li Caixia and her colleagues.

CRITICISM C5: *Providing equipment.* Sequencing equipment and supplies are vital to the genetic testing process. Wee explains how the PRC accounted for 10 per cent of Thermo Fisher's US$20.9 billion sales revenue in 2017 and that it employed some five thousand people there. Wee then cites Chinese government procurement documents that claim the Thermo Fisher sequencers were vital to police DNA testing and there were "no substitutes in China." As well, Thermo Fisher machines were used in the Institute of Forensic Science's patent that differentiates Han Uyghur and Tibetans. She then briefly discusses the announcement by Thermo Fisher that it will stop selling its human identification systems in Xinjiang, quoting Sophie Richardson of Human Rights Watch as saying, "It's an important step, and one hopes that they apply the language in their own statement to commercial activity across China, and that other companies are assessing their sales and operations, especially in Xinjiang" (Wee 2019).

Performance Phase Criticisms

CRITICISM C6: *Completion of these projects advanced repression.* Kidd's coauthoring of the Pakstis et al. (2015a) paper with Li Caixia is strongly criticized in both *New York Times* and NPR pieces. Sophie

Richardson was scathing in her criticism of this cooperation. Referring to the information package on Kidd's cooperation with the Ministry of Public Security I sent her in early 2019, she said, "I was sitting in the Brussels airport reading that packet of information, and I felt physically ill."[6] During the NPR *Planet Money* podcast, Chang asked Alim about Kidd's cooperation:

> CHANG: We asked Alim about Kenneth Kidd, and he said he had heard about him. And even though there is no way that Alim's blood sample was used by Ken, Alim says he's still a victim of a system that Ken helped advance.
> ABDULLAH: As a human being, I can relate to making a mistake or screwing up big-time 'cause everyone does once in a while. But just thinking about it makes me really angry. It's difficult for me to sympathize with him knowing the magnitude of his research and the potential damage his research could do to my people.

C6 DEFENCE: *Cannot predict the future.* When NPR interviewer Ailsa Chang asked Kidd, "Do you feel that you have done absolutely nothing wrong?" he replied, "I have done nothing wrong." Kidd argued that he cooperated with police regularly. The reporters recount Li Caixia's fellowship at Kidd Lab, which culminated in the 2015 AISNP paper (Pakstis et al. 2015a) that included Uyghurs, Kazakhs, and other Xinjiang Indigenous peoples (Chang and Fountain 2019a). Kidd defended his role in the collaboration after NPR cohost Nick Fountain commented, "Which helps China advance this system that tracks, targets and oppresses the Uighurs. Ken says, look; this collaboration – this happened years ago, before much of the world knew what was going on in Xinjiang." Kidd said, "I can't know everything that's going to happen in the future," and tried to disassociate himself from Li Caixia by stating he had stopped collaboration with the Ministry of Public Security.

Sanction Phase Criticisms

CRITICISM C7: *Circulating unethical Uyghur data and findings.* Through publishing the articles, the publishers and reviewers, acting on behalf of the academic discipline, positively sanctioned the research studies by Pakstis et al. (2015a), Li Caixia and her colleagues in Jiang et al. (2018a), and other articles. As well, the inclusion of Li Caixia et al.'s Uyghur data in the ALFRED database, like the scientific articles, is a product of the overall production circuit. Both the *New York Times* and NPR sharply criticized Kidd's inclusion of this Uyghur data, along with that of other research projects involving Uyghurs and other Xinjiang minority peoples. Wee (2019) describes the ALFRED as NIJ funded and her count of 2,143 Uyghurs' data in the database is a type of epideictic rhetoric of quantity to demonstrate the size of the problem and how it also has data from over seven hundred populations. This description is followed by a strong ethical attack by Arthur Caplan, medical ethicist at the New York University School of Medicine, that such data sharing may violate scientific norms because of the lack of clarity over the donation of samples by Uyghurs, stating, "no one should be in a database without express consent." Caplan made a strong ethical rebuke of Kidd with regard to the Institute of Forensic Science claims of consent over the Uyghur subjects: "Honestly, there's been a kind of naïveté on the part of American scientists presuming that other people will follow the same rules and standards wherever they come from." Caplan seems to be referring to the notion that liberal individual ethical norms are universally honoured and acknowledged. Caplan wrongly assumes that Kidd abides by liberal individual ethical norms; the Karitiana and Surui controversies demonstrate that Kidd and other US scientists routinely engage in authoritarian liberal denial of Indigenous peoples' sovereignty and dignity with regard to their own collections of genetic materials (Munsterhjelm 2013; 2014, 46–8, 86).

In the NPR *Planet Money* podcast, Ailsa Chang asked Kidd, "Knowing what you know now about how the Chinese government has treated Uighurs and now treats Uighurs, do you still believe that every genetic sample taken from a Uighur that's referenced in your database or referenced in your research – that each one of those samples was definitely taken with informed consent? Do you really believe that?" To which Kidd responded, "It's impossible to believe that unequivocally. But on the other hand, I have no way of knowing one way or the other" (Chang and Fountain 2019a). Kidd again denies having a responsibility to check the validity of informed consent claims by other researchers.

The publishing process, including peer review of Li Caixia et al.'s articles in prominent journals that completed the global production circuit, was not a major target of criticism in either the *New York Times* or NPR reports on Kenneth Kidd's cooperation, but this process would be in the next controversy.

FACIAL PHENOTYPE SNP PRODUCTION CIRCUIT OF THE TUMXUK PAPERS

Wee's 21 February 2019 article was republished in a number of languages across the world. From a research methodology standpoint, it is a useful quasi-experiment in that the event allows us to determine whether this type of mainstream press article has an impact on subsequent scientific research publications. It did not stop journals from publishing articles coauthored by Li Caixia using the Tumxuk Uyghurs; in April 2019, two articles appeared in prominent journals owned by Springer Nature, one of the largest scientific publishers.[7] The articles (discussed in the previous chapter) were coauthored in part by Liu Fan, Tang Kun, and Li Caixia (on facial shape in *Human Genetics*) and by Liu and Li et al. (on height in *International Journal of Legal Medicine*). Hundreds of articles by PRC security agency scientists using Uyghurs have appeared annually in recent years and these publications indicate that both the publishing industry and the larger international forensic genetics community were still largely indifferent and unwilling to critically assess the ethical claims of this Chinese security agency research on Uyghurs (Moreau 2019). This unwillingness violates the principle of due diligence required in research involving vulnerable populations.

In January 2019, I noticed articles involving Li Caixia and other Ministry of Public Security researchers on 3D forensic genetic phenotyping of the face involving Uyghurs. In February 2019, I spent several days compiling a list of articles (which also included the Xinjiang Uyghur Study and cooperation with the VisiGen Consortium) and did an initial analysis. I then discussed this new set of evidence with Wee Sui-Lee and we decided to write a second article. During the months that followed, I continued my close analysis of the articles and began determining the network of research institutes, sharing my findings and answering Wee's inquiries. The article was published on 3 December 2019. Wee and her coauthor, Paul Mozur, focused on the Tumxuk Uyghur case and the cooperation of Li Caixia with Tang Kun of the Chinese Academy of Sciences-Max Plank Society Partner Institute of Computational Biology and Liu Fan of BIG and Erasmus University, but, as a result, did not deal with the VISAGE and VisiGen networks.

Manipulation Phase Criticisms

This circuit is different from the earlier one with Kidd that involved both the US security apparatus and that of the PRC. In this phenotyping inference SNP (PISNP) assemblage, the PRC government (acting in turn

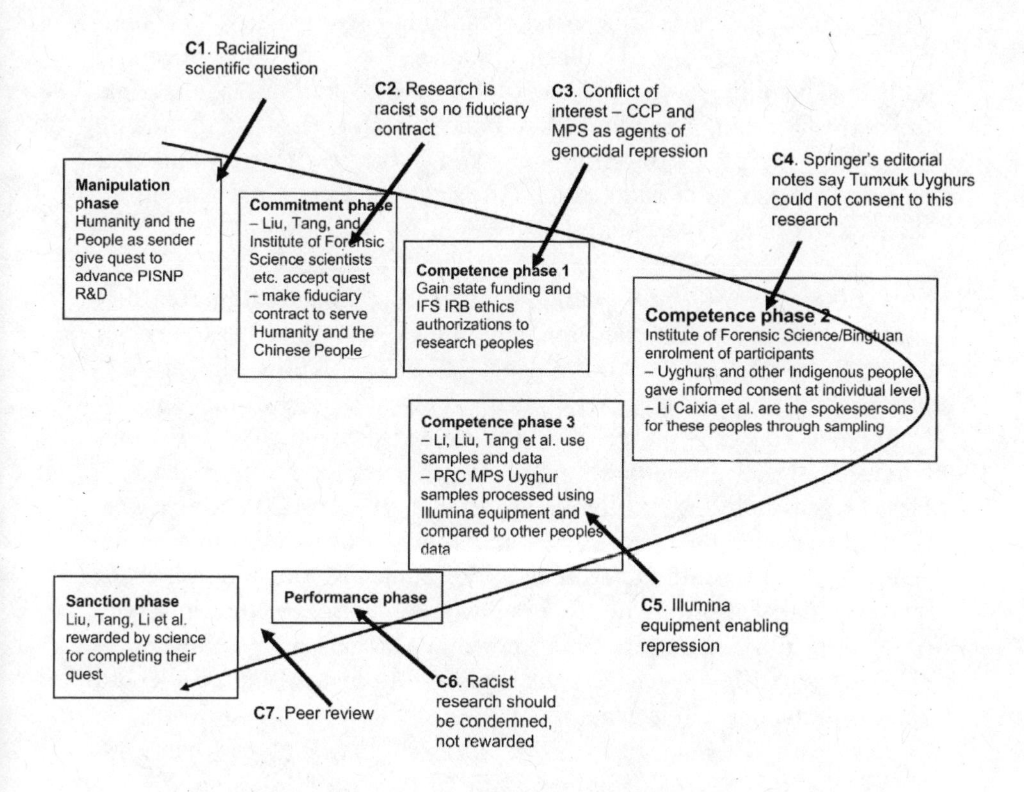

Figure 11.2 | Diagram of how external criticisms destabilized the Institute of Forensic Science's Tumxuk Uyghur phenotyping research organizing narratives.

in the name of the Chinese People and Humanity) is the sender that gives a quest for public security through the development of advanced biometric technologies including phenotyping and ancestry and lineage informative SNP research (se figure 11.2).

CRITICISM C1: *Immature technology.* A set of critical comments were about the still-limited capacities of the technology as "hit or miss" (Wee and Mozur 2019a).

C1 DEFENCE: *It can bring a solution.* Wee and Mozur (2019a) interviewed Peter Claes, a Belgian expert in medical imaging who cooperated with Mark Shriver in the development of the first facial phenotyping technologies. Claes uses the central organizing narrative that asks, what if there is no STR profile? He begins by referring to the PRC's huge STR individual identification databases: "If I were to find DNA at a crime scene, the first thing I would do is to find a match in the 80 million data set." Claes then moves to phenotyping as a viable possibility: "'But what do you do if you don't find a match?' Though the technology is far from accurate, he said, 'DNA phenotyping can bring a solution.'"

CRITICISM C2: *Unethical quest.* The article starts out by arguing that experts and activists are concerned about how China's security agencies are utilizing international assemblages of scientific knowledge production to develop forensic genetic phenotyping as a type of manhunting technology against Uyghurs. They quote me as saying that what the PRC is developing are "essentially technologies used for hunting people." Here I was countering the dominant forensic genetic manhunting narratives of revealing intelligence about the anonymous figures of the insurgent, the terrorist, and the criminal. The article also includes criticisms made by Pilar Ossorio, who teaches law and bioethics at the University of Wisconsin-Madison, about the potential dangers of these technologies being used for genetic dragnets based on phenotyping profile and against particular ethnic groups.

Commitment Phase Criticisms

CRITICISM C3: *Complicity in genocide so no fiduciary contract.* The authors again quote me that "there's a kind of culture of complacency that has now given way to complicity" in scientific research. What I meant here was that, given the extensive media coverage of Xinjiang since 2017 and the emerging international legal norm (jus cogens) against cultural genocide and other crimes against humanity, Western genetic researchers continued cooperation with the Institute of Forensic Science indicated an unwillingness to confront the increasingly genocidal regime in Xinjiang. Ossorio warns against complacency in the development of these technologies: "What the Chinese government is doing should be a warning to everybody who kind of goes along happily thinking, 'How could anyone be worried about these technologies?'" (Wee and Mozur 2019a).

C3 DEFENCE: *Dissociation.* Tang Kun of the Partner Institute engages a disassociation strategy. First, he denies that he was directly involved, instead blaming his graduate student (likely Li Dan) for his name being included as a coauthor on the *Human Genetics* paper. He defends his research interest in the Uyghurs as important subjects for human evolutionary research, also asserting that his sampling of the Xinjiang Uyghur Study subjects was ethical.

C3 DEFENCE: *Tang did not report cooperation.* The Max Planck Society dissociated itself from Tang with a sharp rebuke, stating that he had not informed them of his cooperation with the Institute of Forensic Science and that the Society "takes this issue very seriously" and would carry out an ethics review. This rebuke seeks to disassociate the Max Planck Society from Tang's cooperation with the Institute of Forensic Science.

C3 DEFENCE: *Denial of fiduciary contract.* Erasmus University sought to disassociate itself from Liu Fan's research cooperation with the Ministry of Public Security by denying it had a fiduciary contract to oversee research done by him. They argue that his position at BIG was "totally independent" of his position at Erasmus and "Erasmus added that it could not be held responsible 'for any research that has not taken place under the auspices of Erasmus' by Dr. Liu, even though it continued to employ him" (Wee and Mozur 2019a).

Competence Phase Criticisms

CRITICISM C4: *Ethical oversight is impossible.* Sui-Lee Wee and Paul Mozur (2019a) raise the important issue of how the institutional oversight of the Institute of Forensic Sciences' research on phenotyping of Uyghurs is compromised because of the repressive conditions in Tumxuk and Xinjiang.

C4 DEFENCE: *Must rely on local ethics boards.* The journal *Human Genetics* stated that it had to rely on claims made by scientists that informed consent procedures had been followed under the supervision of local ethics committees. The journal engages in the recognition of the validity of decisions made in other jurisdictions, typical of dispersed recognition of sovereign authority within the graduated sovereignty of research assemblages (Wee and Mozur 2019a). As well, the publisher Springer Nature issued a statement on 2 December 2019 that it would strengthen its oversight of research articles involving vulnerable populations and also placed editorial notes on the Tumxuk papers that had already been published.

C4 DEFENCE: *Outside our jurisdiction.* Graduated sovereignty is evident in the circuits as sovereign-type decision-making is dispersed among different actants in the assemblage and those decisions are accepted by other actants within the assemblage. In this way, we see a familiar pattern of scientists and research institutions attempting to avoid responsibility of due diligence by saying they had no choice but to recognize other countries' and jurisdictions' sovereign decision-making capacities. With Kenneth Kidd, it is evident in his routine acceptance of the ethical and informed consent claims made by Chinese scientists over Uyghur subjects. In this case, Erasmus University emphasized the paramount importance of the independence of the researcher and, though they may be employed by the institution, if funding and ethics oversight was from other institutions, Erasmus had no fiduciary contract to oversee Liu Fan's research. Erasmus University dissociates itself only from Liu Fan's specific research projects that were done without their direct involvement. Their commitment to researcher independence omits consideration of researchers in authoritarian institutions in what is a false assumption of the universalization of liberal individual rights recognition. This avoidance strategy avoids the fact that international networks of cooperation are key to these global assemblages. This avoidance strategy was aided by the fact that Wee and Mozur (2019a) did not directly deal with the Xinjiang Uyghur Study in the VisiGen and VISAGE cooperation (which I had explained to Wee in detail) so they were not able to implicate Erasmus University strongly in the phenotyping research assemblages.

In contrast, the Max Planck Society engaged in a different dissociation process in which they state that Tang had not told them about his cooperation with the police and that it would be reviewed by their ethics committee. This statement is contradicted by the 2014–17 Partner Institute report that clearly stated Tang and his research team's intention to cooperate with the Institute of Forensic Science in forensic applications (see chapter 8). Perhaps the report was not sent to the Max Planck Society and/or no one at the Society bothered to read the report or understood the statement's significance when it was published in 2017.

Tang says that he ended his research cooperation with the Institute of Forensic Science in 2017 because he considered their research and samples collection inept. This dissociation strategy involves denying the capacity of the Institute of Forensic Science to engage in able-to-do and knowing-how-to-do modalities, Tang said, "To be frank, you overestimate how genius the Chinese police is" (Wee and Mozur 2019a). If his negative assessment of the Institute of Forensic Science's scientific competence is true, then it highlights the importance of the continued cooperation of Liu Fan's group at BIG with its leading edge phenotyping to the Institute of Forensic Science, along with that of Partner Institute–trained graduate students.

CRITICISM C5: *The invalidity of informed consent claims*. The willingness of other actants to accept Chinese scientists' informed consent and ethics claims over Uyghur subjects was central to the translation of ethics claims, which then allowed these Uyghur subjects to function as objects of exchange within the circuits of forensic genetic and evolutionary genetic research. Tang dissociated himself from the enrolment process saying he did not know about the sampling of the Tumxuk Uyghurs. He asserted that the Xinjiang Uyghur Study subjects were ethically sampled.

Wee and Mozur travelled to Tumxuk where they filmed a large number of demolished Uyghur houses; Mozur hid the film and later posted it on Twitter (Mozur 2019). The Tumxuk police detained and interrogated them and prevented them from conducting any interviews with local people. They were also forced to delete video, audio, and photos they had taken in Tumxuk. The police effectively stopped the reporters' attempt to verify informed consent. This interference demonstrates that informed consent claims cannot be verified, thereby discrediting the informed consent claim as well. Claims of informed consent must be independently verifiable to be considered valid.

CRITICISM C6: *Illumina provided sequencing equipment*. Wee and Mozur (2019a) discuss how Illumina sequencers were used but the company did not respond to their requests for comment.

Performance Phase Criticism

CRITICISM C7: Research is racist and should be condemned. Wee and Mozur's (2019a) article argued this research is highly unethical and is aiding in the persecution of the Uyghurs and other Turkic peoples in a strong attack on the claim that the phenotyping research assemblage is an ethical contribution to scientific knowledge.

Sanction Phase Criticisms

CRITICISM C8: Circulating unethical Uyghur data and findings. *Human Genetics* was sharply criticized for publishing the article. As a way of disassociating itself, Springer Nature announced that it had added editorial notes of concern to papers involving Uyghur and other Xinjiang minorities published with the involvement of the Institute of Forensic Science and other Chinese security agencies.

This publication was followed by a 4 December 2019 article also by Wee Sui-Lee and Paul Mozur on Yves Moreau and his colleagues conducting a study of papers published in Western journals. Moreau has been working for several years on the ethics issues in Xinjiang, including Thermo Fisher's genetic analyzer sales. His editorial in *Nature* on 3 December 2019 was highly critical of publishers and he stated a literature review had shown hundreds of forensic genetic articles on Uyghurs and other PRC minorities involving researchers affiliated with various Chinese security agencies (Moreau 2019). He and other activist academics have been engaged in a long-term campaign pressuring major publishers over these publications (Moreau n.d.).[8] According to Moreau, "I analysed 529 articles on forensic population genetics in Chinese populations, published between 2011 and 2018 in these journals and others. By my count, Uyghurs and Tibetans are 30–40 times more frequently studied than are people from Han communities, relative to the size of their populations (unpublished data). Half of the studies in my analysis had authors from the police force, military or judiciary. The involvement of such interests should raise red flags to reviewers and editors." Moreau's analysis of the overrepresentation of Uyghurs and other Indigenous groups compared to Han Chinese is similar to what I noted about AISNP research in Jiang et al. (2018a) in chapter 5.

On 5 December 2019, Springer Nature published an editor's note to the Tumxuk facial genes article originally published in *Human Genetics* in April 2019. The editorial note reiterated the PRC jurisdictional authority as authorization and stated there will be investigations and editorial action:

> Concerns have been raised about the ethics approval and informed consent procedures related to the research reported in this paper (Li Yi et al. 2019). The paper includes the following author declarations: "This study was approved by the Ethics Committee of the Institute of Forensic Science of China, and all individuals provided written informed consent. The participants were all volunteers. The consent was discussed in their native language and the signature was in their native language." Editorial action will be taken as appropriate once an investigation of the concerns is complete and all parties have been given an opportunity to respond in full. (*Human Genetics* 2020)

The *International Journal of Legal Medicine* also published a similar editorial note (2020). During 2020, a number of these editorial notes were published for papers involving Chinese security agencies conducting research on Uyghurs. These editorial notes are acknowledgements by corporate publishers of research journals along with scientists of their complacency and the at times complicit disregard for potential human rights issues in forensic genetic research involving Uyghurs and other Turkic peoples.

There was further coverage in a 6 December 2019 article entitled "Science Publishers Review Ethics of Research on Chinese Minority Groups" in *Nature*. The article on changes to policy on research involving vulnerable populations by Springer Nature and Wiley was accompanied by a claim that reiterated the publication's editorial independence of its publisher Springer Nature (Van Noorden and Castelvecchi 2019, 192–3). It recounts as well some of the findings made by Yves Moreau and his colleagues that a disproportionate number of the five hundred papers on the Uyghurs involved Chinese security agency personnel. The article also contains a response from Li Caixia of the Institute of Forensic Science on the two Tumxuk papers: "Both papers state that volunteers gave consent, and that the studies were approved by an ethics committee from the Institute of Forensic Science, which is affiliated with China's police and security authority. 'We are ordinary forensic scientists who carry out forensic research following the scientific research ethics norms,' said Caixia Li of the Institute of Forensic Science in Beijing, a co-author of both papers, in an e-mail to Nature's news team. [S]he said that 'all individuals provided written informed consent'" (Van Noorden and Castelvecchi 2019, 192).

Li reasserted the morality and ethical validity of her institute's research on Uyghurs.

INSTITUTE OF FORENSIC SCIENCE ON THE ENTITY LIST

Within US Congress and government agencies, the legislative process and policy implementation continued.[9] The 21 February 2019 *New York Times* article by Wee was mentioned in legislative hearings held by Congress during 2019 (Congressional Executive Commission on China 2019, 98, 99, 215, 216). On 22 May 2020, the US Department of Commerce issued a press release: "The US Department of Commerce's Bureau of Industry and Security (BIS) announced the impending addition of the People's Republic of China's Ministry of Public Security's Institute of Forensic Science and eight Chinese companies to the Entity

List, which will result in these parties facing new restrictions on access to U.S. technology. These nine parties are complicit in human rights violations and abuses committed in China's campaign of repression, mass arbitrary detention, forced labor and high-technology surveillance against Uighurs, ethnic Kazakhs, and other members of Muslim minority groups in the Xinjiang Uighur Autonomous Region (XUAR)" (US Department of Commerce 2020). Whereas the Commerce Secretary in 2018 had denied the significance of the sales of Thermo Fisher equipment in Xinjiang, now the Department of Commerce was applying American laws on the export of high technology equipment to the Institute of Forensic Science based on the institute's role in Xinjiang (see figure 11.3). This US government ministry enacts its sovereign discretion in the decision-making process over the application of US laws to American companies dealing with the Institute of Forensic Science, an example of extraterritoriality.

This listing took place during a time of escalating tensions between the US and Chinese governments as the Trump administration's jingoism made use of anti-China rhetoric, in part due to the coronavirus pandemic and Trump's lagging political fortunes heading into a presidential election. These tensions were further compounded by the PRC's escalating repression in Hong Kong, which has effectively removed the special administrative region's legal and human rights protections. The listing is significant in that it is supposed to disrupt US companies, particularly Thermo Fisher and Illumina, from supplying advanced biotechnology systems to the Institute of Forensic Science. These restrictions were further elaborated on in a 1 July 2020 Xinjiang Supply Chain Business Advisory that was jointly issued by the US Departments of State, Treasury, Commerce, and Homeland Security. It states that while it is primarily concerned with "private sector entities ... due to potential supply chain concerns," "the scope of this document also includes other entities that may engage with the Xinjiang province, such as academic institutions or those engaged in certain research," including surveillance and genetic databases (US Treasury Department 2020, 2, 5).

PRC OFFICIAL RESPONSES

This assertion of US sovereignty that restricts exports to China led to a strongly worded response from a PRC spokesperson. In his 25 May 2020 press conference, Chinese Ministry of Foreign Affairs spokesperson Zhao Lijiang was asked by an Agence France-Presse reporter (which acts as a manipulation phase): "On Friday the US Department of

Control Policy: End-User and End-Use Based | Supplement No. 4 to Part 744 – page 66

COUNTRY	ENTITY	LICENSE REQUIREMENT	LICENSE REVIEW POLICY	FEDERAL REGISTER CITATION
	Ministry of Public Security's Institute of Forensic Science of China, a.k.a., the following two aliases: -Forensic Identification Center of the Ministry of Public Security of the People's Republic of China; *and* -Material Identification Center of the Ministry of Public Security of the People's Republic of China. No. 18 West Dongbeiwang Road, Haidian District, China; *and* Ministry of Public Security, Xicheng District, Beijing, China; *and* No. 17 Mulidi South Lane, Xicheng District, Beijing, China; *and* No. 5 Qianhai West Street, Tumushuk City, Xinjiang Uighur Autonomous Region (Tumushuk City Public Security Bureau).	For all items subject to the EAR. (See §744.11 of the EAR).	Case-by-case review for ECCNs 1A004.c, 1A004.d, 1A995, 1A999.a, 1D003, 2A983, 2D983, and 2E983, and for EAR99 items described in the Note to ECCN 1A995; presumption of denial for all other items subject to the EAR.	85 FR 34505, 6/5/20.

Figure 11.3 | Image of the Institute of Forensic Science and Tumxuk Bingtuan Lab on the US Entities List.

Commerce said it would sanction relevant Chinese companies and government institute for human rights abuses in Xinjiang. Do you have any response to this?" The commitment phase is already implied because it is a press conference. Lijian's response moves straight to the competence phase in that he seeks to show the US has made an error: "The US, adding relevant Chinese enterprises, institutions and individuals to its 'entity list,' has overstretched the concept of national security, abused export control measures, violated the basic norms governing international relations, interfered in China's internal affairs, and hurt China's interests. China deplores and firmly opposes that" (Ministry of Foreign Affairs 2020). The government spokesperson then asserted the US regulatory actions involved extraterritoriality that infringed on and interfered in China's internal sovereignty and that its war on terror was justified and supported both within Xinjiang and in accordance with Chinese law: "It needs to be highlighted that Xinjiang affairs are purely China's internal affairs which allow no foreign interference. The measures on countering terrorism and deradicalization have been taken to prevent in a fundamental way these two evil forces from taking roots in Xinjiang" (Ministry of Foreign Affairs 2020). The idea of separatism and extremism as evil influences that are seeds that will grow is consistent with the metaphors of contagion and disease discussed in chapter 8. All of these metaphors engage in a type of dehumanization that reduces Uyghur society to a biological process, which makes the violence irrational rather than a result of escalating PRC repression since the 2009 Urumqi Riots.

The spokesperson then claimed the PRC is just in its actions: "They accord with Chinese laws and international practices. They have been proved effective, widely supported by 25 million people of various ethnic groups in Xinjiang, and contributing to the global counter-terrorism cause" (Ministry of Foreign Affairs 2020). He then criticizes the US, saying that it is destabilizing to China's efforts: "The US accusation against China, nothing but absolute nonsense to confound the public, only serves to reveal its vile attempt to disrupt Xinjiang's counterterrorism efforts and China's stability and development." He then concludes with an appeal to the US to change its decision and reasserts China's resolve to protect its sovereignty and its interests: "We urge the US to correct its mistake, rescind the relevant decision, and stop interfering in China's internal affairs. China will continue to take all necessary measures to protect the legitimate rights and interests of the Chinese enterprises and safeguard China's sovereignty, security and development interests."

The PRC spokesperson's response is interesting in the spatiality of its jurisdictional claims of exclusive sovereignty over Xinjiang. US-made and -supplied equipment like the Thermo Fisher 3500 XL are integral parts of the Chinese forensic genetic assemblages. However, this also demonstrates the vulnerability of China to such trade sanctions; US companies and their equipment, US and European Union trained Chinese scientists, and research cooperation have been vital to the development of China's forensic genetic capacities. Their vulnerability is also evident in no-bid purchase contracts by Chinese security agencies in which experts stated that buying genetic testing supplies from Thermo Fisher was necessary because of quality control issues in Chinese made products (Wee 2019).[10]

A similar line of argument is evident in a 27 May 2020 statement published on the Ministry of Public Security website and then picked up by various PRC news websites such as Xinhuanet.org entitled "The Ministry of Public Security's Evidence Appraisal Center issued a statement regarding its inclusion in the US Department of Commerce list of export control entities."

Manipulation Phase

The Institute of Forensic Science has been added to the Entity List: "On May 22, the US Department of Commerce announced that it would include the Physical Evidence Identification Center of the Ministry of Public Security on a list of export control entities" (5月22日，美国商务部宣布将公安部物证鉴定中心列入出口管制实体清单。).

Commitment Phase

"We express our strong dissatisfaction and firm opposition to this, and urge the US Department of Commerce to immediately revoke this erroneous decision" (我们对此表示强烈不满和坚决反对，敦促美国商务部立即撤销有关错误决定。).

Competence Phase

Because it is in support of a type of epideictic/deliberative rhetoric, the competence phase is advocating for a future course of action. It is arguing why it should not be on the Entity List by making a series of claims about its role in China providing forensic evidence for cases and fighting crime. It asserts that it is internationally recognized by peers, a type of ethos statement that relies upon positive evaluation by other countries security agencies.

Subnarratives

Subnarrative One begins with an explanation of the institute's purpose and its record of doing social good: "Our center is a forensic science institution that undertakes on-site investigation, physical identification and scientific and technological research in major cases in China, providing a scientific basis for the determination of criminal facts and court trials, and over the years it has played an important role in fighting crime and maintaining social justice, and is widely recognized by international peers" (Institute of Forensic Science 2020; 我中心属于法庭科学机构，承担我国重大案件的现场勘查、物证鉴定和科技研究等工作，为认定犯罪事实和法庭审判提供科学依据，多年来在打击犯罪、维护社会公平正义中发挥重要作用，得到国际同行的广泛认可).

Subnarrative Two includes the assertion that "the so-called Xinjiang-related issue is not a question of human rights, ethnicity or religion at all, but an issue of anti-terrorism and anti-secession" (所谓涉疆问题根本不是人权、民族、宗教问题，而是反暴恐和反分裂问题。). It rejects framing of Xinjiang in terms of human rights and cultural genocide and attempts to reframe the issue as antiterrorism and national unity. The announcement then makes an argument based on US sanctions as extraterritoriality that violates PRC national sovereignty: "Xinjiang's affairs are purely China's internal affairs, and no country is allowed to interfere in them" (Institute of Forensic Science 2020; 新疆事务纯属中国内政，不容任何国家干涉。). They then accuse the US government of unilateralism and use of domestic law against the institute (extraterritoriality), as well as violating norms of international relations and weakening the civilizational duty to fight terrorism and extremism: "Disregarding the facts, the relevant US authorities have imposed unilateral sanctions on Chinese entities, including forensic scientific institutions, in accordance with their domestic laws, seriously infringed upon the legitimate rights and interests of relevant Chinese institutions and enterprises, and severely undermined the international community's efforts to counter terrorism and extremism, which only harm others and themselves" (美国有关部门罔顾事实，根据其国内法对包括法庭科学机构在内的中国实体实施单边制裁，严重侵犯有关中国机构和企业的合法权益，严重破坏国际社会反恐和去极端化的努力，只会损人害己).

In the final paragraph, the Evidence Center asserts the universal morality of its activities, stating: "Forensic science is an internationally used technical means to fight crime, and it is also an important guarantee for standardized, fair and civilized law enforcement by police in all countries" (Institute of Forensic Science 2020; 法庭科学是国际通用的打击犯罪技术手段，也是各国警察规范公正文明执法的重要保障). It then concludes with how it will resist the sanctions: "Our center will steadfastly improve the level of criminal science and technology, continue international exchanges and cooperation, and provide strong scientific and technological support for maintaining national security and social stability and strengthening international law enforcement cooperation in fighting crime" (我中心将坚定不移地提升刑事科学技术水平，继续开展国际交流与合作，为维护国家安全和社会稳定、加强打击犯罪国际执法合作提供有力科技支撑。).

RETRACTION OF THE TWO TUMXUK PAPERS

During 2020 and 2021, there was increasing pressure on publishers to review papers involving Uyghurs done by scientists affiliated with Chinese security forces. This led to dozens of papers being critically examined at various journals and a number of retractions. In this larger context, Springer Nature journals engaged in a review of the ethics and informed consent of the Tumxuk and a number of other articles involving Uyghurs. Finally, on 30 August 2021, they published a retraction of the 2019 Li Yi et al. article in *Human Genetics* that looked at facial morphology using the Tumxuk Uyghur samples: "The Editors-in-Chief have retracted this Article. Since publication, concerns were raised about the ethics and consent procedures for this study. We requested supporting documentation from the authors, including the application form submitted to the ethics committee and evidence of ethics approval. The documents supplied by the authors contain insufficient information related to the scope of the study for us to remain confident that the protocols complied with our editorial policies or are in line with international ethical standards (World Medical Association Declaration of Helsinki. Ethical principles for medical research involving human subjects)" (Li Yi et al. 2021, 1).

The authors objected to this retraction much as they had to the initial editorial note in December 2019. According to the retraction notice, "Caixia Li stated on behalf of all co-authors that they do not agree to this retraction" (Li Yi et al. 2021). This was then followed by a similarly worded 7 September 2021 retraction in the *International Journal of Legal Medicine* of the Tumxuk Uyghur height paper by Jing et al. (2019) and a similar rebuttal by Li Caixia (Jing et al. 2021). These two retractions were reported on in a number of outlets including the *New York Times* and *South China Morning Post* (Lew 2021; Wee 2021). The retractions were also reported in Holland and this had a number of significant repercussions for the Xinjiang Uyghur Study.

2021: DEFENDING THE XINJIANG UYGHUR STUDY COOPERATION

During 2021, a number of journalists in Holland investigated Liu Fan's cooperation with the Ministry of Public Security and the VisiGen Consortium in their use of the Xinjiang Uyghur Study. The resulting news stories led to heated debates over informed consent claims and the research projects' potential utility in identifying Uyghurs (Janssen 2021). Manfred Kayser of Erasmus University and the VisiGen Consortium defended the use of the Xinjiang Uyghur Study in a 2 August 2021 Amnesty International magazine article on the cooperation (Janssen 2021). In September 2021, there was coverage of Springer Nature's retractions of the two Tumxuk Uyghur articles coauthored by Liu Fan and Li Caixia. This was followed on 6 October 2021 by articles by *Follow the Money* investigative reporters and TV coverage by RTL TV network from the Netherlands about the research cooperation between Erasmus University and Liu Fan that led to a storm of controversy in Holland.

In the following segment, I will analyze this 2021 Dutch media coverage and the public defences of the research by Manfred Kayser and Erasmus University utilizing the same approach of destabilization mapping as the above *New York Times*/NPR coverage. To understand the internal logics of journal publication, I will include email responses I

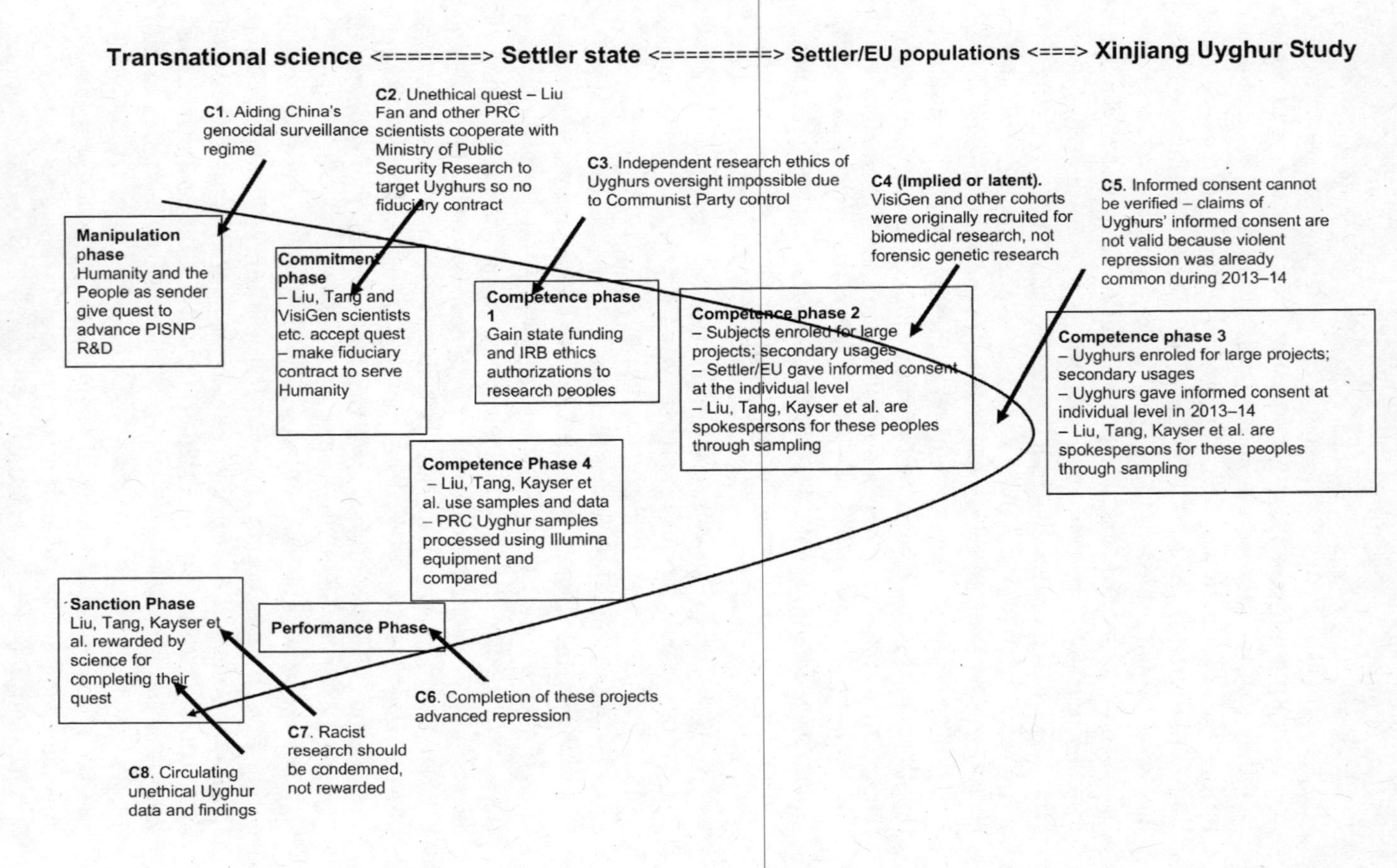

Figure 11.4 | Diagram of the disruption of VisiGen Consortium and Xinjiang Uyghur Study phenotyping research organizing narratives.

received from *eLife* executive editor Andy Collings and *eLife* publisher Michael Eisen, who were responsible for the publication of the Xiong et al. paper in November 2019 (see figure 11.4).[11]

CRITICISM C1: *Aiding a genocidal regime's surveillance.* Yves Moreau is quoted in the 6 October 2021 RTL and *Follow the Money* articles arguing that the use of Uyghurs is a "fundamental error" since it helps improve the capacities of the Chinese security apparatus to differentiate Uyghurs from Chinese (Eikelenboom, de Bruijn and Booij 2021; Kempes and Strijker 2021a). The article by *Follow the Money* reporters discusses how several opposition party MPs were angered by this extensive research cooperation between Holland and China (Eikelenboom, de Bruijn, and Booij 2021). One of the MPs, Hatte Van der Woude, asserted, "We cannot take the risk that there are Dutch fingerprints on instruments of repression" (We kunnen niet het risico nemen dat er Nederlandse vingerafdrukken op onderdrukkingsinstrumenten staan). Another MP, Sjoerd Sjoerdsma, stated, "We must do everything we can to prevent Dutch DNA expertise from being used by the Chinese government as a weapon to suppress Uyghurs" (Eikelenboom, de Bruijn, and Booij 2021; We moeten er alles aan doen om te voorkomen dat Nederlandse dna-kennis door de Chinese overheid als wapen wordt ingezet om Oeigoeren te onderdrukken).

DEFENCE C1: *No harm, no foul.* Some media coverage made the error of directly associating facial phenotype estimation forensic genetic research with facial recognition by surveillance cameras, implying that the two were somehow technically related. This misunderstanding in the media coverage was a problem because it misrepresented the genetic phenotyping research and development and its potential usage and capacities. The error functioned inadvertently like a strawperson argument that allowed Kayser and Erasmus University to deflect some of the criticism about the technological cooperation and claim it was harmless to Uyghurs. Kayser repeatedly defended the use of the Xinjiang Uyghur Study, stating that it could not be used to identify Uyghurs or other ethnic minorities. Erasmus University Medical Center published a 6 October 2021 statement in Dutch and English that asserts, "Various media outlets today suggest that research by Erasmus MC may contribute to the suppression of Uyghurs. This is absolutely incorrect." They argued that the Erasmus Medical Center research collaboration, "involving genetic materials from many thousands of people from around the world ... including Uyghurs, cannot possibly contribute to facial recognition in general."

Furthermore, they argued that the results cannot "be used to recognize Uyghurs on the basis of DNA. And therefore the results cannot contribute to the oppression and exclusion of Uyghurs."

This consequentialist logic of "no harm, no foul" is also at work in *eLife* publisher Michael Eisen's email response to my questions when he argues that the Uyghur samples have little utility (have little or no agency). This no harm no foul logic continues as he argues that the practical significance of the research findings correlating genes with facial morphology was very minimal through a narrow focus on surveillance: "The statements you cite about this paper notwithstanding, I believe is scientifically implausible to believe that these data assisted in surveillance efforts. The degree of genetic correlation with facial morphology, while statistically significant, was small, even within the studied European population, rendering the analysis all but useless in phenotyping individuals" (4 February 2021, email to author).

COUNTER TO DEFENCE C1: *Symbolic power.* Yves Moreau agreed that the research had little practical implications for immediate surveillance but argued that it is part of a larger strategy of control by Chinese security agencies:

> The very fact that DNA is taken from you is an attack on your identity. The symbolic power you can create with that is like magic. The technology helps create symbolic power relations. People who seem to be much smarter than you have collected data, which is processed in an "objective" way by a computer and then something comes out: like an oracle. (Janssen 2021; Moreau legt uit waarom. Hij vergelijkt het met de opkomst van de cartografie. "In de middeleeuwen waren de grenzen tussen landen vaak niet goed gedefinieerd. Pas toen de cartografen het land goed in kaart brachten, kwamen er grenzen die op de meter nauwkeurig bekend waren. Dat was het begin van de natiestaten. Als je kaarten maakt van de genetica van de mens, creëert dat de mogelijkheid om grenzen te trekken. Dat is zorgwekkend.")

CRITICISM C2: *Unethical quest.* Liu and other coauthors cooperate with PRC security agencies involved in the repression of Uyghurs.

DEFENCE C2: *Disassociation.* In a 4 February 2021 email, Michael Eisen dismissed my evidence about research cooperation by trying to disassociate the authors of Xiong et al. (2019) from the Chinese Ministry of

Public Security, stating: "I found your claims that authors of the paper are involved in government surveillance efforts to be highly circumstantial, and based on misinterpretation of the generally accepted meaning of authorship in the scientific community." He dismissed outlines of five articles (Liu et al. 2017; Jing et al. 2019; Li Yi et al. 2018; Li Yi et al. 2019; Pan et al. 2018) I provided him that demonstrated five of the authors of Xiong et al. (2019) – Liu Fan, Li Dan, Xiong Ziyi, Li Yi, and Tang Kun – had directly cooperated with the Ministry of Public Security's Institute of Forensic Science, in part by framing his response in terms of the immediate practical implications for surveillance rather than dealing with the long-term capacity building aspects of this cooperation.

The Erasmus Medical University public statements and its responses to the RTL/*Follow the Money* questions seek to differentiate the VisiGen/VISAGE cooperation from Liu's cooperation with the Ministry of Public Security, stating that Liu was not an employee of the Chinese Ministry of Public Security and that "His research did not focus on Uyghurs, but on mixed Eurasian-Asian DNA. He is also not part of the VisiGen Consortium." (Janssen 2021; Zijn onderzoek was niet gericht op Oeigoeren, maar op gemengd Europees-Aziatisch DNA. Hij is ook geen onderdeel van het VisiGen Consortium.) Both of these statements are disingenuous because the Uyghurs are routinely identified as Eurasian and European/East Asian admixed populations in Ministry of Public Security research, and to say that Liu Fan was not part of the VisiGen Consortium is to deny over a decade of his direct involvement in its research from its very beginnings in the late 2000s, including his joint patent application with Manfred Kayser (discussed in chapter 6).

The Erasmus University statement denied any connection to Liu's cooperation with the Ministry of Public Security despite his Erasmus University affiliation: "Neither the international collaboration partnership nor Erasmus MC have anything to do with the three publications of researcher Dr. Liu, two of which have since been withdrawn, which he performed as part of his main appointment at the Beijing Institute of Genomics Chinese Academy of Sciences" (Erasmus University Medical Center 2021). Furthermore, in their responses to the *Follow the Money* reporters Erasmus University asserted that Liu had discontinued his collaboration with the Institute of Forensic Science, supporting this assertion with Liu's statement on his orcid.org website that he had discontinued research cooperation with the Institute of Forensic Sciences in May 2020 (Siem Eikelenboom, investigative journalist, email to author, 7 October 2021). However, the *Follow the Money* reporters found this claim not to be the case because, a year after this supposed end of cooperation, Liu coauthored a May 2020 *Hereditas* (Beijing) article with Li Caixia and other security agency scientists on facial morphology among Han Chinese (Eikelenboom, de Bruijn, and Booij 2021; Liu Ming et al. 2020). Furthermore, they found that Liu participated in a 23 May 2020 online conference hosted by Southern Medical University in Guangzhou along with Zhao Xingchun, the director of the Institute of Forensic Science's Material Evidence Center in Beijing (where Li Caixia also works) and Li Haiyan of the Guangdong Provincial Public Security Department (whose research presentations using Uyghurs is mentioned in chapter 4) (Eikelenboom, de Bruijn, and Booij 2021; Southern Medical University 2020).

Commitment Phase

CRITICISM C3: *Independent research ethics oversight is impossible.* Ethics review is not valid because research is not independent of the Chinese Communist Party control.

DEFENCE C3: *Consent done according to Chinese law.* The 4 February 2021 email response I received from Michael Eisen, the publisher of *eLife*, shows the logic of no harm, no foul in its publication of Xiong et al. (2019) despite the growing evidence and reports of increasingly genocidal mass repression in Xinjiang. He seeks to justify the inclusion of the Uyghurs stating: "The collection of the Uyghur genetic data was consented in accordance with Chinese law. While we are not obliged as a journal to defer to local laws and regulations, those are the standards we communicate to our authors and they cannot be modified retrospectively. This, of course, does not address the question of whether consent can be obtained from individuals from oppressed populations, which will be discussed below." Given that the article was received in July 2019, the editors cannot claim ignorance of the situation in Xinjiang, which involved charges of crimes against humanity, so Eisen's response that he cannot change standards communicated to the authors is incomprehensible. It is similar then to Kenneth Kidd's comments on not knowing either way about informed consent claims (Chang and Mountain 2019).

In an August 2021 Amnesty International (Netherlands) article, Manfred Kayser argued that the samples had been taken by Fudan scientists "with written consent, before the trouble started in Xinjiang" and that they followed "international standards and are internationally respected scientists" (Janssen 2021; "met schriftelijke geïnformeerde toestemming, in de periode voordat de problemen begonnen in Xinjiang";

"internationale standaarden en zijn internationaal gerespecteerde wetenschappers"). In its 6 October 2021 statement, Erasmus Medical Center reasserts that "The samples and genetic data used by this international partnership were collected by reputable academic institutions following the same internationally used ethical guidelines."

Anthropologist Darren Byler of Simon Fraser University strongly refuted these claims, arguing that no university is separate from the Chinese Communist Party and that the party committees closely scrutinize universities (Eikelenboom, de Bruijn, and Booij 2021). Moreau similarly rejected these claims, stating that it is impossible to conduct independent research in China given the high levels of political control.

CRITICISM C4 (IMPLIED OR LATENT): *Different aims.* While it was not an issue in the press coverage, a latent criticism of the VisiGen and Xinjiang Uyghur Study is that the VisiGen and other cohorts were originally recruited for biomedical or evolutionary research, not forensic genetic research, something I discussed in chapter 8. Toom et al. (2016) criticized the use of biomedical research subjects by Kayser because "forensic applications in a criminal justice context have very different aims and are subject to different ethical regimes of law, science, and practice" (e2).

CRITICISM C5: *Unverified claims.* For a variety of reasons, informed consent cannot be verified.

DEFENCE C5: *Fixing the time-space of donors.* The *eLife* journal editor and Erasmus Medical Center both asserted inclusion of the Xinjiang Uyghur Study in the four papers is justified because the samples were obtained in 2013–14. In a 3 June 2020 email response to me, *eLife* executive editor Andy Collings stated, "We understand that a proportion of participants decided not to provide DNA samples" as evidence of the capacity of the Xinjiang Uyghur Study donors to exercise choice over whether or not to submit blood samples. Similarly, in 2021, RTL and *Follow the Money* reporters asked whether the Erasmus Medical Center had any written documentation that the Xinjiang Uyghur Study donors were not under any form of coercion and were informed that the research would be used for genetic phenotyping. Erasmus MC stated it "had access to the written consents in Chinese and English translation. Our Chinese colleagues told us that this sample collection took place before the political conflicts began." (Siem Eikelenboom, email to author, 7 October 2021; Het Erasmus MC heeft inzage gehad in de schriftelijke toestemmingen in Chinese en Engelse vertaling. Onze Chinese collega's vertelden ons dat deze monsterafname plaatsvond voordat de politieke conflicten begonnen). This claim was reiterated as well in the Erasmus University Medical Center's public statement (2021). When the *Follow the Money* team asked Darren Byler about these claims, Byler argued that most Uyghurs do not have a strong enough knowledge of Mandarin Chinese to fully understand the informed consent process. Therefore, full informed consent would require the process be done in Uyghur. Furthermore, Byler rejected the claim that 2013–14 was before the political conflict began, arguing that state violence against the Uyghurs had begun well before then and had further escalated after 2014 (Eikelenboom, de Bruijn, and Booij 2021).

Once again, the logic of capture within the VisiGen Consortium/Xinjiang Uyghur Study forensic genetic assemblage seeks to firmly fix the Uyghur donors in their time-space of 2013–14 to disassociate them from the ongoing genocide. It also implicitly seeks to disassociate the donors' consenting from the very real probability that a large number of the donors have been or are incarcerated within the gulag of reeducation camps in Xinjiang. The 6 October 2021 Erasmus University Medical Center statement claims that it "has never used Uyghur datasets collected by institutions affiliated with China's Ministry of Security or the Chinese police and disapproves of their use in research." While the Erasmus claim of "affiliated" is correct using a narrow definition of the term, this claim is somewhat contradicted by the relationships established during 2014–15 between the Partner Institute and the MPS Institute of Forensic Science as documented in chapter 9.

Performance Phase

CRITICISM C6: *Completion of these projects advanced repression.* The potential utility of the research findings to repressing Uyghurs was much debated. *eLife* publisher Michael Eisen used a consequentialist ethical logic of no harm, no foul by arguing that the SNP panel had been obtained from the European samples, minimizing the agency of the Uyghur samples, "It is important in this case that the Uyghur data was used as validation, and its inclusion was largely incidental to the main thrust of the work. Indeed, based on the reviews of the paper and my own knowledge of the field, I expect the paper would have been published without it" (email to author, 4 February 2021). Eisen further narrows the framing of the ethical problems in terms of facial recognition surveillance, which is a strawperson argument because this was not, in any way, an aspect of my criticisms: "Indeed, without any data in this paper, it is already possible to identify the genetic ancestry of individuals should governments choose to do so, and that genetic data is not necessary or indeed useful to train surveillance to classify individuals based on images."

CRITICISM C7: *Racism.* The research is racist and should be condemned.

DEFENCE C7: *The research is non-Eurocentric.* In response to an *Erasmus Magazine* article about a Uyghur student who was critical of Erasmus University's affiliation with and cooperation with Liu Fan, the university posted the following response about the necessity of including diverse populations in research to ensure it is non-Eurocentric and thereby inclusive: "This decision has consequences from a scientific point of view, because this population group will not be part of the theoretical research. As scientists, we make every effort to include all relevant ethnic groups so that the knowledge domain contains balanced and comprehensive information. A one-sided (i.e., Eurocentric) focus has been criticised in the past – and rightly so." However, this scientific view of the need for non-Eurocentric research was overridden by Erasmus University's ethical realization: "But the distressing position that Uighurs in China find themselves in, and the social debate surrounding their situation, have compelled us to decide otherwise in this case" (Smaling 2021).

This logic of inclusion involves an appeal to an "inclusion-and-difference-paradigm" in that it seeks to increase the diversity of biomedical research away from predominantly white populations.

Sanction Phase

CRITICISM C8: *Circulating unethical Uyghur data and findings.* In contrast to the positive sanction the researchers received from journal editors and publishers on their quest to help Humanity, external critics negatively sanctioned the research assemblages' use of the Xinjiang Uyghur Study because it helps develop China's surveillance capacities. When Janssen asked Manfred Kayser, "But why don't you use this data anymore?" (Maar waarom gebruiken jullie deze data nu niet meer?), Kayser responded:

> In the VisiGen Consortium, we have now decided not to use these data for the time being. Not because we think they are wrong or because the Chinese scientists are bad. But we do not want to tarnish

> the science of the international consortium with inappropriate publicity. My academic colleagues at Fudan University are not happy about it and see themselves as victims of politics and the media. (Janssen 2021; In het VisiGen Consortium hebben wij inmiddels besloten om deze data voorlopig niet meer te gebruiken. Niet omdat we denken dat ze niet deugen of omdat de Chinese wetenschappers slecht zijn. Maar we willen niet de wetenschap van het internationale consortium aantasten met ongepaste publiciteit. Mijn academische collega's van de Fudan-universiteit zijn er niet blij mee en zien zich als slachtoffer van de politiek en de media.)

The claim they no longer use the data only applies to future research, not its past outputs. Kayser and the VisiGen Consortium and their colleagues have already integrated the paper's findings into their own EU-funded VISAGE research and development. As chapter 8 shows, the data from Pośpiech et al. (2018) on human hair shape has been integrated into the VISAGE Working Package for police phenotype prediction for use in Europe (VISAGE 2019a, 5, 2019b, 34). The findings from Xiong et al. (2019) on facial morphology phenotypes, while not included in the working package, was still included as a deliverable under the VISAGE grant (VISAGE 2019a, 17). The Xinjiang Uyghur Study data continues to make a difference for the VISAGE Consortium and the VisiGen Consortium.

The outcome of the destabilization of the cooperation between the VisiGen Consortium and the Xinjiang Uyghur Study has been more mixed. None of the four papers has an editorial note (Liu Fan et al. 2018; Pośpiech et al. 2018; Wu Sijie et al. 2018; Xiong et al. 2019), which suggests the editorial panels and publishers have no problem with the inclusion of the Xinjiang Uyghur Study donors – a fact that is also made evident in Michael Eisen's responses to my criticisms. Therefore, it is unlikely any of them will retract the articles for the time being. However, the use of the Xinjiang Uyghur Study in genetic research cooperation with the VisiGen Consortium has ceased. Furthermore, the intense political and media scrutiny of Erasmus University's employment of Liu Fan and linkages to other PRC researchers who have cooperated with the Ministry of Public Security means this long-term assemblage, dating back to the late 2000s, has, in some regards, either ceased or been restricted.

As we have seen time and time again in this book, forensic genetics as a discipline has proven incapable of meaningful ethical and political governance. These four Xinjiang Uyghur Study papers demonstrate the failures of conventional ethics procedures including ethical self-evaluation protocols, informed consent, and institutional review boards to properly and adequately consider racializing forensic genetic research involving vulnerable peoples. As part of its Horizon 2020 project, the European Union produced a considerable set of ethical research materials including guidelines and practical implementation guides. The four papers violate the following criteria from the European Commission's February 2015 "Guidance note – Potential misuse of research ethics," which covers all Horizon 2020 research projects (European Commission 2015). The guidance note defines the problem: "The term 'potential misuse of research' refers to research involving or generating materials, methods, technologies or knowledge that could be misused for unethical purposes" (1). It advocates that there be attention to the problem of misuse of research "from the very beginning" and, in addition to the research project planning, "they should also assess if their research could be used to serve unethical purposes, and examine if there are potential risks beyond the lifetime of the project." It expressly states that research must consider a number of forms of "research most vulnerable to misuse" including those that involve "the development of surveillance technologies that could result in negative impacts on human rights and civil liberties" and "research on minority or vulnerable groups and research involving the development of social, behavioural or genetic profiling technologies that could be misapplied for stigmatisation, discrimination, harassment or intimidation" (1–2). The VisiGen Consortium and the VISAGE Consortium did not think these criteria applied to their research involving the Xinjiang Uyghur Study, adamantly denying that there was anything unethical or misguided in using these samples.

Continuing this pattern of denial, Erasmus Medical Center filed a complaint with Netherlands Press Council (Raad voor de Journalistiek) against *Erasmus Magazine*. The magazine published the article by Elmer Smaling (2021), discussed above, that dealt with an Erasmus University Uyghur student's criticism of Erasmus University's denial of any responsibility for the research cooperation of Liu Fan with the Chinese Ministry of Public Security (Gunneweg 2021). The Uyghur student, using the pseudonym of Meryem to protect herself and her family, said: "But surely it should not be so simple to throw the Uyghurs under the bus because of business interests? Every scientist should always act humanely, and that has not happened here." According to the July 2022 Netherlands Press Council decision, this statement by the Uyghur student constituted an opinion and therefore Erasmus Medical Center should have had the right to respond with its own opinion and by not doing so the article's author violated journalistic standards.

In this decision, it is evident that Erasmus Medical Center once again engaged in equivocation with its repetition of the red herring argument that research cannot advance facial recognition of Uyghurs. According to the decision,

> This basic research, which used genetic material from many thousands of people from all over the world, including Uyghurs, cannot possibly contribute to facial recognition. The research is not aimed at it and the findings can in no way lead to the selection of minorities by means of DNA. So neither to recognition, suppression and exclusion of Uyghurs. Not at this time and not in the near and conceivable future. It would also go against our core values. (Netherlands Press Council 2022; Dit fundamenteel onderzoek, waarvoor genetisch materiaal van vele duizenden mensen vanuit de hele wereld werd gebruikt, waaronder Oeigoeren, kan onmogelijk bijdragen aan gezichtsherkenning. Het onderzoek is er niet op gericht en de bevindingen kunnen op geen enkele manier leiden tot het selecteren van minderheden middels DNA. Dus evenmin aan herkenning, onderdrukking en uitsluiting van Oeigoeren. Niet op deze termijn en ook niet in de nabije en voorstelbare toekomst. Het zou bovendien ingaan tegen onze kernwaarden.)

The Erasmus Medical Center then engages in further equivocation when it states, "Erasmus MC has never used Uighur datasets collected by institutions affiliated with the Chinese Ministry of Security or the Chinese police and disapproves of their use in research" (Het Erasmus MC heeft nooit gebruik gemaakt van Oeigoeren-datasets die verzameld zijn door instellingen die zijn gelieerd aan het Chinese ministerie van Veiligheid of de Chinese politie en keurt het gebruik ervan in onderzoek af). Erasmus Medical Center's claims are contradicted by a number of facts:

- BIG has a joint research center with the Institute of Forensic Science, founded in December 2015, as discussed in Chapter 9 (Beijing Institute of Genomics 2016, 96–7).
- Ye Jian has been a BIG committee member since 2015 (Beijing Institute of Genomics 2016, 5; 2017, 5; 2018, 11; 2019, 9; 2020, 6; 2021, 6).
- The Partner Institute has cooperated with the Institute of Forensic Science since 2014 (Chinese Academy of Sciences-Max Planck Society 2017, 106, 204).

- Statements are made in the BIG annual reports (2019, 89; 2020, 35) about how the Xinjiang Uyghur Study papers involving Liu Fan advanced BIG's cooperation with the Institute of Forensic Science.

The Erasmus Medical Center's complaint also focused on a minor error by the article's author – that the Medical Center had not sanctioned Liu and had not even discussed the issue with him – to which the Medical Centre asserted that it had discussed the issue with Liu but had not considered it necessary to sanction him. They also complained that they had not been provided the opportunity to respond, though *Erasmus Magazine* included a response shortly thereafter (Gunneweg 2022). The July 2022 conclusion of the hearing (attended by lawyers from Erasmus Medical Center and *Erasmus Magazine*) states that Erasmus Medical Center had been wronged because it "was mentioned in one breath with controversial scientific research conducted by one of its researchers and has been severely disqualified as a result. The article contains a mix of facts and opinions, which are not always distinguished in a way that is clear to a reader" (quote from Gunneweg 2022; Netherlands Press Council 2022). Legally and ethically, the Erasmus Medical Center's claims that the *Erasmus Magazine* article violated its rights to representation in the media is a highly selective and ultimately petty complaint. These claims of violations of journalist standards pale in comparison to the violations of research ethics and human rights standards of this research cooperation:

- Autonomy. There is an utter denial of autonomy, such as the impossibility of enroled Uyghurs ever exercising their right to withdraw from the Xinjiang Uyghur Study.
- Non-malificence (do no harm). The scientific cooperation of Liu Fan with the Institute of Forensic Science advanced PRC forensic genetic phenotyping inference targeting Uyghurs.
- Beneficience (do good). This research does not benefit the Uyghurs in any way.
- Justice. This research does nothing good for the Uyghurs, it only benefits the researchers and their respective security apparatuses. Furthermore, it helps the PRC security apparatus develop phenotyping inference technologies at a time when this apparatus is actively involved in the repression of the Uyghurs and other Indigenous peoples in Xinjiang.

CONCLUSION

The international research assemblages of the mid-2010s involving Kidd Lab and Li Caixia at the Institute of Forensic Science are no more. However, each assemblage continues on their particular research trajectory with its own internal hierarchies. Kenneth Kidd and his colleagues continue to utilize the Karitiana and other Indigenous peoples in their genetic research, as they have for three decades. For example, Kidd coauthored a paper entitled "Population Relationships Based on 170 Ancestry SNPs from the Combined Kidd and Seldin Panels," published by *Nature* in December 2019, that used the Kidd Lab's collection of Karitiana, Surui, Ticuna, and many other Indigenous people (Pakstis et al. 2019). The PRC connections continue in this paper with coauthor Li Hui of Fudan University, who tested four populations, "Kazakhs, Inner Mongolians, Khamba Tibetans, and BaimaDee [classed as a Tibetan subgroup]," at his lab in Shanghai (2). Kidd and Li Hui have coauthored other papers such as Li Hui et al. (2009) and Pakstis et al. 2015 (along with Li Caixia). Li Hui is noted for his research into the origins of the Chinese people, which has been criticized for its nationalist overtones (McLaughlin 2016).

Ignoring Kidd's request to stop using the samples of DNA extract he gave her, Li Caixia and her colleagues published further research in 2020 in the Chinese journal *Hereditas* that also made use of the Karitiana and other Indigenous peoples in conjunction with further Uyghur subjects in a paper on the seventy-four AISNP panel (Liu Yang et al. 2020, 302–3) and filed another patent application, CN111073888A, on the seventy-four AISNP panel (Liu, Ma, and Li 2020, 27, 68). The disruption of their international assemblage is part of the larger fracturing of PRC–US relations that has limited forensic genetics to their respective security apparatuses. In effect, the once seamless research cooperation among these three major security assemblages is now fragmented.

In response to media scrutiny, the Kidd Lab ALFRED database has removed some of the Uyghur data supplied by Li Caixia and her colleagues. However, Li Caixia et al.'s Karitiana, Ticuna, and Surui data, along with that of some forty populations from DNA extract supplied by Kidd Lab, remain on the ALFRED at the time of writing (October 2021) (ALFRED 2019h, 2019i, 2019j). As well, the ALFRED, at the time of writing, still includes the twenty-seven and seventy-four AISNP panels developed and submitted by Li Caixia and her colleagues (ALFRED 2019a). This highly selective data removal shows how the data produced by forensic genetic assemblages tends to remain mobile, stable, and combinable with other data sets unless directly challenged (Latour 1987).

Li Caixia and her colleagues have likely made use of the Tumxuk Uyghurs in another paper. This paper seeks to justify its research on male pattern baldness among "Eurasian" males in terms of being able to help psychological well-being: "Male pattern baldness is the most common type of progressive hair loss, which has a considerable negative impact on personal image and mental health" (Pan et al. 2020, 1079). They tested a set of over six hundred SNP markers associated with male pattern baldness on "684 Eurasian males in China" with an average age of 35.6, which likely makes it the Tumxuk Uyghurs, as I suggest in chapter 10 (Pan et al. 2020, 1072). The Institute of Forensic Science researchers conclude that "The prediction model constructed here may help understand the genetic mechanism, early diagnosis and prevention of MPB in East Asian populations" (1079). It is unlikely that the Institute of Forensic Science is particularly concerned with male pattern baldness's "negative impact on personal image and mental health" on Uyghur men's self-esteem (1079). They are Ministry of Public Security forensic geneticists and they do mention its potential use in forensic science, which is likely their main purpose (1077).

Similarly, the VISAGE Consortium (without Liu Fan) published a paper that tested their VISAGE Ancestry panel on Diversity Project samples including the Karitiana and Surui and other Indigenous peoples (de la Puente et al. 2021). They made use of data from the 1000 Genomes Project, which had "completed whole-genome-sequencing of the full HGDP-CEPH panel in 2019 soon after our in-house genotyping had been completed, so we compiled the complete dataset from 1000 Genomes and used the overlapping sample data to measure" the massive parallel sequencing effectiveness of their VISAGE basic tool set of ancestry and appearance prediction genetic markers (4). An international team with Wellcome Trust, European Union, and UK government funding "sequenced DNA extracted from the lymphoblastoid cell lines of the HGDP-CEPH panel on Illumina HiSeq X[10] machines and incorporated data from a subset of samples that had been previously sequenced" (Bergström et al. 2020, 9–10). Now the Diversity Project's Karitiana, Surui, Uyghur, and other Indigenous peoples samples are mapped in some of the highest resolution possible, which is a significant addition to the scientific commons resource for future forensic genetic human hunting research and development, as shown by how the VISAGE Consortium made use of the data.

Kayser coauthored a 2021 article with Liu Fan, Xiong Ziyi, and Li Yi (who have all, in turn, directly cooperated with the Institute of Forensic Science). However, this continued cooperation on phenotyping shifted to mice in an article entitled "The effects of Tbx15 and Pax1 on facial and other physical morphology." This paper uses some of the genetic markers identified in the study by Xiong et al. (2018) that involved the Xinjiang Uyghur Study subjects. In an early version of the Xiong et al. paper submitted to the journal *Cell* (the article was not published), the scientists used CRISPR to turn off ("knock out") gene expression of two genes, TBX15 and PAX1, in mice embryos that led to physical defects in mice's facial structure (Xiong et al. 2018, 5, 41). The 2021 paper used the same methodology and had similar findings. The authors state that "general agreement between our findings in knockout mice with those from previous GWASs suggests that the functional evidence we established here in mice may also be relevant in humans, which is of value in developmental biology, evolutionary biology, human genetics, medical genetics, and forensic genetics" (Qian et al. 2021, 7). In effect, through a shift to animal studies, this paper builds on the findings of the earlier studies of Xiong et al. (2018, 2019) and advances forensic genetic research with researchers who have directly cooperated with the Ministry of Public Security.

This chapter demonstrates how external pressure focusing on tenuous informed consent and ethical oversight claims can destabilize these large, well-funded research assemblages. However, it is the larger political hegemonic competition between the US and China and the growing international outrage over Chinese repression in Xinjiang that really increased the strength of the external assemblages opposing the forensic genetic research assemblages. The reconfigured research assemblages of Kidd et al., Li Caixia et al., and the VISAGE Consortium nonetheless continued utilizing Indigenous peoples as part of their routine research. As well, the shift to animal studies by Qian et al. (2021) could be a way to avoid attention while still allowing forensic genetic research cooperation to continue between Erasmus University's Manfred Kayser and Liu Fan and other Chinese researchers. Additional future studies of the disruption of other research networks involving Uyghurs will provide a more detailed and broad view of how these international research assemblages were destabilized and transformed.

CONCLUSION

Indigenous Peoples in Racializing Security Assemblages

Trading the Other is a vast industry based on the positional superiority and advantages gained under imperialism. It is concerned more with ideas, language, knowledge, images, beliefs and fantasies than any other industry. Trading the Other deeply, intimately, defines Western thinking and identity.

Linda Tuhiwai Smith, *Decolonizing Methodologies: Research and Indigenous Peoples*

As we have seen in the assemblages analyzed in this book, population and forensic genetic scientists as violence technology workers have been very productive in their roles, coproducing the subjectivity of Indigenous peoples as violence technology resources. The international assemblages (both long- and short-term production networks) analyzed in this book are moments in the reproduction and technological expansion of US, European Union, and Chinese security apparatuses. A forensic genetic research article is a primary unit of production and accumulation that enacts a nomos of appropriation, production, and distribution spanning numerous political jurisdictions (Schmitt 2003). These are narratively organized cycles of accumulation that articulate a sequence of events into a coherent article. This sequence begins in the spaces of transnational science, it then moves to the specific research assemblage, with its researchers and their laboratories, its research ethics committee overview and approval, its funding from national and international organizations, and its cooperation with policing agencies and biotechnology firms involved in the development of these new forensic genetic technologies. In these assemblages, scientists enact their subjectivities as scientists by representing Indigenous peoples as research subjects that are their objects of research, not vice versa. As resources, Indigenous peoples do not represent themselves or have any actual say in the research. The forensic genetic assemblages' organizing narratives hierarchically order into a coherent whole these disparate actants within their respective time-spaces. These assemblages of time-spaces involve a hierarchy of recognition and nonrecognition as scientists are recognized as legitimate users of these genetic samples within the time-spaces of the assemblage, with the resulting articles added to their curriculum vitae, included in funding reports as examples of "deliverables," incorporated into forensic genetic analysis systems like the Illumina MiSeq FGX, and so on. In contrast, Indigenous peoples exist in a state of capture as resources used to produce knowledge within these assemblages based upon nonrecognition and exclusion from contemporary norms over secondary usage and related regulatory and legal frameworks (Reardon and TallBear 2012). Crucially then, the forensic genetic research assemblage is also a form of governance in that a specific set of hierarchical relationships between scientists and Indigenous peoples are enacted, relationships that originate from colonial and settler colonial violence (L.T. Smith 1999).

In the case studies analyzed in this book, the scientists are reacting to the call to action by specifying how the particular technology of ancestry inference can help in hunting criminals. This scientific knowledge functions as an object of exchange and value to be pursued by the scientists on behalf of Humanity. Racial categories and classifications involve the interaction of synecdoche and metonymy. Racial categories involves synecdoche in that terms like European, Eurasian, and East Asian use taxonomies to categorize and classify populations in which the Geographical Region stands FOR the People of that Region. The common forensic genetic classification of Uyghurs as Eurasians classifies them as a subcategory that combines European and Asian, which moves from two more comprehensive to one less comprehensive composite category, and relies on a synecdochic relation of genus TO species. Synecdoche is at work here in that it seeks to classify. Metonymy is significant in the various traits ascribed to these categories and classifications such as SNP data on allele frequency, eye colour frequency, hair colour, and other physical traits.

The manipulation phase's universal and specific premise leads to a conclusion in the commitment phase, in which the authors as receivers of the quest submit to serving Humanity through their research projects, thus synecdochically acting as a member (hyponym or species) of the larger category of scientist (hypernym or genus) (Greimas 1971, 103–4). The manipulation phase as universal scientific knowledge and the commitment phases with the scientists' specific projects are articulated through a synecdochic transfer from genus TO species that articulates the manipulation and commitment phases of the forensic genetic narrative schema. This set of synecdoche-based transfers between taxonomic levels (genus TO species and species TO genus) are enacted through a series of enthymemes and syllogisms that define and coproduce not only the scientists but also the Uyghurs. The authors use the Uyghurs as objects of exchange and do so by moving from the general category of human populations to regional continental groupings like European, East Asian, and Eurasian as a blending of these two larger racial populations, which is then used to define Uyghur as Eurasian in a genus TO species taxonomy. These transfers in the forensic genetic articles' competence phase become species TO genus in that the specific findings of genetic research on Uyghurs and other Indigenous peoples of Xinjiang become equivalent to knowledge about Eurasian populations, which also allows for the defining of proportions, for example, defining Kazakhs as more East Asian and Uyghurs as the most European. Scientists and others within the forensic genetic time-space accept this quest to help by conducting genetic research in their area of expertise.

While genetic testing for individual identification, lineage, and phenotype does have other applications and hence guiding narratives, such as in paternity lawsuits, citizenship claims, or mass casualty events like plane crashes, these are secondary in the manhunt organizing narratives of forensic genetic assemblages. Rather, the close analysis of articles in this book shows how forensic genetics is dominated by the manhunt narrative with its central relationship between the Hunter, who is authorized government acting on behalf of the People and/or Humanity with tracking and capturing and/or killing the Hunted. Within the assemblages of forensic genetics, scientists have used Indigenous peoples as violence technology research and development resources in various productive subjectivities (actant roles):

- "Resources" of genetic samples, cell lines grown in "unlimited amounts," and data that represent Indigenous peoples and are de facto property within the assemblage.
- Proxies of a people or the peoples of a particular region. For example, the Karitiana and Surui cell lines represent these peoples but also and in other configurations function as representative of South American Indigenous peoples or "Native Americans." The synecdochic metonymic chain of cell lines FOR People FOR peoples of a region is central to the agency of these proxies.
- Hunted proxies. Keating et al. (2013) tested the Identitas identification system on several hundred Indigenous peoples' samples and cell lines as part of a globally representative sample, Budowle and David Smith et al. tested the Illumina FGx forensic system on Yavapai,

> and Li Caixia et al. tested their ancestry SNP panels on Uyghurs and other Turkic peoples and also on Indigenous peoples' DNA extracts provided by Kidd Lab.

Scientific assemblages involving settler colonial institutions including universities have engaged in the capture of Indigenous peoples through genetic sampling, immortalization to make cell lines, and cryogenic storage. Genetic research articles are vital units of knowledge production and accumulation, so the narrative schema approach developed in this book pays attention to the differential relationships between the disparate actants in their respective time-spaces. This approach highlights the differential sets of rights and obligations between scientists and Indigenous peoples within the field of forensic genetics and their respective larger security apparatuses. These hierarchies are graduated and involve scientists as highly privileged entrepreneurial violence workers (actants) able to organize assemblages across multiple forms of time-space including transnational science, national government security agencies and funding agencies, university institutional review boards, genetic sequencer companies, and scientific publishers (Ong 2006; Munsterhjelm 2014). In contrast, many of the Indigenous peoples sampled decades ago, such as the Karitiana and Surui, are likely now deceased, but cell lines and samples taken from them by scientists have continued to serve science as genetic materials that exist as human objects of exchange (boundary objects) as resources grown, distributed, and processed in these scientific assemblages (Castiglione et al. 1995, 1448; Munsterhjelm 2014, 2015; Reardon and TallBear 2012).

When forensic genetic researchers within their assemblages of capture cooperate with security agencies' assemblages of capture, there is a meeting of two petty sovereigns: that of the police and that of the scientists and their research and development assemblages (Bourne, Johnson, and Lisle 2015, 310). And it is through this meeting and cooperation that forensic genetic global assemblages involve a complex set of graduated sovereignty decisions in the articulation of the disparate actants in their respective organizational time-spaces (Ong 2006). Scientists are active participants in their respective security apparatuses (Bourne, Johnson, and Lisle 2015). For the scientists who translate the manhunt organizing narratives into global assemblages of forensic science, their research and development processes anticipate decisions of the police during manhunts and, in some cases, provide direct aid to police agencies in investigations. They are trying to provide timely genetic intelligence by exploiting the density of information within genetics that has been made possible by advances in sequencing technologies and mass data techniques.

Ong (2006) argues for "a rethinking of sovereignty not as a container concept but rather as a political order produced by an assemblage of administrative strategies" (98). In terms of analyzing forensic genetic assemblages, these assemblages' administrative strategies are coordinated through translations of dominant metanarrative schemas of fighting terrorism and crime to improve the biopolitical security of populations that provide the basis of interaction between disparate agents across numerous jurisdictions. This shared commitment to fighting terrorism and crime is translated into their different respective organizational agendas and, as a shared set of goals, helps coordinate the interaction between biotech companies, state security agencies, genetic researchers, and research funders. However, if these organizing narratives are rendered incoherent or contradictory, external actants will either ignore or actively oppose them and existing assemblage members may withdraw from the assemblage. Such passive or active rejection thereby defines the parameters or borders and boundaries of the assemblage; something that occurred with the disruption of Li Caixia et al.'s AISNP and PISNP assemblages research on Uyghurs during 2019 and the US government's embargo on technology transfers to and cooperation with the Institute of Forensic Science in 2020. Therefore, to empirically study the sovereign decision-making process within assemblages, we must look at the contingent and overdetermined assemblages that include state agencies but also exceed the state, as opposed to dominant ideas of the state as a container or as irrelevant under some oversimplified globalization hypotheses (Sassen 2006). In particular, within the research assemblages, we must consider how the dispersed sovereign decisions over informed consent practices, ethics review, legislation, regulatory regimes, and funding made and then recognized by actants situated in other jurisdictions is central to the international production of forensic genetic research. When forensic genetic researchers within their assemblages of capture cooperate with security agencies' assemblages of capture, there is a meeting of two petty sovereigns that of the police but also that of the scientist (Bourne, Johnson, and Lisle 2015, 310, 314). It is through this meeting and cooperation that forensic genetic global assemblages involve complex sets of graduated sovereignty decisions in the articulation of the disparate actants in their respective time-spaces (Ong 2006). As violence technology workers, involved scientists are active participants in their respective security apparatuses (Bourne, Johnson, and Lisle 2015). For the scientists translating the manhunt organizing narratives into global assemblages of forensic science, their research and development processes anticipate decisions of the police during manhunts, and in some cases provide direct aid to police agencies (Bourne, Johnson, and Lisle 2015). They are trying to provide timely genetic intelligence by exploiting the density of information within genetics that has been made possible by advances in sequencing technologies and mass data techniques.

IMPACT OF INDIGENOUS PEOPLES

The direct and indirect impacts of Indigenous peoples' resistance to and human rights criticisms of forensic genetics' ability to represent Indigenous peoples within its assemblages has been significant and should not be understated. Kenneth Kidd and his colleagues mentioned the political problems of gaining research cooperation from "populations" and the limitations this requirement poses on global research cooperation (see, for example, Soundararajan et al. 2016, 31). The research assemblages' reliance on pre-1990s samples shows how these assemblages have been shaped in their reach and scope by the resistance of increasingly strong global Indigenous networks and the growing recognition of Indigenous peoples' rights and sovereignty as reflected in documents such as the UN Declaration of the Rights of Indigenous Peoples (2007). In doing an analysis of genetic research, it is necessary to consider the heterogeneous contingency of these assemblages and in particular how they are vulnerable to resistance and external pressure at various points and how the effects of resistance can persist. The effects of this resistance are evident in the much attenuated size and scope of the Human Genome Diversity Project, from its proponents' original ambitions of tens of thousands of cell lines involving five to seven hundred populations to its 1064 cell lines representing about fifty-two Indigenous and non-Indigenous populations.

From 2017 onwards, there was growing international criticism and condemnation of China's crimes against humanity in Xinjiang, which, in 2019, finally began disrupting research assemblages:

- Corporate entities like Thermo Fisher announced it would not sell to PRC security agencies and Springer Nature began investigations of informed consent and ethics violations.
- In the context of the escalating conflicts between the US and the PRC over Xinjiang, US government agencies enacted embargoes against Chinese security agencies.

- Scientists begin to disassociate themselves from these research assemblages, including Kenneth Kidd.

The fragmenting of the research assemblages followed the respective security apparatus, so we might expect continued cooperation between European Union and American researchers, and PRC researchers will become increasingly isolated. However, Kenneth Kidd's and Manfred Kayser's respective continued cooperation with PRC scientists suggests these assemblages will adopt low-key cooperation and so continue to advance forensic genetics across these three major security apparatuses of the United States, the European Union, and China. There is a peremptory norm (jus cogens), a fundamental principle of international law that genocide is a crime against Humanity that demands a strong decisive principled response (Finnegan 2020). Given this peremptory norm, it is important ask why such an internal response was absent within forensic genetics assemblages until faced with strong external criticism in 2019 over research involving and targeting Uyghurs, and why has there been a continued denial by some researchers and institutions of any culpability. This resistance to criticism is exemplified by Erasmus Medical Centre's denials, including filing of a complaint with the Netherlands Press Council about not being allowed the opportunity to respond in an article about a Uyghur student's criticisms of Erasmus University's responses to Liu Fan's cooperation with the Institute of Forensic Science and how journals like *Forensic Science International Genetics* have not retracted any of the institute's articles involving Uyghurs and other Indigenous peoples in Xinjiang.

NEW ETHICS NEEDED

The objectification of Indigenous peoples rests upon wilful ignorance in which scientific norms suppress strong questioning or inquiry about claims of informed consent and negative impacts upon Indigenous peoples. Such wilful ignorance continues even when these claims are publicly discredited, such as the strong withdrawals of informed consent made by Karitiana and Surui leaders in a 2007 *New York Times* article (Rohter). There have also been numerous criticisms of such usage of the Karitiana and Surui in academic journals, as well as discussions in the Brazilian Parliament and actions by the Brazilian government. If the Karitiana and Surui withdrawal of informed consent had also been recognized by Kenneth Kidd and his colleagues, Kidd would not have been able to transfer the DNA extract to Li Caixia at the Institute of Forensic

Science. The recognition of Indigenous peoples' sovereignty and attendant rights and dignity would have prevented this transfer. Though Kenneth Kidd was credited as the source of the Coriell Cell Repository's Karitiana and Surui cell lines, Kidd Lab with its collection of samples and cell lines from Indigenous peoples went unnoticed in Rohter's 2007 *New York Times* coverage. In effect, the article did not provoke significant public debate and inquiry over Kidd Lab's continued use of the cell lines. It was only in 2019, with the rising evidence and condemnation of the mass incarceration of the Uyghurs, that Kidd's networks were finally disrupted. However, the disruption was restricted to Kenneth Kidd's direct research cooperation with the Chinese Ministry of Public Security involving Uyghurs. There has been no destabilization of the Kidd Lab collection of cell lines and samples obtained from Indigenous peoples; Kidd and his colleagues continue to use the Karitiana and Surui in their AISNP and other research. These complex interactions of settler state security and science institutions, genetic researchers, and companies follow a pattern similar to that identified by Reardon and TallBear:

> In short, Euro-American law and science operate within and act to enforce dominant social formations. Further, power inequities exist between these formations and indigenous peoples. As a result, indigenous peoples' efforts to reclaim rights to their resources and identities through dominant legal and regulatory mechanisms are likely to continue to fail. These mechanisms are mediated at every turn by power relations shaped by histories of racism and colonialism, and it is these relations that must be addressed if we are to recognize and respond to the problems created by the constitution of whiteness as property by both the law and the life sciences. (Reardon and TallBear 2012, S242)

These conditions also apply to the hierarchies between Han Chinese–dominated settler state security agencies and the Uyghurs and other Indigenous peoples in China.

The graduated and dispersed enactment of the sovereign decisions of rights and obligations was performed throughout the global assemblage of a typical research project (Ong 2006). The involved scientists and their respective institutions recognized these enactments of enrolment and ethics review by the various actants in the assemblage. This mutual acceptance of legitimacy is centrally important to the underlying political economy that structures the hierarchies of recognition of claims of informed consent. The circulation of Uyghurs as objects of exchange in the assemblages of forensic genetics evolutionary research relied on the various actants' shared acceptance of Chinese scientists' and PRC institutional review board's ethics claims about the enrolment of Uyghurs. The workings of informed consent and institutional review boards were sovereign enactments in which designated PRC institutions made decisions over these processes. In this way, the Uyghurs as research subjects and as representatives of the Uyghur people as an entity were stable, mobile, and combinable (Latour 1987). They circulated metonymically as samples and cell lines within the PRC, from Xinjiang to laboratories of BIG and Institute of Forensic Science in Beijing or the Partner Institute and Fudan University in Shanghai, and as data within the PRC and internationally within research assemblages that spanned Xinjiang, the PRC, Western research institutions, genetic equipment manufacturers, and publishers. It was only when the legitimacy of these ethics claims was publicly brought into question by assemblages of human rights organizations and activists through influential Western media outlets and US Congress that what had been routine acceptance became problematic.

Forensic genetic assemblages are guided in part by neoliberal market logic and rationality (governmentality), including the use of patent applications, competitive grant seeking, and active long-term cooperation with the corporate sector (e.g., Kidd's fifty-five AISNP panel integrated into Thermo Fisher and Illumina forensic genetic systems). However, the foundational importance of the state sovereignty rationale is clear in the 2019–20 disruption of the Chinese Ministry of Public Security's AISNP and PISNP assemblages by activists and, eventually, US government agencies and legislation imposing embargoes and restrictions. US government agencies have argued that China is engaged in crimes against Humanity and that the US government is thereby required to act in accordance with international laws banning genocide, whereas Chinese government agencies have countered that it is engaged in legitimate efforts against terrorism, extremism, and separatism.

It was through their shared use of the manhunt narrative that involved scientists, ethics review boards, funding agencies, police and security agencies, corporations, and other actants organized circuits of scientific knowledge production that began in the transnational discourses of forensic genetics. In this narrative schema, scientists submit to a quest given to them to use genetic research to improve the capacity of police and security agencies to estimate the appearance of suspects and their ancestral origins. In their claims of having the informed consent of Indigenous peoples, scientists constructed these Indigenous peoples as also submitting to these quests for scientific knowledge for the good of

Humanity. Yet, despite these moral claims, they readily cooperated in research on and about Uyghurs and other Turkic peoples without regarding the larger context of Chinese settler colonialism in Xinjiang. Ethics oversight boards, journal editors, and peer reviewers have and will continue to fail in their oversight because, as is the norm in genetic research, they accept the moral and ethical claims of the scientists –made clear by the continued acceptance of decades old claims of informed consent over the Karitiana and Surui and other Indigenous peoples' genetic materials. This book has demonstrated that forensic genetic assemblages in the name of protecting Humanity and the People have relentlessly exploited Indigenous peoples as productive "resources" in "unique collections," which raises fundamental ethical, legal, and political questions for the growing debates over the implementation and use of these deeply racializing manhunting technologies.

Notes

INTRODUCTION

1 For example, following a major scandal over the misuse of blood samples from the Nuu-chah-nulth First Nation of Vancouver Island, Canada's Tri-Council Funding Agency adapted the concept of "DNA on loan" in which scientific researchers are effectively only borrowing samples from Indigenous communities and control ultimately remains with these communities (Arbour and Cook 2006). Following years of Taiwan Aboriginal lobbying and a number of scandals, in 2011, Taiwan's Parliament passed legislation protecting Aboriginal rights in research that have strong financial penalties and bans from future research funding (Munsterhjelm 2014, 218–20).

2 Though there were various attempts to use things like blood typing before the 1980s, Jeffreys's invention of DNA fingerprinting led to the scientifically credible entry of genetics into the courtroom.

3 As well, the controversial practice of familial searching involves investigators looking for individuals who share a large number of markers to see if an unidentified sample might be related to somebody already in the database. I will not deal with this issue because it did not occur in the case studies covered in this book.

4 A team of fourteen US legal scholars in a 2013 *amicus curiae* (a briefing by parties not directly involved in a case) submitted to the US Supreme Court on forensic genetic evidence argue that as "DNA technology has evolved, almost all new techniques have been implemented without express legislative permission or judicial oversight" (Garrett and Murphy 2013, 40).

5 A recent paper by Ahuriri-Driscoll, Tauri, and Veth (2021) on Maori views of forensic genetics is an exception. As well, there has been work done on Latin American Indigenous peoples and forensic genetic identification of victims of repressive regimes such the Guatemalan government's mass murder of Maya during the Guatemalan Civil War (see, for example, Smith and García-Deister 2021). However, I will not deal with those assemblages since they do not intersect with the US-, European Union–, and China-centred assemblages that are the focus of this book.

6 Some readers may be familiar with Gilles Deleuze and Félix Guattari's concept of the assemblage of capture in regard to the state in their book *A Thousand Plateaus*. My analysis will not utilize their conceptualization because it is very theoretically dense and not applicable to my mid-level actor-network theorizing and analysis.

7 Police are also honoured as sacred dead. For example, in the province of Ontario, numerous bridges and public facilities are named after police who died in the line of duty under the Highway Memorials for Fallen Police Officers Act, 2002, S.O. 2002, c. 26 – Bill 128. The bill was recently amended to include officers who have taken their own lives by Bill 24, Highway Memorials for Fallen Police Officers Amendment Act (In Memory of Officers Impacted by Traumatic Events), 2021.

8 For example, the May 2017 Human Identification Symposium sponsored by Thermo Fisher Scientific had the theme of "Increasing Security, Solving Crime" (Lackey 2017). This slogan is featured on the conference brochure along with an image of a young white woman hugging a young white man whose face is unseen, while, near the bottom, a superimposed nighttime image of a backlit silhouette of a male in an ominous pose on a grassy area suggests a stalker or rape scenario (Applied Biosystems 2017). This scenario was repeated in 2018 with the theme "Seeking Answers Solving Crime," which features a female silhouette in the foreground on a pedestrian bridge with a male silhouette in the background.

9 Victims' rights activist Jayann Sepich gave a keynote presentation to the 2017 Human Identification Solutions Conference entitled "DNA Databases: Solving Crimes and Saving Lives … A Mother's Story" (ellipsis in original). Sepich became an advocate for the increased use of forensic genetic testing after her daughter was brutally raped and murdered in 2003. The presentation begins with a number of slides of Sepich's family photos, including several photos of her daughter. Some slides have white or yellow print against a black background and contain statements that

support routine genetic testing of suspects in violent crimes such as "DNA is truth" or the names of victims followed by, "These young women did not have to die."

10 In 2000, the US Department of Homeland Security began the Universal Adversary Program, which tried simply summarizing this disparate range of enemies as "the universal adversary" in various possible scenarios (Neocleous 2015).

11 Neocleous (2013; 2014, 14) argues that pacification involves a continuum of violence such that war and policing should not be distinguished as some sort of dichotomy because both war power and police power violently impose and maintain the relations of capitalist accumulation: "On the one hand, we need to grasp the exercise of the police power in constant war against the 'enemies of order.' Police treatises, texts, speech and action never cease telling us of the constant police wars being fought against the disorderly, unruly, criminal, indecent, disobedient, disloyal and lawless. On the other hand, we also need to grasp the ordering capabilities of the war power. 'War is the motor behind institution and order,' as Foucault puts it. Just as the police power is not reducible to the police but depends on a whole range of technologies to form the social order, so the war power is not reducible to the military but depends on a whole range of technologies to do the same" (Neocleous 2014, 14). In forensic genetics, there are extensive back and forths of technologies between military and police agencies like the use of mass biometric profiling (including DNA profiles based on the CODIS system) by the US military as part of its counterinsurgencies in Afghanistan and Iraq, with centralized databases used and shared by various US government agencies including the US military, FBI, and Homeland Security (Bell 2013, 469–70, 477; Ryan 2009, 60–1, 65; Wikoff and Holmes 2009, 7–9).

12 In the development of actor network theory, Bruno Latour and Michael Callon drew extensively on the work of Greimas, including the concept of actants and enrolment, so Greimas's narrative schema can be readily adapted to science and technology (STS) approaches (see, for example, Callon 1986).

13 Philosopher Judith Butler explains how problem definitions affect practices (2002, 216): "What it suggests is that certain kinds of practices which are designed to handle certain kinds of problems produce, over time, a settled domain of ontology as their consequence, and this ontological domain, in turn, constrains our understanding of what is possible. Only with reference to this prevailing ontological horizon, itself instituted

through a set of practices, will we be able to understand the kinds of relations to moral precepts that have been formed as well as those that are yet to be formed."

14 In this book, I utilize all caps "FOR," from cognitive linguistics notation, to indicate metonymic relationships between concepts (Radden and Kövecses 1999).

CHAPTER ONE

1 There were various disputes over the Amazon region among the region's settler states. The last major shift in borders occurred in 1904 when Brazil annexed the State of Acre under the Treaty of Petrópolis. This treaty concluded a conflict in which a local independence movement made up of Brazilian rubber tappers (who were the majority of the population), with support from the governor of the State of Amazonas, fought against rule by the Bolivian government from 1899 to 1903. This included declaring the Republic of Acre, which was short-lived. In the Treaty, Brazil gained 191,000 km² of territory from Bolivia. In exchange, it gave up a small amount of territory in Mato Grasso State, provided Bolivia two million British pounds in compensation, and promised to build a railway that would allow landlocked Bolivia access to the Atlantic Ocean. Brazil has not had any military confrontations with neighbouring settler states since then.

2 The early biopolitical management of the Brazilian settler colonial population in the Amazon region included the provision of some basic health, education, and transportation infrastructure and administration (Rodrigues and Kalil 2021, 12).

3 However, by the late 1960s and early 1970s, the accelerating scale of environmental destruction and displacement and genocide of many Indigenous peoples in the Brazilian Amazon region were becoming a focus of significant international criticism and lobbying (Garfield 2001, 143–4; Lewis and McCullin 1969). Published in 1967, a thirty-volume report by a Brazilian public prosecutor named Jader de Figueiredo Correia chronicled the role of the Indigenous Protection Service in crimes committed against Indigenous peoples from 1940 to 1967, including misappropriation of funds, illegal logging and mining, and murders and massacres, such as the bombing of a Cinta Larga Indigenous village with dynamite dropped from a small plane (Lewis and McCullin 1969). Rather than suppress the report, the dictatorship released it in an attempt to show that it was committed to "racial democracy" and to show the crimes committed against

Indigenous peoples by the populist democratic governments that had preceded the dictatorship (Garfield 2001, 143–4). However, instead, the dictatorship received intensified criticism.

4 The English word capture derives from the Latin *captus*, which is the past participle of *capere* "to take, hold, seize."

5 *Online Etymology Dictionary*, s.v. "vanish," accessed 19 January 2023, https://www.etymonline.com/word/vanish.

6 The Diversity Project's use of *squander* has similarities to how early theorists of international law like Grotius and political philosophers like John Locke sought to justify European claims over Indigenous peoples lands and resource as waste because Indigenous peoples were not using their territories in a productive manner according to self-serving European imperial standards (Neocleous 2014).

7 In a 1993 email listserv response, Cavalli-Sforza and Henry Greely of Stanford Law School defended the Diversity Project against Indigenous peoples' efforts to stop it: "Today we read with great concern a message on this network seeking to enlist support to stop the Human Genome Diversity (HGD) Project. That message conveyed factually incorrect and grossly misleading information about the proposed Project and, more importantly, was deeply wrong in its stance toward that proposed Project" (Cavalli-Sforza and Greely 1993). Later in the same message they state: "The proposed HGD Project is not and will not be a commercial venture. It is thought the chance that this research will lead to the devlopment of commercially valuable products is very remote. Our current planning anticipates that this unlikely event is not impossible and so we are seeking to ensure that, should the cell lines have commercial value, the benefits will flow back to the sampled populations." The history of forensic genetic research and development and the attendant use of Indigenous peoples' genetic materials from the Diversity Project and Kidd Lab has since proven these assertions wrong.

8 Early genetic researchers' widespread belief in a salvage paradigm was crucial to their view of Indigenous peoples as a premodern primitive Other. This belief became entrenched in a series of projects and assemblages that developed in the post–World War II period, informed by a type of logic of the emergency in which scientists sought to sample Indigenous peoples before they were acculturated and/or disappeared as entities (de Chadarevian 2015; Radin 2017; Reardon 2005). In the 1950s, within the US national security apparatus, the recent rapid development of nuclear weapons and atomic energy led to extensive concerns over the potential impacts of radiation on human genetics. These concerns led to

calls for a sustained and systematic effort to understand these effects, with an abstract from a 1955 World Health Organization (WHO) meeting calling for "a world-wide scientific study of radiation induced genetic effects from the standpoint of the world population," and, by the late 1950s, a study group was formed to consider this issue (quoted in de Chadarevian 2015, 367). In this way, radiation's potential effects on human genetics were becoming increasingly viewed as a public health issue (369). From the early 1960s onward, Neel was deeply involved in this research in conjunction with the Brazilian geneticist Francisco Salzano in what eventually became the International Biological Program (Radin 2017, 101–17).

9 For example, Meriwether in a 1999 article entitled "Freezer Anthropology: New Uses for Old Blood" states that between 1967 and 1977 Neel took some 9,800 samples from Indigenous peoples in the Amazon region (121; see also Radin 2017, 157–60).

10 The expedition involved researchers from a number of different disciplines, who took turns, so Neel and his colleagues participated in the second phase in July and August 1976. James Neel began research on Indigenous peoples in the Amazon region in conjunction with Salzano in the early 1960s (Radin 2017, 101–17).

11 Blood group typing was a standard way of doing genetic research in the 1970s. DNA-based techniques were still very limited during that time.

12 The FUNAI did approve Neel's sampling but only after they refused Neel's request to sample five more isolated peoples (Radin 2017, 146).

13 Other examples include the fur trade in North America, camphor in Taiwan, ivory in the Congo, and gold in California.

14 The rubber trade in Brazil collapsed after 1912 due to cheap supplies from plantations in British colonies in Malaya and elsewhere in Asia (grown from rubber seeds smuggled out of Brazil in 1876). There was a brief boom during World War II, with the Japanese occupation of Southeast Asia (1942–45), when the Brazilian government, in an effort to provide rubber supplies for the American war effort, conscripted labourers, generally impoverished men from the northeast known as "rubber soldiers," who were then abandoned by the Brazilian government after the war. Thousands died of disease (Tully 2011, 83–4, 326–7).

15 Antônio Moraes's birth date is the source of some contradiction in the various sources ranging from as early as 1905 to 1919. He was baptized in Porto Vehlo in 1957 and the baptism registry entry describes him as being around forty (Vander Velden 2010). He is said to have died between 1965 and 1968.

16 Though the data was introduced under seal, meaning it was not to be published or distributed until this article in *Human Biology* was published, it soon began circulating among defence lawyers.

17 Kidd and Friedlaender coauthored a paper in 1971 using Friedlaender's genetic data from Bougainville (Friedlaender et al. 1971). However, after the 1971 paper, they did not cooperate again until the 1980s when they became involved in work on population variation, discussions that contributed to the Diversity Project (Friedlaender and Radin 2009).

18 Yet incredibly, particularly given the fact that Kenneth Kidd had already given Nasioi cell lines to the Diversity Project and Coriell Cell Repositories (the latter began selling them in the 1990s), Friedlaender nonetheless asserted that "the open-ended nature of the cell lines ultimately meant that they were difficult to control, so the Kidds laid down strict rules on who could get the DNA harvested from their cell lines – basically, no profit-making research ventures. The cell lines themselves remained under their (and Cavalli's) control" (Friedlaender and Radin 2009, 150).

Friedlaender's assertions were later contradicted by how Kenneth Kidd and his colleagues used the Nasioi cell lines in testing the Identitas forensic genetic system (Keating et al. 2013, 562, 565) and in developing the fifty-five AISNP panel used in Illumina and Thermo Fisher forensic genetic testing systems (Pakstis et al. 2015a, 269; 2015b, 4). Kidd would later give Nasioi DNA extract to Li Caixia, which was used in Institute of Forensic Science patents, such as CN107419017B (Liu Jing, Li Caixia et al. 2020, paras 46, 80).

19 Epideictic rhetoric is used to define the manipulation phase because it is a call to action, which seeks to express the significance of the larger quest to which the scientists willingly submit.

20 The metaphors of "gene tree" and later of "bottleneck" rest on the metonymy of genetics FOR ancestry (Velasco-Sacristán 2010). Mitochondrial DNA is passed from mother to child, so we do not receive any from our fathers. It was the first part of the human chromosome to be decoded and contains about 16,500 base pairs that code for thirty-seven genes. In contrast, nuclear DNA contains around 3 billion base pairs and codes for some twenty to twenty-five thousand genes. The estimate of "1000 fold more genes" was based upon pre–Human Genome Project estimates that were significantly higher than the number of genes (twenty to twenty-five thousand) the Project eventually revealed.

21 HLA haplotypes refers to human leukocyte antigens, which were an early form of testing frequently used in population genetic research on

migrations and heredity. Haplotypes are groups of genes that tend to be inherited together.

22 The Karitiana were sampled as part of a larger project by American and Brazilian researchers on disease immunity involving some twenty-five hundred individuals from twenty different Indigenous peoples in Brazil (Atwood 1988).

23 It also appears that Epstein-Barr virus immortalization may cause other changes in cell lines that have been cultured for a long time. There is evidence that some display chromosomal instability, including abnormal karyotype, particularly aneuploid, which means they have an abnormal number of chromosomes (Danjoh et al. 2013; Volleth et al. 2020).

24 The data from the seven VNTR loci that were tested would later play a role in what came to be termed the DNA Wars.

25 As Gerard Steen argues, "metonymy is important for narrative ... As a mode of narration; then it plays a structural role in narrative. Thus the chronological and causal sequencing of distinct events plot has been analyzed as so many metonymic moves by the narrator, taking the addressee from one situation to another with the situations constituting contiguous parts evoking a larger all that is left unexpressed. A similar metonymic function has perpetuated to the narrator switching from characters to settings to events themselves" (Steen 2005, 307).

26 Nussbaum (1995) identifies these traits in the 1990s feminist debates over the depiction of women in pornography. Not all practices have to be present for objectification to occur.

27 In July 2018, the *Guardian* newspaper carried a report about the last surviving member of an uncontacted Indigenous people in Rondonia state. FUNAI monitored the man described as being in his fifties and in good health. He was the last survivor of a group of six, the rest of whom were murdered by settlers in the 1990s. The man refused gifts of machetes and other items left in the vicinity of his hut by FUNAI officials and so the officials did not directly contact him. The reports were accompanied by a short video of the man cutting a tree with an axe (Phillips 2018). According to a 2022 BBC report: "The man, whose name was not known, had lived in total isolation for the past 26 years. He was known as Man of the Hole because he dug deep holes, some of which he used to trap animals while others appear to be hiding spaces. His body was found on 23 August in a hammock outside his straw hut. There were no signs of violence" (Buschschlüter 2022).

28 A crucial qualification to the distribution of these cell lines is the claims that they are derived from ethically approved samples, willingly and

knowingly given and claimed to be ethically approved. This 1995 mention of ethics may reflect the changing research environment after 1992–93 when the controversies over the Diversity Project and patenting of genetic research on Indigenous people began, because such a mention of ethics is completely absent in the 1991 paper by J.R. Kidd et al.

Han Chinese are the majority population in China. The Druze are considered by some as a religious minority in Israel, Lebanon, and Syria. Yet given the contested definitions of Indigenous peoples, the Druze might be considered as Indigenous to the region.

CHAPTER TWO

1 The Hardy-Weinberg equilibrium proposes that frequencies of alleles and genotype will remain constant across generations provided there is no influence by the forces of evolution, such as bottlenecks and founder effects. Its idealized assumptions include random mating, no migration, and an infinite population size. It calculates the proportions of alleles in a given population. For example, a simple case with two alleles $(p + q)^2 = 1$ or $p^2 + 2pg + q^2 = 1$.

2 VNTRs do not code for genes that produce proteins so they are not subject to positive selection and therefore more variable between individuals. However, recent research indicates that VNTR is not really "junk DNA" but likely has functions such as regulation of transcription in which DNA is converted into RNA (Marshall et al. 2021).

3 For example, one of these loci D2S44, located on Chromosome 2, has a wide range of sizes, "from less than 871 to more than 5,686 base pairs" (National Research Council 1996, 21). The large number of alleles (variants) for each VNTR location and their high variability between individuals made them useful in individual identification. Segments are cut out of the genome using enzymes (e.g., HaeIII) before and after the particular VNTR sequences. The length of these fragments was then measured, which allows for an estimate of the number of times that each particular VNTR repeats. For example, the repeat unit length of D2S44 is thirty-one base pairs long and the lengths of fragments ("fragment sizes") ranged from under 700bp (about 22 repeats) up to 8,500bp (more than 274 repeats) (Butler 2010; National Research Council 1996, 74). In practice, measuring the exact number of repeats was difficult (if not impossible) so the FBI divided up the length range of repeats into thirty-one bins ("fixed bins"). Adjacent bins with small numbers were combined, so for example Bin 1 for D2S44 was from 0 to 871 base pairs while Bin 25 for

D2S44 was 5,686 base pairs or more (National Research Council 1996, 20).

4 In the Castro case, the judge allowed only exclusionary usage and so limited the prosecution's use of genetic evidence to try to show the blood on the watch was not Castro's. Furthermore, the judge did not allow inclusionary use by the prosecution trying to show the blood was that of the victims (Aronson 2007, 74).

5 This setback contributed to the National Academy of Sciences forming a study group to standardize forensic genetic methodologies.

6 A 2017 summary from Cold Spring Harbor Laboratory of the meeting's importance states, "This meeting had the most significant outcomes of any meeting at the Banbury Center: In part as consequence of this meeting, there were substantial changes in the conduct of DNA fingerprinting" (1).

7 The Frye Test derives from a 1923 appeal of a murder conviction, *Frye versus US*, in which the District of Columbia Court rejected the scientific validity of polygraphs (lie detectors). The opinion reads, "Just when a scientific principle or discovery crosses the line between the experimental and demonstrable stages is difficult to define. Somewhere in this twilight zone the evidential force of the principle must be recognized, and while the courts will go a long way in admitting experimental testimony deduced from a well-recognized scientific principle or discovery, the thing from which the deduction is made must be sufficiently established to have gained general acceptance in the particular field in which it belongs."

8 In June 2018, I received an estimate from my university library of over US$9,000 for copying the *Yee* case materials located in the National Archives in Chicago.

9 In contrast, in the United States judiciary, a voir dire hearing refers to the questioning process during jury selection.

10 A "match window" is typically plus/minus 2.5 per cent the number of basepairs of the allele. A match would be declared providing the match window was within the bin. Cellmark and Lifecodes utilized floating bins, which varied depending on the length of the allele. For an explanation of the differences see Aronson (2007) and National Research Council (1996, 5–14, 5–18).

11 Originally sentenced to twenty-five years in jail, Legere has since been legally deemed a dangerous offender who is likely to reoffend and will spend the rest of his life in prison.

12 The 1,431 possible pairs (r=2) from the fifty-four Karina (n=54) is calculated using the following formula: $n!/(r!(n-r)!) = 54!/(2!(54-2)!)$.

13 Hicks was found guilty and sentenced to death in 1989. He was eventually executed in 2000 by lethal injection.
14 Kidd's Yale University webpage states he received two awards for his efforts in the 1990s: a "'Profile in DNA Courage' award" from the "United States National Institute of Justice (2000)" and a "Recognition of Your Efforts During Our Decade of DNA, 1988–1998" from the "U.S. Federal Bureau of Investigation (1998)" (Yale University 2022).
15 For example, in trial by jury, the jury functions metonymically as representing the will of the People.

CHAPTER THREE

1 Many 9/11 victims were identified by taking DNA samples from personal items such as toothbrushes or combs (Biesecker et al. 2005).
2 The NSF data is from https://www.nsf.gov/awardsearch/simpleSearchResult?queryText=+Kidd+Yale. Awards 9408934, 9632509, and 9912028 were to Judith Kidd as primary investigator and the other nine were to Kenneth Kidd as primary investigator.
3 The 00965588 abstract posits wide ranging usages for genetic variation: "One area of particular relevance to history, historical demography, linguistics, medical anthropology, forensic anthropology, and ethnic studies, among others, is the gene frequency variation that exists among human populations."
4 The ALFRED was originally funded through grants from the US NSF and the US Public Health Service. For example, a US$2,652,755 NSF grant (NSF #0096588) that ran from 2001 to 2008 was entitled "A Genetic Database for Anthropology" (https://www.nsf.gov/awardsearch/showAward?AWD_ID=0096588&HistoricalAwards=false).
5 I composed this map with geographic information systems (GIS) software using a file from the ALFRED database named "alfredPops.zip," a text file dated 19 October 2013 that provides a set of coordinates of the maximum and minimum longitude and the maximum and minimum latitude of over seven hundred populations (ALFRED n.d.a.). These coordinates if mapped would create a sort of quadrilateral region for each population. However, mapping all of these quadrilaterals at once would render the map unprintable (though it would work on an interactive online map). Figure 3.2 uses a midpoint based on the averages of these maximum and minimum longitude and latitude points. As well, the GIS program I am using (ArcGIS) does not print every population name in areas of the map with a higher density of populations, so such regions will have a reduced number of

population names. A number of these populations are aggregates of numerous populations, such as "Asian Mixed," which covers several populations, and "Amerindians," which included Indigenous peoples from throughout North and South America. As well, Greenlanders and Inuit who live in coastal regions appear in the centre of Greenland's icesheet. The 2013 list includes 722 populations so it is an underestimate of the full number of populations in the ALFRED database.

6 In contrast, a typical scientific research paper's manipulation phase is situated in the transnational biotechnology time-space that goes through the cycle returning in the sanction phase to the transnational biotechnology time-space (Munsterhjelm 2014).

7 Some might disagree with the inclusion of the Samaritans but they have been established for thousands of years in the same area and now only number less than a thousand people.

8 Fst or Fixation index is a statistical process in population genetics used to measure differentiation between and within populations based on genetic polymorphism data.

9 Their paper may, however, be challenged or even discredited later, a form of negative sanction (Latour 1987, 91–2).

10 Later, he argues about the quality of SNPs in legal practice: "To the degree that SNPs identified from our studies are brought before the courts, this work that we have published and the data we have deposited into ALFRED provide a firm scientific basis for their acceptability" (Kidd 2015b, 23).

CHAPTER FOUR

1 This paper was funded in part by two US NIJ grants totalling $1,505,056 to Kidd entitled "Population Genetics of SNPs for Forensic Purposes" (2004-DN-BX-K025 and 2007-DN-BX-K197) (Li Hui et al. 2009, 937).

2 The Tibetan Autonomous Region is 1.228 million km^2 or around 12 per cent of the PRC (while adjacent Tibetan regions of Qinghai and Sichuan provinces are also large areas).

3 The PRC has inflicted radioactive contamination in the region of the Lop Nor nuclear testing zone in the eastern Xinjiang. From 1964 to 1996, the Chinese military conducted all of its open-air and underground nuclear testing at the Lop Nor site (some forty tests including atomic and hydrogen bombs) (Alexis-Martin 2019; Millward 2007, 310).

4 Xi'an Jiaotong University is another major centre of forensic genetic research on Uyghurs and other minorities. Zhu Bofeng is an important

figure there. For example, he is a coauthor of Wei et al. (2016b) along with members of the Ministry of Public Security. However, this important node will not be covered in this book.

5 According to Ye Jian's biography entry for the 2015 conference: "In 2000, she was in the Pennsylvania State University as a visiting scholar for one year" (International Conference on Genetics 2015c, 2). In 2002, she was lead author on a paper with Mark Shriver and Peter Underhill entitled "Melting Curve SNP (McSNP) Genotyping: A Useful Approach for Diallelic Genotyping in Forensic Science."

6 One might expect this university to be located in the Tibet region, but instead it is located in the city of Xianyang, near Xi'an in Shaanxi province in eastern China. Xi'an was the first capital of a unified Chinese empire under the first emperor Qin Se Huang, and it is where his tomb is located, including its famous army of terracotta warriors.

7 They also make use of "Casework samples derived from the Department of Public Security, Guangdong Province, and the Institute of Forensic Sciences, Ministry of Public Security, Beijing" from Kyrgyzstan, Nigeria, and Uganda, which would raise questions about informed consent (Jia et al. 2014, 1).

8 Kidd also made a trip to the Huazhong University of Science and Technology in Wuhan in early March 2013 where he gave lectures on forensic genetic markers for phenotyping estimation and ancestry inference (Huazhong University 2013). This report features several photos of his visit.

9 Guangzhou has several thousand African residents who work as traders and run small businesses, etc.

10 The article states, "Received: 3 December 2015 / Accepted: 10 February 2016 / Published online: 1 March 2016," so the 741 Uyghurs from Urumqi were sampled before December 2015.

11 In Table S1: Data and sample information for Wei et al. (2016b) there is a change in wording as they omit Kidd and Budowle as the sources of the DNA extract samples and simply state "Our lab" (2016c), which seems to indicate they are increasingly viewing these DNA extract samples as their own property.

12 They use the Chinese term 尼格罗, pronounced *Nígéluó* in Mandarin Chinese, which is a transliteration of the English word.

13 This error rate is given significant attention in the discussion, so they argue that ancestral components are not sufficient and likelihood ratios have to be considered as well in ancestry inference. They discuss how to deal with Uyghur examples in particular.

14 In this map, I have combined two sets of Salar samples that were submitted by Li Caixia's lab to the ALFRED and included in Jiang et al. (2018): SA004729W (n=92) plus SA004730O (n=115) since they are likely from the same area.

15 The Chinese word Zhǒngzú (种族) can be translated as either race or ethnicity. However, because Zhang et al. (2019) use the large population categories of East Asian, European, and African, I think race is the more accurate translation here.

16 Those with existing passports had to return them to the government, and, to obtain a new one, they would have to submit a DNA sample with their application.

CHAPTER FIVE

1 The idiom "large mixed living and small clusters" means that the Chinese population as a whole is characterized by large populations that have a mix of ethnic groups, for example, in large urban areas. However, at the local or regional level, there are areas where particular minorities predominate.

2 CN112011622B uses a different technical approach to infer ancestry from the other patents and applications since it makes use of forty-two insertion-deletion polymorphisms (InDels) rather than SNPs.

3 I make use of these images from Chinese patents based on Article 5 of the Copyright Law of the People's Republic of China, which states, "This Law shall not be applicable to: (1) laws; regulations; resolutions; decisions and orders of state organs; other documents of legislative, administrative and judicial nature; and their official translations;"

4 Based on an analysis of genetic research published in *Nature Genetics*, Panofsky and Bliss (2017) argue that a range of classification logics like race, religion, continental, country, ethnicity, and language are frequently used and combined, with such ambiguity being advantageous because it allows translation among disparate actors, which is the basis of research coordination.

CHAPTER SIX

1 The webpage's "Recent site changes" states it was created on 12 October 2010 (VisiGen Consortium 2010b).

2 The patent is entitled "Method for Characterising Variability in Telomere DNA by PCR."

3 STR refers to short tandem repeat polymorphisms, which is one of the original forensic genetic techniques for individual identification techniques. The FBI's "Combined DNA Index System" (CODIS) system analyzes the allele at twenty different loci and stores these in national databases. Because every human cell is diploid (has one allele from each parent), it actually tests for forty alleles, twenty on each diploid (Federal Bureau of Investigation 2020).

CHAPTER SEVEN

1 Tang Kun and Guo Jianya filed Chinese patent application CN105740851A in 2016 for the facial mapping process.
2 The 2013 BMC Bioinformatics Article is covered by the following licence, which allows commercial use provided there is proper attribution: https://creativecommons.org/licenses/by/2.0/.
3 Mark Shriver's project was funded by the US NIJ and began in the early 2000s (Fullwiley 2014, 808–11). Shriver was an early collaborator in the development of the controversial Parabon Nanolabs genetic profiling service, which has been marketed to police agencies since 2014 (809).
4 A reporter writing on the paper's findings used the phrase "mitochondrial Eve," which has since become one of the foundational mythic metaphors of modern genetic ancestral genealogy.
5 There are two researchers from Xinjiang Medical University, Agu Hashan (阿古哈山) and Nurmamat Bahaxar (努尔买买提•巴哈夏尔), who are from Xinjiang Turkic minorities (I am not sure which ones); however, these two researchers are not involved in any further papers.
6 As well, the number eight is considered auspicious and lucky in Chinese culture.
7 They used a process called an array, which is a well-established and widely used chip technology that allows testing for a large number of targeted genetic markers. This array process has several major cost advantages because sequencing of whole genomes is still quite expensive ($800–$1,000 per genome at the time of writing in May 2020 using costly sequencers like the Illumina X10 system) and produces a massive amount of data (Illumina 2016, 2017). In contrast, a single array could be done for $50–$100 on cheaper equipment and requires less time and resources.
8 The scientists used imputation to create a shared set of 8,100,752 SNPs.

CHAPTER EIGHT

1 In his influential book, *The Normal and the Pathological*, the French philosopher Georges Canguilhem contends that "The normal is not a static or peaceful, but a dynamic and polemical concept" such that "every value must be earned against an anti-value," which sets up an oppositional relation in defining the norm: "To set a norm (*normer*), to normalize, is to impose a requirement on an existence, a given whose variety, disparity, with regard to the requirement, present themselves as a hostile, even more than an unknown, indeterminant" (Canguilhem 1966, 146).

2 Instituted in 2008, this PRC government program recruits scientists under the age of forty from prestigious foreign laboratories and institutions with generous salaries, research grants, and housing allowances as well as ensuring local residence permits for spouses and children under the country's *hukou* household registration system (Recruitment Program n.d.).

3 The resulting data was then imputed using 1000 Genomes Phase 3 data to find another 6,414,304 ungenotyped SNPs. In this way, each Uyghur subject was translated through a metonymic chain into a set of 7,308,821 SNPs and an accompanying hair shape classification of straight, wavy, or curly data ready for analysis.

4 Another entry was for a paper by Zhao Shilei et al. (2019), who are all affiliated with BIG, on ancestry informative marker SNP identification: "November 1. We developed a computational method (AIMSNP-Tag) that can efficiently analyze large samples of genomic sequences to identify highly-informative SNP panels for ancestry inference. By using this method, we constructed a series of SNP panels for distinguishing individuals with different continental, spatial and minor group origins, which are useful in forensic and medical genetic studies" (Beijing Institute of Genomics 2019, 89).

5 Of the SNPs that were genotyped using the Illumina array chip, 810,648 were used to impute another 6,414,304 SNPs to create a 7,224,952 SNP set for each Uyghur subject (Wu et al. 2018, 13).

6 *Transcription* is the first stage of gene expression when DNA is transcribed into messenger RNA (mRNA). The second stage of gene expression is *translation*, when mRNA is translated into a protein.

7 This SNP data was further imputed with the 1000 Genomes Phase Three data to yield 6,414,304 imputed SNPs that, "were combined with the 810,648 genotyped SNPs for further analyses" (Xiong et al. 2019, 17).

8 I asked the QIMR Berghofer Human Research Ethics Committee (HREC) Secretariat whether the QIMR informed consent procedures specifically mentioned use in forensic genetics to research subjects. Their response seems to indicate the consent procedures do not specifically deal with use in forensic genetic research: "Prof Nick Martin's contributed genome data collected from various genetic studies conducted at QIMR Berghofer over a period of time. Prof Martin collected samples with informed consent, which includes sharing of genome data with collaborators and for unspecified research projects, subject to the new proposed research having ethical approval. Data collected by Prof Martin have been deposited into genome consortiums, which was utilised in at least 2 of the 3 studies. For example, EUROFORGEN-NoE Consortium – Pośpiech et al. 2018 and International Visible Trait Genetics (VisGen) Consortium – Xiong et al. 2019. When data is deposited into the consortium, the consortium undertakes the role of the data custodian, and requires ethics approval to grant access to the data for researchers to conduct new studies" (Eva Baxter, email to author, 30 September 2021).

CHAPTER NINE

1 In a 2015 article in *Forensic Science International: Genetics*, Manfred Kayser advocated the searching of biomedical databases as a legitimate usage in police investigations (Toom et al. 2016, e3). A group of researchers on forensic genetic ethics sharply criticized this advocacy of using medical records "to identify a pool of potential suspects. Such trawling breaches patient confidentiality and trust, two aspects that are central to the idea and work of biomedical science and healthcare."
2 Other grants will include 2017YFC0803503, which differentiates forensic bodily fluid mixtures (see, for example, Pang et al. 2020), while 2017YFC0803505 covers computer vision and facial recognition (see, for example, Fan et al. 2020, 13).
3 Like Liu Fan, Wang did his advanced studies abroad, completing a PhD at University College London in 2009 followed by a joint appointment as a postdoc at Harvard, both of which he completed in 2012. He then moved to the Partner Institute in 2012 (Shanghai Institute of Nutrition and Health n.d.).
4 Tang Kun also did advanced studies abroad. He completed a PhD at the National University of Singapore in 2004. He then did a postdoctoral fellowship from 2004 to 2007 at the Max Planck Institute of Evolutionary Anthropology in Germany under the supervision of Mark Stoneking. He

returned to China and joined the Partner Institute in 2008 (https://web.archive.org/web/20190526160152/http://www.picb.ac.cn/picb/peopleeng.jsp?ntype=pi&ID=11).

5 According to a 2016 Illumina brochure, the US$10 million system promises "Maximum throughput and lowest cost for population-scale whole-genome sequencing" and claims that "The HiSeq X Ten System, a set of 10 HiSeq X instruments, is the first and only platform to deliver a $1000 human genome, generating tens of thousands of high-quality, high-coverage genome sequences" (Illumina 2016, 1).

6 The program provides a competitive salary and a range of benefits including housing support, a one million renminbi starting bonus, and access to Chinese government research grants (Recruitment Program n.d.). As well, "Awardees with Chinese citizenship will be free to settle down in any city of their choice and will not be restricted by his or her original residence registry," which refers to the *hukou* registration system that controls residence status within the country (Recruitment Program n.d.). *Hukou* has been termed an internal social class hierarchy within China since those without local residence status cannot send their children to local schools, are denied access to various government benefits, and can be subjected to forced eviction and other restrictions.

CHAPTER TEN

1 Archive.org has some of the content from Uighurbiz.net. The 26 July 2013 story is still available; however, unfortunately, several articles cited by the UHRP report were not archived. Nonetheless, the available content provides a significant record of the way Uyghur activists used the public forum for dissent, activism, and organizing. The UHRP report has several photos from Uighurbiz.net of the destruction of Mehet Imin's home, including one of front-end loader razing the house and another of the ruins burning.

2 The terahertz body scanners tested in Tumxuk were produced by the Beijing Aerospace Yilian Science and Technology Development Company, which is a subsidiary of the China Aerospace Science and Technology Corporation, a major defence and aerospace production company that manufactures a range of technologies, including missiles for the Chinese military and launch vehicles for the Chinese space program. A China Aerospace Science and Technology Corporation press release claimed this testing in Tumxuk was part of the protection of the 19th Peoples Congress (2017).

3 Another indication of Liu Fan's close working relationship with the Institute of Forensic Science is his software copyright with Peng Fuduan, Jiang Li, and Li Caixia, No. 2018SR922822 entitled "A Software of Intercontinental Source Inference Calculator Based on 27 SNPs" (彭付端，江丽，李彩霞，刘凡，《基于27个SNP基因型的洲际族群来源推断计算器软件》), which is likely related to the twenty-seven AISNP panel developed by Li Caixia and her colleagues as discussed in chapter 4 (Beijing Institute of Genomics 2018, 50).

4 Liu Fan's Erasmus University curriculum vitae from June 2017 states that he received a number of grants, including a Chinese Ministry of Science and Technology (MOST) grant 2017YFC0803501 entitled "Key technology of the suspect characteristics of precise description and identification" project, and describes him as co-owner of grant (Liu 2017). This title is a direct English translation of the above MOST program, so his grant fits within this larger program. This grant is cited in various papers Liu coauthored with the VisiGen Consortium including Xiong et al. (2018) and Pośpiech et al. (2018), discussed earlier. The other grant holder is likely Li Caixia since the grant is cited in several papers including Hao et al. (2018), Jiang et al. (2018a, 2018b), Liu et al. (2018), Li Yi et al. (2018), and Zhao Wenting et al. (2017), in which Li Caixia is a coauthor but Liu Fan is not. This grant is also cited in a series of papers on the new population samples, likely taken in 2017–18, from 715 Uyghurs in the city of Tumxuk in western Xinjiang (Li Yi et al. 2018).

5 Ectodermal is a general term that covers external physical features such as ears, nose, lips, skin, and so on.

CHAPTER ELEVEN

1 As part of my research during that time, I was regularly checking the Coriell Cell Repository website's catalog of populations. On 15 May 2015, I noticed that the Karitiana and Surui cell lines and DNA extract had been removed, but I found that there was still a Google cache of the Coriell catalog webpage listing the Karitiana samples dated 25 March 2015 (https://catalog.coriell.org/0/sections/BrowseCatalog/SDesc_List.aspx?PgId=249&s_id=Karitiana&coll=GM), which suggests the withdrawal occurred in April or early May 2015.

2 This press release is part of an information package that Yves Moreau sent to the Office of the High Commissioner for Human Rights (Moreau n.d.)

3 I cooperated with Ailsa Chang in the preparation of the National Public Radio reports.

4 Byler (2018) did field research in Xinjiang during the 2010s and has been an outspoken critic of the PRC government's repression in Xinjiang.

5 The archive.org 21 February 2019 capture of Budowle's University of North Texas biography had this sentence on his list of affiliations: "member of the Academic Committee, Key Laboratory of Forensic Genetics at the Institute of Forensic Science of the Ministry of Public Security in Beijing, China." This sentence is deleted in the 22 February 2019 version on archive.org (https://www.unthsc.edu/bios/budowle/).

6 Richardson mentioned this incident to me in a 20 February 2019 email.

7 Li Caixia and her colleagues also published another paper about their AISNP research in *Forensic Science International: Genetics* on 10 May 2019. The paper included Uyghur subjects as well as a large number from Kidd Labs (Ren et al. 2019, 147).

8 These important efforts lie outside the scope of this book, which is intended as a qualitative case study of prominent figures and institutions rather than an exhaustive survey of forensic genetic research on PRC minorities.

9 Marco Rubio's bill, the Uyghur Human Rights Policy Act, worked its way through the Congressional hearings process and was finally passed into law in 2020.

10 The spokesperson Zhao Lijian is well known for his "wolf warrior" diplomatic style (Huang 2020). *Wolf Warrior* is the title of a 2015 film and its 2017 sequel, Rambo-type action films about an elite Chinese special forces operator fighting US mercenaries, which were major successes in China, particularly the sequel.

11 Given the publication in November 2019 of the article by Xiong et al., I contacted the editor, Andy Collings, in February 2020 and, in March 2020, provided a twenty-three page information kit detailing the linkages between Liu Fan and other coauthors with the Ministry of Public Security, the problematic ethics of the article, and how Liu et al. (2018) was cited in the 2019 BIG annual report as advancing the Ministry of Public Security's goals. Reporters from RTL and *Follow the Money* in the Netherlands did their own investigations into the VisiGen Consortium usage of the Xinjiang Uyghur Study along with the Tumxuk Uyghur papers (I was not involved), the results of which were published in September 2021 and October 2021. After the publications, I contacted them and they shared their questions and Erasmus University Medical Centre's responses with me.

References

Note: Wherever possible I have made use of permanent web links from archive.org. In keeping with Chinese naming conventions, Chinese names are formatted as surname followed by given name to prevent confusion between Chinese and Western articles.

Agamben, Giorgio. 1998. *Homo Sacer: Sovereign Power and Bare Life.* Stanford, CA: Stanford University.

Ahuriri-Driscoll, Annabel, Juan Tauri, and Johanna Veth. 2021. "Māori Views of Forensic DNA Evidence: An Instrument of Justice or Criminalizing Technology?" *New Genetics and Society* 40 (3): 249–66. DOI: 10.1080/14636778.202 0.1829463.

Alexis-Martin, Becky. 2019. "The Nuclear Imperialism Necropolitics Nexus: Contextualizing Chinese Uyghur Oppression in Our Nuclear Age." *Eurasian Geography and Economics* 60 (2): 152–76.

ALFRED. n.d. "The ALlele FREquency Database." Yale University School of Medicine. https://web.archive.org/web/20220608120901/https://alfred.med.yale.edu/alfred/index.asp.

– n.d. "ALFRED Populations file." Allele Frequency Database, Yale University School of Medicine. https://web.archive.org/web/20150914163152/http://alfred.med.yale.edu/alfred/FlatFileFormat/alfredPops.zip.

– 2018. "Welcome to the New FROG-kb." Allele Frequency Database, Yale University School of Medicine. https://web.archive.org/web/20181004154119/http://frog.med.yale.edu/FrogKB.

– 2019a. "AISNP Sets." Allele Frequency Database, Yale University School of Medicine. https://web.archive.org/web/20211107024139/https://alfred.med.yale.edu/alfred/AISnpSets.asp.

– 2019b. "Population Information: Karitiana, PO000028K." Allele Frequency Database, Yale University School of Medicine. https://web.archive.org/web/20211025231945/https://alfred.med.yale.edu/alfred/recordinfo.asp?UNID=PO000028K.
– 2019c. "Population Information: Mbuti, PO000006G." Allele Frequency Database, Yale University School of Medicine. https://web.archive.org/web/20220724175320/https://alfred.med.yale.edu/alfred/recordinfo.asp?UNID=PO000006G.
– 2019d. "Population Information: Surui, SA004725S." Allele Frequency Database, Yale University School of Medicine. https://web.archive.org/web/20220724175512/https://alfred.med.yale.edu/alfred/recordinfo.asp?UNID=PO000014F.
– 2019e. "Population Information: Ticuna, PO000027J." Allele Frequency Database, Yale University School of Medicine. https://web.archive.org/web/20220810233309/https://alfred.med.yale.edu/alfred/recordinfo.asp?UNID=PO000027J.
– 2019f. "Population Information: Uyghur, PO000399V." Allele Frequency Database, Yale University School of Medicine. https://web.archive.org/web/20190308063937/https://alfred.med.yale.edu/Alfred/recordinfo.asp?UNID=PO000399V.
– 2019g. "Population Information: Yavapai, PO000893U." Allele Frequency Database, Yale University Medical School. https://web.archive.org/web/20180326022933/https://alfred.med.yale.edu/alfred/recordinfo.asp?UNID=PO000893U.
– 2019h. "Sample Information: Karitiana, SA004725S." Allele Frequency Database, Yale University School of Medicine. https://web.archive.org/web/20211025231945/https://alfred.med.yale.edu/alfred/recordinfo.asp?UNID=PO000028K.
– 2019i. "Sample Information: Surui, SA004727U." Allele Frequency Database, Yale University School of Medicine. https://web.archive.org/web/20211212015423/https://alfred.med.yale.edu/alfred/sampleDescrip.asp?sampleID=%274727%29%27.
– 2019j. "Sample Information: Ticuna, SA004728V." Allele Frequency Database, Yale University School of Medicine. https://web.archive.org/web/20200823202445/https://alfred.med.yale.edu/alfred/sampleDescrip.asp?sampleID=%274728%27%29.
Anand, Dibyesh. 2019. "Colonization with Chinese Characteristics: Politics of (In)Security in Xinjiang and Tibet." *Central Asian Survey*, 38 (1): 129–47.
Anderson, Christopher. 1991. "DNA Fingerprinting Discord." *Nature* 354:500.

Applied Biosystems. 2017. HIDS Vienna Conference Brochure. Big Think Marketing Agency (UK). https://web.archive.org/web/20170610192428/http://hidsvienna.ebrochure.bigthinkagency.co.uk/files/assets/common/downloads/publication.pdf.

Aradau, Claudia, and Rens Van Munster. 2007. "Governing Terrorism Through Risk: Taking Precautions, (Un)Knowing the Future." *European Journal of International Relations* 13 (1): 89–115. https://doi.org/10.1177/1354066107074290.

Arbour, Laura, and Doris Cook. 2006. "DNA on Loan: Issues to Consider when Carrying Out Genetic Research with Aboriginal Families and Communities." *Community Genetics* 9: 153–60.

– 2007. "Domestic War: Locke's Concept of Prerogative and Implications for US Wars Today." *Polity* 39:1–28.

Aronson, Jay D. 2007. *Genetic Witness: Science, Law, and Controversy in the Making of DNA Profiling*. New Brunswick, NJ: Rutgers University Press.

Atwood, Roger. 1988. "Virus in Amazon Indians May Hold Clues to AIDS Origin." *Reuters News*, 20 May 1988. Factiva Database Document lba0000020011203dk5k01q01.

Australian Strategic Policy Institute. 2022. Xinjiang Data Project. Australian Strategic Policy Institute. https://xjdp.aspi.org.au/.

Baird, Duncan Martin, Nicola Jane Royle, and Alec John Jeffreys. 2001. "Method for Characterising Variability in Telomere DNA by PCR." US patent 6235468. https://web.archive.org/web/20230112155821/https://patentimages.storage.googleapis.com/8e/d6/e6/1b546f37afa3e4/US6235468.pdf.

Ballantyne, Jack, George Sensabaugh, and Jan A. Witkowski. 1989. "Preface." In *DNA Technology and Forensic Science*, edited by Jack Ballantyne, George Sensabaugh, and Jan A. Witkowski, xi–xiii. Cold Spring Harbor, NY: Cold Spring Harbor Laboratory.

Barker, Joanne. 2004. "The Human Genome Diversity Project." *Cultural Studies* 18 (4): 571–606.

Bartram, Isabelle, Tino Plümecke, and Susanne Schultz. 2021. "Genetic Racial Profiling: Extended DNA Analyses and Entangled Processes of Discrimination." *Science & Technology Studies* 35 (3): 44–69. DOI: 10.23987/sts.101384. https://web.archive.org/web/20220620070208if_/https://sciencetechnologystudies.journal.fi/article/download/101384/65867/213339.

Beijing Institute of Genomics. 2016. *2015 Annual Report*. Beijing: Beijing Institute of Genomics. https://web.archive.org/web/20200626182911/https://www.naozhouzhen.com/zt/jyzsnb/201612/P020161202609978042577.pdf.

– 2017. *2016 Annual Report*. Beijing: Beijing Institute of Genomics. https://web.archive.org/web/20200626184601/http://www.big.ac.cn/zt/jyzsnb/201701/P020170126471477832268.pdf.

– 2018. *2017 Annual Report*. Beijing: Beijing Institute of Genomics. https://web.archive.org/web/20200626184804/http://www.big.ac.cn/zt/jyzsnb/201802/P020180202588900853152.pdf.

– 2019. *2018 Annual Report*. Beijing: Beijing Institute of Genomics. https://web.archive.org/web/20191208214752/http://www.big.ac.cn/zt/jyzsnb/201903/P020190328408341082136.pdf.

– 2020. *2019 Annual Report*. Beijing: Beijing Institute of Genomics. https://web.archive.org/web/20200430043128/http://www.big.cas.cn/zt/jyzsnb/202004/P020200410618970867240.pdf.

– 2021. *2020 Annual Report*. Beijing: Beijing Institute of Genomics. https://web.archive.org/web/20210705204442if_/http://www.big.cas.cn/ztbd/nb/202102/P020210207508251141016.pdf.

– n.d. "Faculty: Liu Fan." Beijing Institute of Genomics. https://web.archive.org/web/20160728160935/http://sourcedb.big.cas.cn/yw/pl/fs/201504/t20150416_4338247.html.

Beijing Institute of Genomics Datacenter. 2019a. "Accession SAMC051815, Sample Name uyg1." Beijing: Beijing Institute of Genomics Datacenter. https://web.archive.org/web/20190304042308/http://bigd.big.ac.cn/biosample/browse/SAMC051815.

– 2019b. "The Candidate Gene Analysis for Human Facial Morphology in Eurasian Population." Beijing: Beijing Institute of Genomics Datacenter. https://web.archive.org/web/20190304030626/http://bigd.big.ac.cn/bioproject/browse/PRJCA001171.

– 2019c. "Genome Variation Map, Data Submission." Beijing: Beijing Institute of Genomics Datacenter. https://web.archive.org/web/20190311123949/http://bigd.big.ac.cn/gvm/getSample?dataId=152.

– 2019d. "Title Predicting Adult Height from DNA Variants in a European-Asian Admixed Population." Beijing: Beijing Institute of Genomics Datacenter. https://web.archive.org/web/20190402024106/http://bigd.big.ac.cn/bioproject/browse/PRJCA001236.

Bell, Colleen. 2006. Surveillance Strategies and Populations at Risk: Biopolitical Governance in Canada's National Security Policy." *Security Dialogue* 37 (2): 147–65. https://doi.org/10.1177/0967010606066168.

– 2013. "Grey's Anatomy Goes South: Global Racism and Suspect Identities in the Colonial Present." *The Canadian Journal of Sociology* 38 (4): 465–86.

Bergström, Anders, Shane A. McCarthy, Ruoyun Hui, Mohamed A. Almarri, Qasim Ayub, Petr Danecek, Yuan Chen, et al. 2020. "Insights into Human

Genetic Variation and Population History from 929 Diverse Genomes." *Science* 367 (6484). https://doi.org/10.1126/science.aay5012.

Berube, Margery, ed. 2005. *Webster's II New College Dictionary*. Boston: Houghton Mifflin.

Bidcenter.com.cn. 2017. "第三师五十一团法制教育培训中心建设项目 [工程施工]" [Construction of Legal Education and Training Center of the 51st Regiment of the Third Division (Engineering Construction)]. Bidcenter.com.cn (PRC commercial website). https://web.archive.org/web/20180617184622/https://www.bidcenter.com.cn/newscontent-36968273-1.html.

Bierwiaczonek, Bogusław. 2020. "Figures of Speech Revisited: Introducing Syntonymy and Syntaphor." In *Figurative Meaning Construction in Thought and Language*, edited by Annalisa Baicchi, 225–51. Amsterdam: John Benjamins Publishing Company.

Biesecker, Leslie G., Joan E. Bailey-Wilson, Jack Ballantyne, Howard Baum, Frederick R. Bieber, Charles Brenner, Bruce Budowle, et al. 2005. "DNA Identifications after the 9/11 World Trade Center Attack." *Science* 310 (5751): 1122–3.

Bigo, Didier. 2008. "Security, Exception, Ban and Surveillance." In *Theorizing Surveillance: The Panopticon and Beyond*, edited by David Lyon, 46–68. Cullompton, UK: Willan Publishing.

Bingtuan Satellite Television. 2017. "兵团去极端化法治宣讲团深入三师重点连队、中小学校、法治教育转化中心宣讲" [The Bingtuan De-Extremeification Legal Rule Publicity Team went Deep into the Key Companies of the Three Divisions, Primary and Secondary Schools, and the Rule of Law Education Transformation Center.] Bingtuan Satellite Television story on Sohu.com (PRC). https://web.archive.org/web/20180528112750/http://www.sohu.com/a/141338747_233637.

Black, Francis L. 1991. "Reasons for Failure of Genetic Classifications of South Amerind Populations." *Human Biology* 63 (6): 763–74.

Bliss, Catherine. 2012. *Race Decoded: The Genomic Fight for Social Justice*. Stanford, CA: Stanford University Press.

Borofsky, Robert. 2005. *Yanomami: The Fierce Controversy and What We Can Learn from It*. Berkeley: University of California Press.

Bourne, Mike, Heather Johnson, and Debbie Lisle. 2015. "Laboratizing the Border: The Production, Translation and Anticipation of Security Technologies." *Security Dialogue* 46 (4): 307–25.

Bowcock, Anne and Luca Cavalli-Sforza. 1991. "The Study of Variation in the Human Genome." *Genomics* 11: 491–8.

Braatz, Timothy. 2003. *Surviving Conquest: A History of the Yavapai Peoples*. Lincoln: University of Nebraska Press.

Brdar, Mario. 2015. "Metonymic Chains and Synonymy." FLUMINENSIA 27 (2): 83–101.
Brophy, David. 2016. *Uyghur Nation: Reform and Revolution on the Russia-China Frontier.* Cambridge, MA: Harvard University Press
Brown, Matthew. 2017. "China Moves to Expand DNA Testing in Muslim Region." *Associated Press*, 16 May 2017. https://web.archive.org/web/20180915131850/https://apnews.com/17fc345187e74e72a527f1c2c5b1c3ca.
Browne, Simone. 2015. *Dark Matters: On the Surveillance of Blackness.* Chapel Hill, NC: Duke University Press.
Budowle, Bruce. 2015a. "Perspectives on the Future of Forensics Genetics – HIDS 2015." Human Identification Solutions Conference, Madrid, Spain, 3 March. YouTube video. https://www.youtube.com/watch?v=uVIl1XR56t8.
– 2015b. "Perspectives on the Future of Forensic Genetics." Human Identification Solutions Conference, Madrid, Spain, 3 March. https://web.archive.org/web/20190530172015/https://www.thermofisher.com/content/dam/LifeTech/Documents/PDFs/1-1_CO014018_Future%20Madrid%20March%202015%20Light%20Bruce%20Budowle_FINAL_March%202015.pdf.
Buschschlüter, Vanessa. 2022. "'Man of the Hole': Last of His Tribe Dies in Brazil." *BBC News.com*, 29 August. https://web.archive.org/web/20220901045625/https://www.bbc.com/news/world-latin-america-62712318.
Butler, John M. 2010. *Fundamentals of Forensic DNA Typing.* San Diego, CA: Elsevier Academic Press.
Butler, Judith. 2002. "What Is Critique? On Foucault's Virtue." In *The Political*, edited by David Ingram, 212–26. Malden, MA: Blackwell.
Butler, Judith, and Athena Athanasiou. 2013. *Dispossession: The Performative in the Political.* Cambridge, UK: Polity Press.
Byler, Darren T. 2018. "Spirit Breaking: Uyghur Dispossession, Culture Work and Terror Capitalism in a Chinese Global City." PhD diss., University of Washington.
– 2022. *Terror Capitalism: Uyghur Dispossession and Masculinity in a Chinese City.* Durham, NC: Duke University Press.
Cage Report. 2023. "Verogen, Inc." Commercial & Government Entity Report (US) website. https://web.archive.org/web/20230115180159/https://cage.report/DUNS/080948871.
Callon, Michel. 1986. "Some Elements of a Sociology of Translation: Domestication of the Scallops and the Fisherman of St. Brieuc Bay." In *Power, Action and Belief: A New Sociology of Knowledge?* edited by J. Law, 196–233. London: Routledge.

Campbell, Joseph. 2004. *The Hero with a Thousand Faces*. Princeton, NJ: Princeton University Press.

CANDELA. 2011. "General Memorandum of Understanding (GMOU) (Version 1 – July 2011)." Cayetano Heredia University (Peru). https://web.archive.org/web/20161021140517/http://www.upch.edu.pe/upchvi/durin/host/internacionales/CONSORTIUM%20CANDELA.pdf.

– 2020. "CANDELA: Consortium for the Analysis of the Diversity and Evolution of Latin America." University College London. https://web.archive.org/web/20200809190214/https://www.ucl.ac.uk/biosciences/node/18043/home-link.

Canguilhem, Georges. (1966) 1978. *On the Normal and the Pathological*. Dordrecht: Reidel.

Carlson, Jean M., and John Doyle. 2002. "Complexity and Robustness." *Proceedings of the National Academy of Sciences* (US) 99:2538–45.

Castiglione, C.M., A.S. Deinard, W.C. Speed, G. Sirugo, H.C. Rosenbaum, Y. Zhang, D.K. Grandy, et al. 1995. "Evolution of Haplotypes at the DRD2 Locus." *American Journal of Human Genetics* 57:1445–56. https://web.archive.org/web/20191112074208/https://europepmc.org/backend/ptpmcrender.fcgi?accid=PMC1801414&blobtype=pdf.

Cavalli-Sforza, L.L. 2005. "The Human Genome Diversity Project: Past, Present, and Future." *Nature Review Genetics* 6 (April): 333–40.

Cavalli-Sforza, L.L, A.C. Wilson, C.R. Cantor, R.M. Cook-Deegan, and M.C. King. 1991. "Call for a Worldwide Survey of Human Genetic Diversity: A Vanishing Opportunity for the Human Genome Project." *Genomics* 11:490–1.

Cavalli-Sforza, L.L., and Henry T. Greely. 1993. "Human Genome Diversity Project – Organizers' Response." Nativenet email list, 8 July 1993. https://web.archive.org/web/20080830091346/http://nativenet.uthscsa.edu/archive/nl/9307/0046.html.

Chacko, Priya, and Kanishka Jayasuriya. 2018. "A Capitalising Foreign Policy: Regulatory Geographies and Transnationalised State Projects." *European Journal of International Relations* 24 (1): 82–105. https://doi.org/10.1177/1354066117694702.

Chakraborty, Ranajit, and Kenneth K. Kidd. 1991. "The Utility of DNA Typing in Forensic Work." *Science*, 254:1735–9

Chamayou, Grégoire. 2012. *Manhunts: A Philosophical History*. Princeton, NJ: Princeton University Press.

Chang, Ailsa, and Nick Fountain. 2019a. "Episode 924: Stuck in China's Panopticon." *Planet Money* podcast, National Public Radio, 5 July 2019. https://web.archive.org/web/20200411072344/https://www.npr.org/transcripts/738949320.

– 2019b. “How Americans – Some Knowingly, Some Unwittingly – Helped China’s Surveillance Grow.” *All Things Considered* news program, National Public Radio, 18 July 2019. https://web.archive.org/web/20190726040644/https://www.npr.org/2019/07/18/743211959/how-americans-some-knowingly-some-unwittingly-helped-chinas-surveillance-grow.

Charles, Dan. 1992. “Courtroom Battle over Genetic Fingerprinting.” *New Scientist* 134 (1817): 10. https://web.archive.org/web/20160413022812/https://www.newscientist.com/article/mg13418171-700-courtroom-battle-over-genetic-fingerprinting/.

Chiappino, Jean. 1975. *The Brazilian Indigenous Problems and Policy; The Example of the Aripuana Indigenous Park*. International Working Group for Indigenous Affairs. https://web.archive.org/web/20200702225834/https://www.iwgia.org/images/publications/0196_19Aripuana.pdf.

China Aerospace Science and Technology Corporation. 2017. “航天科技集团公司以航天技术全力保障党的十九大胜利召开” [China Aerospace Science and Technology Group Co., Ltd. is Making Every Effort to Assure the Victory of the 19th National Congress of the Communist Party of China with Aerospace Technology], press release, 18 October 2017. https://web.archive.org/web/20200920074621/http://www.spacechina.com/n25/n2014789/n2014804/c1755637/content.html.

China CNTC International Tendering Co., Ltd. 2017. “招标文件. 项目名称：公安部物证鉴定中心援建新疆兵团第三师公安局DNA实验室设备、耗材和实验室综合管理系统采购项目招标编号” [Bidding Documents. Project Name: The Ministry of Public Security’s Material Evidence Appraisal Center assisted in the construction of the DNA laboratory equipment, consumables and laboratory comprehensive management system procurement project of the Public Security Bureau of the Third Division of Xinjiang Corps]. PRC Ministry of Finance, bidding documents TC170F629. https://web.archive.org/web/20200609205543/http://www.ccgp.gov.cn/oss/download?uuid=7F4E64F9834C5BACDD75BD10478CCE.

China Postdoctoral Science Foundation. 2016. “公安部物证鉴定中心2016年博士后招收简章. 来源：发布日期：2016年5月30日” [2016 Postdoctoral Recruitment.” Ministry of Public Security’s Physical Evidence Identification Center, 30 May 2016]. https://web.archive.org/web/20190125165120/http://www.chinapostdoctor.org.cn/WebSite/program/Info_Show.aspx?InfoID=1e8d0f4e-f9ec-48ed-a167-e14440a16ce1.

Chinese Academy of Sciences – Max Planck Society Partner Institute of Computational Biology. 2007. *Research Report 2005–2007*. Shanghai: Partner Institute of Computational Biology. http://web.archive.org/web/20170627125256/http://www.picb.ac.cn/picb/report2007.pdf.

– 2009. *Research Report 2005–2009*. Shanghai: Partner Institute of Computational Biology. http://web.archive.org/web/20170627140502/http://www.picb.ac.cn/picb/report2009.pdf.

– 2012. *Research Report 2009–2012*. Shanghai: Partner Institute of Computational Biology. http://web.archive.org/web/20170627132146/http://www.picb.ac.cn/picb/report2012.pdf.

– 2014. *Research Report 2012–2014*. Shanghai: Partner Institute of Computational Biology. http://web.archive.org/web/20170627143602/http://www.picb.ac.cn/picb/report2014.pdf.

– 2017. *Research Report 2014–2017*. Shanghai: Partner Institute of Computational Biology. https://ia904700.us.archive.org/15/items/picb-research-report-2014-2017/PICB%20Research%20Report%202014-2017.pdf.

Chinese National Intellectual Property Administration. 2020. "Patent Law of the People's Republic of China." Chinese National Intellectual Property Administration. https://web.archive.org/web/20200628115308/http://english.sipo.gov.cn/lawpolicy/patentlaws regulations/915574.htm.

Chu Jiayou. 2001. "Chinese Human Genome Diversity Project: A Synopsis." In *Genetic, Linguistic and Archaeological Perspectives on Human Diversity in Southeast Asia*, edited by Li Jin, Mark Seielstad, and Chunjie Xiao, 95–100. New Jersey: World Scientific.

Chu Jiayou, Huang W., Kuang S.Q., Wang J.M, Xu J.J., Chu Z.T., Yang Z.Q., et al. 1998. "Genetic Relationship of Populations in China." *Proceedings of the National Academy of Sciences of the United States of America* 95 (20): 11763–8.

Chun, W. H. K. 2009. "Introduction: Race and/as Technology; or, How to Do Things to Race." *Camera Obscura: Feminism, Culture, and Media Studies* 24 (1): 6–35.

Clarke, Michael. 2021. "Settler Colonialism and the Path toward Cultural Genocide in Xinjiang." *Global Responsibility to Protect* 13 (1): 9–19. https://doi.org/10.1163/1875–984X-13010002.

Cold Spring Harbor Laboratory. 2017. "Banbury Center Highlights: DNA Technology and Forensic Science." https://www.cshl.edu/wp-content/uploads/2017/05/BanburyCenter_1988_DNA-technology-forensic-science.pdf.

Coleman, Howard, and Eric Swenson. 1994. *DNA in the Courtroom: A Trial Watcher's Guide*. Seattle, WA: GeneLex Press.

Committee to Protect Journalists. n.d. "Ilham Tohti." https://web.archive.org/web/20220528040728/https://cpj.org/data/people/ilham-tohti/.

Conant, E. 2009. "Terror: The Remains of 9/11 Hijackers." *Newsweek*, 2 January 2009. https://web.archive.org/web/20140529160255/http://www.newsweek.com/terror-remains-911-hijackers-78327.

Congressional Executive Commission on China. 2019. *Annual Report 2019*. U.S. House of Representatives, 18 November 2019. https://web.archive.org/web/20200808050238/https://www.cecc.gov/sites/chinacommission.house.gov/files/CECC%202019%20Annual%20Report.pdf.

Cooren, François. 2000. *The Organizing Property of Communication*. Amsterdam: John Benjamins Publishing Company.

Cooren, François, and Gail T. Fairhurst. 2004. "Speech Timing and Spacing: The Phenomenon of Organizational Closure." *Organization* 11:793–824.

Cooren, François, and James R. Taylor. 1997. "Organization as an Effect of Mediation: Redefining the Link between Organization and Communication." *Communication Theory* 7 (3): 219–60.

– 2000. "Association and Disassociation in an Ecological Controversy: The Great Whale Case." In *Technical Communication, Deliberative Rhetoric, and Environmental Discourse: Connections and Directions*, edited by Nancy W. Coppola and Bill Karis, 171–90. Stanford, CT: Ablex Publishing Corporation.

– 2006. "Making Worldview Sense, and Paying Homage, Retrospectively, to Algirdas Greimas." In *Communication as Organizing: Empirical and Theoretical Explorations in the Dynamic of Text and Conversation*, edited by Francois Cooren, James R. Taylor, and Elizabeth Van Every, 115–38. Mahwah, NJ: Lawrence Erlbaum Associates.

Coulthard, Glen S. 2014. *Red Skins White Masks: Rejecting the Colonial Politics of Recognition*. Minneapolis: University of Minnesota Press.

Cover, Robert M. 1986. "Violence and the Word." Faculty Scholarship Series. Paper 2708. https://web.archive.org/web/20220107072314if_/https://openyls.law.yale.edu/bitstream/handle/20.500.13051/2050/Violence_and_the_Word.pdf?sequence=2&isAllowed=y.

Crosby, Andrew, and Jeffrey Monaghan. 2012. "Settler Governmentality in Canada and the Algonquins of Barriere Lake." *Security Dialogue* 43 (5): 421–38.

– 2016. "Settler Colonialism and the Policing of Idle No More." *Social Justice* 43 (2): 37–57.

Cultural Survival. 2004. "Brazil: Indians' Genetic Material Sold on Internet." *Cultural Survival*. https://web.archive.org/web/20180920011929/http://www.culturalsurvival.org/news/brazil-indians-genetic-material-sold-internet.

Cyranoski, David. 2017. "China Expands DNA Data Grab in Troubled Western Region." *Nature* 545 (7655): 395–6.

Danjoh, I., R. Shirota, T. Hiroyama, and Y. Nakamura. 2013. "Dominant Expansion of a Cryptic Subclone with an Abnormal Karyotype in B Lymphoblastoid Cell Lines During Culture." *Cytogenet Genome Research* 139 (2): 88–96. doi:10.1159/000343757.

de Chadarevian S. 2015. "Human Population Studies and the World Health Organization." *Dynamis* 35 (2): 359–88. doi:10.4321/s0211-95362015000200005.

de la Puente, María, Jorge Ruiz-Ramírez, Adrián Ambroa-Conde, Catarina Xavier, Jacobo Pardo-Seco, Jose Álvarez-Dios, Ana Freire-Aradas, et al., on behalf of the VISAGE Consortium. 2021. "Development and Evaluation of the Ancestry Informative Marker Panel of the VISAGE Basic Tool." *Genes* 12 (8): 1284.

Dean, Mitchell. 1999. *Governmentality: Power and Rule in Modern Society*. London: Sage Publication.

– 2007. *Governing Societies: Political Perspectives on Domestic and International Rule*. Berkshire, CA: Open University Press.

Dean, Mitchell, and Kaspar Villadsen. 2016. *State Phobia and Civil Society: The Political Legacy of Michel Foucault*. Stanford, CA: Stanford University Press.

Debrix, François, and Alexander D. Barder. 2009. "Nothing to Fear but Fear: Governmentality and the Biopolitical Production of Terror." *International Political Sociology* 3 (4): 398–413.

Denroche, Charles. 2015. *Metonymy and Language: A New Theory of Linguistic Processing*. London: Routledge.

Dillon, Michael. 2008. "Security, Race and War." In *Foucault on Politics, Security and War*, edited by Michael Dillon, and Andrew W. Neal, 166–96. London: Palgrave Macmillan. https://doi.org/10.1057/9780230229846_9.

du Plessis, Gitte. 2015. "Hunting as Techniques of Governing: Chamayou's Manhunts, and Fassin's Enforcing Order." *Theory & Event* 18 (2). muse.jhu.edu/article/578642.

Duster, Troy. 2006. "Explaining Differential Trust of DNA Forensic Technology: Grounded Assessment or Inexplicable Paranoia?" *J Law Med Ethics* 34 (2): 293–300.

– 2015. "A Post-Genomic Surprise." *The British Journal of Sociology* 66:1–27.

Early, John D., and John F. Peters. 2000. *The Xilixana Yanomami of the Amazon: History, Social Structure, and Population Dynamics*. Gainesville: University Press of Florida.

Eikelenboom, Siem, Annebelle de Bruijn, Dorine Booij. 2021. "Hoe Nederlandse onderzoeksinstituten de Chinese politiestaat helpen" [How

Dutch research institutes are helping the Chinese police state]. *Follow the Money* investigative journalism website (Holland), 6 October 2021. https://web.archive.org/web/20221213214852/https://www.ftm.nl/artikelen/nauwe-banden-erasmus-en-chinese-politie.

eLife. 2021. "Novel Genetic Loci Affecting Facial Shape Variation in Humans" (webpage for Xiong et al.). https://web.archive.org/web/20210411152343/https://elifesciences.org/articles/49898.

England, Ryan, and Sallyann Harbison. 2020. "A Review of the Method and Validation of the MiSeq FGx™ Forensic Genomics Solution." *WIREs Forensic Sci.* 2 (1): e1351.

Epstein, Charlotte. 2007. "Guilty Bodies, Productive Bodies, Destructive Bodies: Crossing the Biometric Borders." *International Political Sociology* 1 (2): 149–64. https://doi.org/10.1111/j.1749-5687.2007.00010.x.

Erasmus University Medical Center. 2021. "DNA Research by Erasmus MC Cannot Contribute to Suppression of Uyghurs." 6 October 2021. https://web.archive.org/web/20211014225812/https://amazingerasmusmc.nl/biomedisch/dna-onderzoek-erasmus-mc-kan-niet-bijdragen-aan-onderdrukking-oeigoeren/.

Esposito, Roberto. 2010. *Communitas: The Origin and Destiny of Community*. Translated by Timothy Campbell. Stanford, CA: Stanford University Press.

EuroForGen. 2016. "Seminar Series on Genetics, Technology, Security and Justice at Northumbria University 2015–2017." (European Union website). https://web.archive.org/web/20161107041512/https://www.euroforgen.eu/news/seminar-series-on-genetics-technology-security-and-justice-at-northumbria-university/.

– 2019a. "About EUROFORGEN-NoE." (European Union website). https://web.archive.org/web/20190708042006/https://www.euroforgen.eu/.

– 2019b. "Erasmus MC University Medical Center Rotterdam." (European Union website). https://web.archive.org/web/20190708070445/https://www.euroforgen.eu/the-group/consortium/erasmus-mc-university-medical-center-rotterdam/.

European Commission. 2015. "Explanatory Note on Potential Misuse of Research." European Commission. https://web.archive.org/web/20200810200840/https://ec.europa.eu/research/participants/portal/doc/call/h2020/h2020-drs-2015/1645162-explanatory_note_on_potential_misuse_of_research_en.pdf.

– 2017. "European Forensic Genetics Network of Excellence." European Commission. https://web.archive.org/web/20220522151632/https://cordis.europa.eu/project/id/285487.

– 2022. "Horizon 2020: Visible Attributes through Genomics: Broadened Forensic Use of DNA for Constructing Composite Sketches from Traces." https://web.archive.org/web/20230306002634/https://cordis.europa.eu/project/id/740580.
– n.d. "Secure Societies – Protecting Freedom and Security of Europe and Its Citizens." https://wayback.archive-it.org/12090/20220124144231/https://ec.europa.eu/programmes/horizon2020/en/h2020-section/secure-societies-%E2%80%93-protecting-freedom-and-security-europe-and-its-citizens.
Fairhurst, Gail T. 2007. *Discursive Leadership: In Conversation with Leadership Psychology*. Thousand Oaks: Sage.
Fan Zhenfeng, Hu Xiyuan, Chen Chen, Wang Xiaolian, and Peng Silong. 2020. "Facial Image Super-Resolution Guided by Adaptive Geometric Features." *EURASIP Journal on Wireless Communications and Networking* no. 1: 1–15.
Fanon, Frantz. 1963. *The Wretched of the Earth*. New York: Grove Press.
– 1967. *Towards an African Revolution*. New York: Grove Press.
Faye, David J. 2004. "Bioprospecting, Genetic Patenting and Indigenous Populations." *The Journal of World Intellectual Property* 7:401–28. https://doi.org/10.1111/j.1747–1796.2004.tb00213.x.
Federal Bureau of Investigation. 2020. "Frequently Asked Questions on CODIS and NDIS." US Federal Bureau of Investigation. https://web.archive.org/web/20200812212250/https://www.fbi.gov/services/laboratory/biometric-analysis/codis/codis-and-ndis-fact-sheet.
– 2021. "Laboratory Services." US Federal Bureau of Investigation. https://web.archive.org/web/20210813085752/https://www.fbi.gov/services/laboratory.
Finnegan, Ciara. 2020. "The Uyghur Minority in China: A Case Study of Cultural Genocide, Minority Rights and the Insufficiency of the International Legal Framework in Preventing State-Imposed Extinction." *Laws* 9 (1): 1. https://doi.org/10.3390/laws9010001.
Fondation Jean Dausset. n.d.a. "HGDP-CEPH Human Genome Diversity Cell Line Panel." Fondation Jean Dausset-CEPH. https://web.archive.org/web/20131213213147/http://www.cephb.fr/en/hgdp/diversity.php.
– n.d.b. "Geographic Origins." Fondation Jean Dausset-CEPH. https://web.archive.org/web/20131030203646/http://www.cephb.fr/en/hgdp/table.php.
Foucault, Michel. 2003. *Society Must Be Defended: Lectures at the College De France, 1975–76*. Edited by Mauro Bertani and Alessandro Fontana, translated by David Macey. New York: Picador.
Friedlaender, Jonathan, and Joanna Radin. 2009. *From Anthropometry to Genomics: Reflections of a Pacific Fieldworker*. Bloomington, IN: Universe.
Friedlaender, J.S., L.A. Sgaramella-Zonta, K.K. Kidd, L.Y. Lai, P. Clark, and R.J. Walsh. 1971. "Biological Divergences in South-Central Bougainville:

An Analysis of Blood Polymorphism Gene Frequencies and Anthropometric Measurements Utilizing Tree Models, and a Comparison of These Variables with Linguistic, Geographic, and Migrational 'Distances.'" *American Journal of Human Genetics* 23 (3): 253–70.

Frye v. United States. 1923. 293 F. 1013 (d.c. Cir. 1923). Quotation from Harvard Law School. https://web.archive.org/web/20180103133331/http://www.law.harvard.edu/publications/evidenceiii/cases/frye.htm.

Fudan University. 2017. "Jin Li." Fudan University. https://web.archive.org/web/20180724040202/http://www.fudan.edu.cn/entries/view/965/.

Fullwiley, Dorothy. 2014. "The Contemporary Synthesis: When Politically Inclusive Genomic Science Relies on Biological Notions of Race." *Isis* 105:803–14.

– 2015. "Race, Genes, Power." *The British Journal of Sociology*, 66:36–45.

Galis, Vasilis, and Francis Lee. 2014. "A Sociology of Treason: The Construction of Weakness." *Science, Technology, & Human Values* 39 (1): 154–79.

Gao Shan, Ho Shan, and Jilil Kashgari. 2016. "China: Xinjiang Residents Must Give DNA, Voice-Print for Passports." *Radio Free Asia* (US government funded organization), 8 June 2016. https://web.archive.org/web/20170103103730/https://www.refworld.org/docid/5760fc35c.html.

Garfield, Seth. 2001. *Indigenous Struggle at the Heart of Brazil: State Policy, Frontier Expansion, and the Xavante Indians, 1937–1988*. Durham, NC: Duke University Press.

Garrett, Brandon L., and Erin Murphy. 2013. "Brief of 14 Scholars of Forensic Evidence as Amici Curiae Supporting Respondent." No. 12 – 207 in The Supreme Court of the United States. State of Maryland, petitioner, v. Alonzo Jay King, Jr., Respondent. https://web.archive.org/web/20150912124309/http://www.law.virginia.edu/pdf/news/md_v_king.pdf.

Giannelli, Paul C. 1993. "Junk Science: The Criminal Cases." *The Journal of Criminal Law & Criminology* 84 (1): 105–28.

– 1997. "The DNA Story: An Alternative View." *The Journal of Criminal Law and Criminology* 88 (1): 380–422.

– 2011. "Daubert and Forensic Science: The Pitfalls of Law Enforcement Control of Scientific Research." *University of Illinois Law Review* 2011 (1): 53–90.

Gibbs, Raymond W. 1999. "Speaking and Thinking with Metonymy." In *Metonymy in Language and Thought*, edited by Klaus-Uwe Panther and Günter Radden, 60–76. Amsterdam: John Benjamins.

Greitens, Sheena Chestnut, Myunghee Lee, and Emir Yazici. 2019. "Counterterrorism and Preventive Repression: China's Changing Strategy in Xinjiang." *International Security* 44 (3): 9–47.

Gunneweg, Wieneke. 2022. "Netherlands Press Council Issues Conclusion on Dispute between Erasmus MC and Erasmus Magazine." *Erasmus Magazine*, Erasmus University (Netherlands). https://web.archive.org/web/20220815014343/https://www.erasmusmagazine.nl/en/2022/07/15/netherlands-press-council-issues-conclusion-on-dispute-between-erasmus-mc-and-erasmus-magazine/.

Guo Jianya, Mei Xi, and Tang Kun. 2013. "Automatic Landmark Annotation and Dense Correspondence Registration for 3D human Facial Images." *BMC Bioinformatics* 14: article 232. https://web.archive.org/web/20220809110202if_/https://bmcbioinformatics.biomedcentral.com/track/pdf/10.1186/1471-2105-14-232.pdf.

Guo Jing, Tan Jingze, Yang Yajun, Zhou Hang, Hu Sile, Agu Hashan, Nurmamat Bahaxar, et al. 2014. "Variation and Signatures of Selection on the Human Face." *Journal of Human Evolution* 75:143–52.

Hao Weiqi, Liu Jing, Jiang Li, Han Jun-Ping, Wang Ling, Li Jiu-Ling, Ma Quan, Liu Chao, Wang Hui-Jun, and Li Caixia. 2019. "Exploring the Ancestry Differentiation and Inference Capacity of the 28-plex AISNPs." *International Journal of Legal Medicine* 133:975–82.

Harry, Deborah. 1995. "The Human Genome Diversity Project: Implications for Indigenous Peoples." Indigenous Peoples Council on Biocolonialism (US). https://web.archive.org/web/20210325055012/http://www.ipcb.org/publications/briefing_papers/files/hgdp.html.

– 2009. "Indigenous Peoples and Gene Disputes." *Chicago-Kent Law Review* 84 (1): article 8. https://web.archive.org/web/20200322043446/https://scholarship.kentlaw.iit.edu/cgi/viewcontent.cgi?referer=&httpsredir=1&article=3640&context=cklawreview.

– 2011. "Biocolonialism and Indigenous Knowledge in United Nations Discourse." *Griffith Law Review* 20 (3): 702–28.

Harvey, David. 2003. *The New Imperialism*. New York: Oxford University Press.

Heger, Monica. 2015. "Illumina Launches MiSeq FGx for Forensic Applications." *GenomeWeb*, 21 Jan 2015. https://www.genomeweb.com/business-news/illumina-launches-miseq-fgx-forensic-applications.

Her Majesty the Queen v. Allan Joseph Legere. Voir Dire, Vol VIII, pp. 1–277, 7–8 May, 1991. University of New Brunswick Law School. https://web.archive.org/web/20221012042253/https://www.unb.ca/fredericton/law/library/_resources/pdf/legal-materials/allan-legere/voir_dire/legerevoirdire_volviii_may7-8_opt.pdf.

– Voir Dire, Vol. XI, 14–15 May, 1991. University of New Brunswick Law School. https://web.archive.org/web/20180429004157/https://www.unb.ca/fredericton/law/library/_resources/pdf/legal-materials/allan-legere/voir_dire/legerevoirdire_volxi_may1415_opt.pdf.

– Voir Dire, Vol. XII, 16–17 May, 1991. University of New Brunswick Law School. https://web.archive.org/web/20160920162443/https://www.unb.ca/fredericton/law/library/_resources/pdf/legal-materials/allan-legere/voir_dire/legerevoirdire_volxii_may1617_opt.pdf.

– Voir Dire, Vol XIII, 27 May 1991. University of New Brunswick Law School. https://web.archive.org/web/20180429001155if_/http://www.unb.ca/fredericton/law/library/_resources/pdf/legal-materials/allan-legere/voir_dire/legerevoirdire_volxiii_may27_opt.pdf.

– Voir Dire, Vol XIV, 6–7 June, 1991. University of New Brunswick Law School. https://web.archive.org/web/20180429001206/https://www.unb.ca/fredericton/law/library/_resources/pdf/legal-materials/allan-legere/voir_dire/legerevoirdire_volxiv_june67_opt.pdf.

Her Majesty the Queen v. Allan Joseph Legere. 1991. Vol. XVIII Pages 4,421 to 4,673 incl. 21 and 22 October 1991. University of New Brunswick Law School. https://web.archive.org/web/20160720195001/http://www.unb.ca/fredericton/law/library/_resources/pdf/legal-materials/allan-legere/trial_transcripts/020.pdf.

– Vol. XIX Pages 4,674 to 4,823 incl. 23–24 October 1991. University of New Brunswick Law School. https://web.archive.org/web/20180429004659/https://www.unb.ca/fredericton/law/library/_resources/pdf/legal-materials/allan-legere/trial_transcripts/021.pdf.

– Vol. XX Pages 4,957 to 5,180 incl. 28 and 29 October 1991. University of New Brunswick Law School. https://web.archive.org/web/20180429004706/https://www.unb.ca/fredericton/law/library/_resources/pdf/legal-materials/allan-legere/trial_transcripts/022.pdf.

– Vol. XXII. 2 and 3 November 1991. University of New Brunswick Law School. https://web.archive.org/web/20180429004720/https://www.unb.ca/fredericton/law/library/_resources/pdf/legal-materials/allan-legere/trial_transcripts/024.pdf.

Hopman, Roos. 2020. "Opening Up Forensic DNA Phenotyping: The Logics of Accuracy, Commonality and Valuing." *New Genetics and Society* 39 (4): 424–40.

Huang Lanlan. 2020. "Young Chinese Idolize FM Spokespersons, Welcome 'Wolf Warrior' Diplomats." *Global Times* (PRC government news website), 21 May 2020. https://web.archive.org/web/20200804092247/https://www.globaltimes.cn/content/1189118.shtml.

Huang Yue. 2021. Vocational Training Center Graduates Tell Personal Stories, Refute 'Genocide' Accusations. *China Global Television Network (CGTN)*, 13 June 2021. https://web.archive.org/web/20221103221404/https://news.cgtn.com/news/2021-06-13/Vocational-training-center-graduates-refute-genocide-accusations-113IDQ14VI6/index.html.

Huazhong University. 2013. "美国耶鲁大学遗传学教授Kenneth K. Kidd应邀到法医学系讲学" [Kenneth K. Kidd, Professor of Genetics at Yale University, was invited to give lectures in the Department of Forensic Medicine]. Huazhong University (China), 18 June 2013. https://web.archive.org/web/20181204164323/http://fayixi.tjmu.edu.cn/info/1047/1113.htm.

Huggins, Martha Knisely, Mika Haritos-Fatouros, and Philip G. Zimbardo. 2002. *Violence Workers: Police Torturers and Murderers Reconstruct Brazilian Atrocities*. Berkeley: University of California Press.

Hughes-Stamm, Sheree. 2016. "高通量测序在高度降解样本和爆炸后样本法医鉴定中的应用" [Sheree Hughes-Stams: Application of High-Throughput Sequencing in Forensic Identification of Highly Degraded Samples and Post-Blast Samples]. YouTube video, 29 June 2016. https://web.archive.org/web/20190130004018/https://www.thermofisher.com/cn/zh/home/clinical/clinical-translational-research/Forensic-Science-New-Technology-Applicaiton-Summit-2016/speech01.html.

HUGO Committee for Human Genetic Diversity 1991. *A Project for the Study of Human Diversity with Special Attention to Vanishing Human Populations*. US National Institutes of Health. https://web.archive.org/web/20221211190230if_/https://collections.nlm.nih.gov/ext/document/101584632X153/PDF/101584632X153.pdf.

Human Genetics. 2020. "Editors' Note to: EDAR, LYPLAL1, PRDM16, PAX3, DKK1, TNFSF12, CACNA2D3, and SUPT3H Gene Variants Influence Facial Morphology in a Eurasian Population." *Human Genetics* 139:273. https://doi.org/10.1007/s00439–019–02097–3.

Human Rights Watch. 2017. "China: Police DNA Database Threatens Privacy." Human Rights Watch. https://web.archive.org/web/20200404034847/https://www.hrw.org/news/2017/05/15/china-police-dna-database-threatens-privacy.

– 2017. "China: Minority Region Collects DNA from Millions." Human Rights Watch. https://web.archive.org/web/20180305233032/https://www.hrw.org/news/2017/12/13/china-minority-region-collects-dna-millions.

Humes, E. 1992. "The DNA Wars: Touted as an Infallible Method to Identify Criminals, DNA Matching Has Mired Courts in a Vicious Battle of Expert Witness." *Los Angeles Times*, 29 November 1992. https://web.archive.org/web/20230112212208/https://www.latimes.com/archives/la-xpm-1992-11-29-tm-2666-story.html.

Identitas. 2016a. "ISAS: Predicted Phenotypes." Identitas. https://web.archive.org/web/20160401220313/http://www.identitascorp.com/ISAS.html.

– 2016b. "Sample ISAS Report." Identitas. https://web.archive.org/web/20160603001021if_/http://identitascorp.com/uploads/Sample_report_form.pdf.

Ikram, M. Arfan, Guy G.O. Brusselle, Sarwa Darwish Murad, Cornelia M van Duijn, Oscar H. Franco, André Goedegebure, Caroline C.W. Klaver, et al. 2018. "The Rotterdam Study: 2018 Update on Objectives, Design and Main Results." *European Journal of Epidemiology* 32 (9): 807–50.

Illumina. 2011. "Illumina Enters Collaborative Research Agreement with UNTHSC for Next-Generation Sequencing in Forensic DNA Analysis." Illumina. 4 October 2011. https://web.archive.org/web/20221225174618/https://emea.illumina.com/company/news-center/press-releases/2011/1613146.html.

– 2015a. "Targeted Next Generation Sequencing for Forensic Genomics." Illumina. https://web.archive.org/web/20150321053841/http://www.illumina.com/content/dam/illumina-marketing/documents/products/appspotlights/app_spotlight_forensics.pdf.

– 2015b. "Towards Validation and Implementation of the MiSeq FGx Forensic Genomic System." 2015 International Symposium on Human Identification. https://web.archive.org/web/20160628100727/https://www.ishinews.com/toward-validation-and-implementation-of-the-miseq-fgx-forensic-genomics-system/.

– 2016. "HiSeq X™ Series of Sequencing Systems." Illumina. https://web.archive.org/web/20180723044515/https://www.illumina.com/content/dam/illumina-marketing/documents/products/datasheets/datasheet-hiseq-x-ten.pdf.

– 2017. "Illumina Investor Presentation May 8, 2017." Illumina. https://web.archive.org/web/20170712210638/https://www.illumina.com/content/dam/illumina-marketing/documents/company/investor-relations/investor_presentations/illumina_investor_presentation.pdf.

– 2019. "Infinium OmniZhongHua-8 Kit." Illumina. https://web.archive.org/web/20170720094653/https://www.illumina.com/products/by-type/microarray-kits/infinium-omni-zhonghua.html.

Ince, Onur Ulas. 2018. "Between Equal Rights: Primitive Accumulation and Capital's Violence." *Political Theory* 46 (6): 885–914.

Information Office of the State Council. 2014. "The History and Development of the Xinjiang Production and Construction Corps (full text)." (PRC government website.) https://web.archive.org/web/20200806185332/http://english.www.gov.cn/archive/white_paper/2014/10/05/content_281474992384669.htm.

Institute of Forensic Science. 2020. "公安部物证鉴定中心就被美国商务部列入出口管制实体清单事发表声明" [The Ministry of Public Security's Physical Evidence Identification Center issued a statement on its inclusion in the US Department of Commerce list of export control entities]. (PRC government website.) https://web.archive.org/web/20200904211102/http://www.gov.cn/xinwen/2020-05/28/content_5515460.htm.

International Conference on Genetics. 2015a. "Sponsors (conference website)." International Conference on Genetics, Xi'an, China, 23–25 October 2015. Beijing Institute of Genomics, Chinese Academy of Sciences. https://web.archive.org/web/20160313095942/http://icgchina.org/sponsors.

– 2015b. "Scientific Program (conference website)." International Conference on Genetics, Xi'an, China, 23–25 October 2015. Beijing Institute of Genomics, Chinese Academy of Sciences. https://web.archive.org/web/20160313095530/http://icgchina.org/scientific-program.

– 2015c. "Abstracts for Speakers: E 37-Jian Ye.docx." In *The Conference Handbook*. International Conference on Genetics, Xi'an, China, 23–25 October 2015. Beijing Institute of Genomics, Chinese Academy of Sciences. https://web.archive.org/web/20160402152436if_/http://icgchina.org/2015handbook.rar.

– 2015d. "International Conference on Genetics Group Photo." International Conference on Genetics, Xi'an, China, 23–25 October 2015. Beijing Institute of Genomics, Chinese Academy of Sciences. https://web.archive.org/web/20160402150852/http://icgchina.org/icg-photo.jpg.

International Journal of Legal Medicine. 2020. "Editors' Note to: Predicting Adult Height from DNA Variants in a European-Asian Admixed Population." *International Journal of Legal Medicine* 134 (2): 393. https://doi.org/10.1007/s00414-019-02224-9.

James, Adrian. 2013. *Examining Intelligence-Led Policing Developments in Research, Policy and Practice*. Hampshire, UK: Palgrave MacMillan.

Janssen, Gerard. 2021. "Waarom China DNA van de Oeigoeren verzamelt" [Why is China collecting DNA from the Uyghurs]. *Amnesty International* (Netherlands), 2 August 2021. https://web.archive.org/web/20220119173611/https://www.amnesty.nl/wordt-vervolgd/waarom-china-dna-van-de-oeigoeren-verzamelt.

Jen, Erica. 2005. "Introduction." In *Robust Design: A Repertoire of Biological, Ecological, and Engineering Case Studies*, edited by Erica Jen, 1–6. New York: Oxford University Press.

Jia Jing, Wei Yiliang, Qin Cuijiao, Hu Lan, Wan Lihua, and Li Caixia. 2014a. "Developing a Novel Panel of Genome Wide Ancestry Informative Markers for Biogeographical Ancestry Estimates." *Forensic Science International: Genetics* 8 (1): 187–94.

– 2014b. "Appendix A: Supplementary data for Developing a Novel Panel of Genome Wide Ancestry Informative Markers for Biogeographical Ancestry Estimates." *Forensic Science International: Genetics* 8 (1).

Jiang Li, Li Caixia, Zhao Wenting, Liu Jing, and Huang Meisha. 2020. "一种在27个群体中识别中国青藏高原藏族群体个体的方法和系统" [Method and System for Recognizing Tibetan Population Individuals on Qinghai-Tibet Plateau of China Among 27 Groups]. People's Republic of China patent CN107400713B, priority date 18 August 2017; Issued 30 June 2020. https://web.archive.org/web/20230114170033/https://patentimages.storage.googleapis.com/55/55/f5/ba70ccb6cee04e/CN107400713B.pdf.

Jiang Li, Li Caixia, Zhao Wenting, Liu Jing, and Zhao Lei. 2017. "种对男性个体进行y-snp分型的方法和系统" [Male Individuals Y-SNP Typing Method and System]. People's Republic of China patent application CN108342489A, priority date 23 January 2017. https://web.archive.org/web/20230114170941/https://patentimages.storage.googleapis.com/10/fb/76/8f46fb9fa3f61c/CN108342489A.pdf.

Jiang Li, Sun Qifan, Ma Quan, Zhao Wenting, Liu Jing, Zhao Lei, Ji Anquan, Li Caixia. 2017. "Optimization and Validation of Analysis Method Based on 27-Plex SNP Panel for Ancestry Inference." *Hereditas* (Beijing) 39 (2):166–73. https://web.archive.org/web/20200802151628/http://www.chinagene.cn/CN/article/downloadArticleFile.do?attachType=PDF&id=5130.

Jiang Li, Wei Yi-Liang, Zhao Lei, Li Na, Liu Tao, Liu Hai-Bo, Ren Li-Jie, et al. 2018a. "Global Analysis of Population Stratification using a Smart Panel of 27 Continental Ancestry-Informative SNPs." *Forensic Science International: Genetics* 35:e10–e12.

– 2018b. "Table S2: Geographic Locations and Further Information on Population Subjects." Supplemental material for the article "Global Analysis of Population Stratification Using a Smart Panel of 27 Continental Ancestry-Informative SNPs." *Forensic Science International: Genetics* 35:e10–e12.

Jing Xiaoxi, Yanan Sun, Wenting Zhao, Xingjian Gao, Mi Ma, Fan Liu, and Caixia Li. 2019. "Predicting Adult Height from DNA Variants in a European-Asian Admixed Population." *International Journal of Legal Medicine* 133 (6): 1667–79.

– 2021. "Retraction Note: Predicting Adult Height from DNA Variants in a European-Asian Admixed Population." *International Journal of Legal Medicine* 135:2151.

Junqueira, Carmen, and Betty Mindlin. 1987. *The Aripuana Park and the Polonoroeste Programme*. Copenhagen: International Work Group for Indigenous Affairs. https://web.archive.org/web/20200724134102/https://www.iwgia.org/images/publications/0165_59_The_Aripuana_Park.pdf.

Kahn, Jonathan. 2015. "'When Are You from?' Time, Space, and Capital in the Molecular Reinscription of Race." *The British Journal of Sociology* 66:68–75.

Kam, Stefanie, and Michael Clarke. 2021. "Securitization, Surveillance and 'De-extremization' in Xinjiang." *International Affairs* 97 (3): 625–42.
Kanindé Associação de Defesa Etnoambiental, Organização Metareilá do Povo Indígena Paiter, and Betty Mindlin. 2018. "Surui Paiter." Povos Indígenas no Brasil. https://web.archive.org/web/20220422061547/https://pib.socioambiental.org/en/Povo:Surui_Paiter.
Kaye, David H. 2010. *The Double Helix and the Law of Evidence.* Cambridge, MA: Harvard University Press.
Kayser, Manfred, and Peter de Knijff. 2011. "Improving Human Forensics through Advances in Genetics, Genomics and Molecular Biology." *Nature Reviews Genetics* 12:179–92.
Kayser, Manfred, Liu Fan, and Albert Hofman. 2011. "Method for Prediction of Human Iris Color." US patent application 20110312534A1. https://web.archive.org/web/20200809182604/https://patentimages.storage.googleapis.com/fe/b2/72/cb2f4ae94837e2/US20110312534A1.pdf.
Keating, Brenden, Aruna T. Bansal, Susan Walsh, Jonathan Millman, Jonathan Newman, Kenneth Kidd, Bruce Budowle, et al., on behalf of the International Visible Trait Genetics (VisiGen) Consortium. 2013. "First All-In-One Diagnostic Tool for DNA Intelligence: Genome Wide Inference of Biogeographic Ancestry, Appearance, Relatedness and Sex with the Identitas Version 1 Forensic Chip." *International Journal of Legal Medicine* 127 (3): 559–72.
Kempes, Maaike, and Roland Strijker. 2021a. "Nederland doet samen met China DNA-onderzoek: 'Fundamenteel fout'" [The Netherlands is Conducting DNA Research Together with China: 'Fundamental Error']. *RTL News* (Netherlands), 6 October 2021. https://web.archive.org/web/20211009083234/https://www.rtlnieuws.nl/nieuws/artikel/5258161/nederland-china-dna-oeigoeren-mensenrechten.
– 2021b. "Nederlandse kennis over DNA mag niet in handen komen van China" [Dutch Knowledge about DNA Must Not Come into the Hands of China]. *RTL News* (Netherlands), 6 October 2021. https://web.archive.org/web/20211009020836/https://www.rtlnieuws.nl/nieuws/artikel/5258513/dna-china-wetenschap-nederland-oeigoeren-d66-vvd-samenwerking.
Kidd, Judith R., Francis Lee Black, Kenneth M. Weiss, Ivan Balázs, and Kenneth K. Kidd. 1991. "Studies of Three Amerindian Populations Using Nuclear DNA Polymorphisms." *Human Biology* 63:775–94.
Kidd, Judith R., Andrew J. Pakstis, and Kenneth K. Kidd. 1993. "Global Levels of DNA Variation." Proceedings of the Fourth International Symposium on Human Identification (Promega Corporation). Yale University School of Medicine. https://web.archive.org/web/20120130202554/http://medicine.yale.edu/labs/kidd/www/302.pdf.

Kidd, Kenneth K. 2008. "Final Report: Population Genetics of SNPs for Forensic Purposes." National Institute of Justice Grant #2004-DN-BX-K-25. United States Department of Justice. https://web.archive.org/web/20170224000550/https://www.ncjrs.gov/pdffiles1/nij/grants/223982.pdf.

– 2011. "Population Genetics of SNPs for Forensic Purposes (Updated)." National Criminal Justice Reference Service (US). https://web.archive.org/web/20170131035021/https://www.ncjrs.gov/pdffiles1/nij/grants/236433.pdf.

– 2015a. "Ancestry, Mixtures, Relationships – All with MPS." Green Mountain DNA Conference, 27 July 2015. Vermont State Department of Public Safety. https://web.archive.org/web/20160720234223/http://vfl.vermont.gov/sites/pslab/files/pdfs/conference/Kidd.pdf.

– 2015b. "Further Development of SNP Panels for Forensics." Report for National Institute of Justice Grant 2010-DN-BX-K225. https://web.archive.org/web/20160720212018/https://www.ncjrs.gov/pdffiles1/nij/grants/249548.pdf.

– 2015c. "Forensic Resource/Reference on Genetics Knowledge Base (FROG-kb)." National Criminal Justice Reference Service, US Department of Justice. https://web.archive.org/web/20170617085808/https://www.ncjrs.gov/pdffiles1/nij/grants/249549.pdf.

Kidd, Kenneth K., Judith R. Kidd, Andrew J. Pakstis, and William C. Speed. 2012. "Better SNPs for Better Forensics: Ancestry, Phenotype, and Family Identification." Poster presented at the National Institute of Justice Annual Meeting, Arlington, VA. https://web.archive.org/web/20160720211146/http://medicine.yale.edu/lab/kidd/publications/NIJposter2012_Minihaps_237328_174718_29491.pdf.

Kidd, Kenneth K., Andrew J. Pakstis, William C. Speed, Elena L. Grigorenko, Sylvester L.B. Kajuna, Nganyirwa J. Karoma, Selemani Kungulilo, et al. 2006. "Developing a SNP panel for Forensic Identification of Individuals." *Forensic Science International: Genetics* 164 (1): 20–32.

Kidd, Kenneth K., William C. Speed, Andrew J. Pakstis, Manohar R. Furtado, Rixun Fang, Abeer Madbouly, Martin Maiers, Mridu Middha, Françoise R. Friedlaender, and Judith R. Kidd. 2014. "Progress Toward an Efficient Panel of SNPs for Ancestry Inference." *Forensic Science International: Genetics* 10:23–32.

Kidd, Kenneth K., William C. Speed, Andrew J. Pakstis, and Judith R. Kidd. 2011. "The Search for Better Markers for Forensic Ancestry Inference." 22nd International Symposium on Human Identification. https://web.archive.org/web/20200731192712/https://pdfs.semanticscholar.org/dc8a/cb54a157048248b940e0e1e7b6726ad78c9b.pdf.

Kidd, Kenneth K., William C. Speed, Sharon Wootton, Robert Lagace, Reina Langit, Eva Haigh, Joseph Chang, and Andrew J. Pakstis. 2015. "Genetic

Markers for Massively Parallel Sequencing in Forensics." *Forensic Science International: Genetics* 5 (Supplement Series):e677–e679.

Kidd Lab. 2017. "47 Population Samples Routinely Studied at Kidd Lab." Yale University. https://web.archive.org/web/20160419212424/http://medicine.yale.edu/lab/kidd/research/populations/.

Kirkwood, Karen A. 2017. "Letter Denying Responsibility in Response to HRW Request on Xinjiang." Human Rights Watch. https://web.archive.org/web/20191026185011/https://www.hrw.org/sites/default/files/supporting_resources/thermo_fisher_response_.pdf.

Koops, B.J., and M. Schellekens. 2008. "Forensic DNA Phenotyping: Regulatory Issues." *The Columbia Science and Technology Law Review* 9:158–202. doi:10.2139/ssrn.975032.

Kowal, Emma, and Joanna Radin. 2015. "Indigenous Biospecimen Collections and the Cryopolitics of Frozen Life." *Journal of Sociology* 51 (1): 63–80. https://doi.org/10.1177/1440783314562316.

Kruger, E. 2013. "Image and Exposure: Envisioning Genetics as a Forensic-Surveillance Matrix." *Surveillance & Society* 11 (3): 237–51.

Labaton, Stephen. 1990. "DNA Fingerprinting Showdown Expected in Ohio." *The New York Times*, 22 June 1990.

Lackey, Angie. 2017. "Increasing Security, Solving Crime – Join Us at HIDS 2017 in Vienna." Thermo Fisher Scientific. https://web.archive.org/web/20200823142635/https://www.thermofisher.com/blog/behindthebench/increasing-security-solving-crime-join-us-at-hids-in-vienna-may-16-17/.

Landin, Rachel M. 1989. "Kinship and Naming among the Karitiana of Northwestern Brazil." MA thesis, University of Texas at Arlington.

Latour, Bruno. 1987. *Science in Action: How to Follow Scientists and Engineers through Society*. Cambridge, MA: Harvard University Press.

Leibold, James. 2020. "Surveillance in China's Xinjiang Region: Ethnic Sorting, Coercion, and Inducement." *Journal of Contemporary China* 29 (121): 46–60.

Leibold, James, and Danielle Xaiodan Deng. 2015. "Segregated Diversity: Uyghur Residential Patterns in Xinjiang, China." In *Inside Xinjiang: Space, Place and Power in China's Muslim Far Northwest*, edited by Anna Hayes and Michael Clarke, 122–48. London: Routledge.

Leverhulme Trust. 2010. "Newsletter April 2010." Leverhulme Trust. https://web.archive.org/web/20200907212156/https://www.leverhulme.ac.uk/sites/default/files/Apr2010.pdf.

Lew, Linda. 2021. "China's Genetic Profiling Research Faces Pushback from Academic Journals over Ethics Concerns." *South China Morning Post* (Hong Kong), 12 September 2021. https://web.archive.org/

web/20210912044930/https://www.scmp.com/news/china/science/article/3148333/chinas-genetic-profiling-research-faces-pushback-academic.

Lewis, Norman, and D. McCullin. 1969. "Genocide – From Fire and Sword to Arsenic and Bullets, Civilization Has Sent 6 Million Indians to Extinction." *Sunday Times Magazine* (UK), 23 February 1969.

Lewontin, R.C., and D.L. Hartl. 1991. "Population Genetics in Forensic DNA Typing." *Science* 254 (5039): 1745–50.

Li Caixia, Han Junping, Zhao Lei, and Jiang Li. 2019. "一种对未知来源个体进行非、东亚、欧洲群体来源分析的方法和系统" [A Method and System for Analysing the Origin of Unknown Individuals from African, East Asian and European Groups]. Peoples Republic of China patent CN112011622B, filed 29 May 2019; issued 2 December 2022. https://web.archive.org/web/20230114174013/https://patentimages.storage.googleapis.com/8a/b4/e7/0e5757ad55ce8c/CN112011622B.pdf.

Li Caixia, Hu Lan, and Wei Yiliang. 2013. "获得人种特异性位点的方法和人种推断系统及其应用" [Method for Obtaining Race Specific Loci and Race Inference System and Application Thereof]. People's Republic of China patent application CN101956006B, filed 27 August 2010; issued 16 October 2013. https://web.archive.org/web/20230114163234/https://patentimages.storage.googleapis.com/97/d8/d5/a3da1208bc05fa/CN101956006B.pdf.

Li Caixia, Jiang Li, Sun Qifan, Zhao Wenting, and Liu Jing. 2016a. "一种对未知来源个体进行十个群体来源分析的方法和系统" [Method and System for Analyzing Ten Population Sources for Individuals of Unknown Origin]. People's Republic of China patent application CN105861654A, priority date 5 April 2016. https://web.archive.org/web/20230114171323/https://patentimages.storage.googleapis.com/c6/5b/e0/cd439a39fbc74c/CN105861654A.pdf.

Li Caixia, Andrew J. Pakstis, Jiang Li , Wei Yi-Liang , Sun Qi-Fan Sun, Wu Hong , Ozlem Bulbul, Wang Ping, Kang Longli, Judith R. Kidd, and Kenneth K. Kidd. 2016b. "A Panel of 74 AISNPs: Improved Ancestry Inference within Eastern Asia." *Forensic Science International: Genetics* 23:101–10.

Li Caixia, Wei Yiliang, Hu Lan, Ji Anquan, Jia Jing, and Li Wanshui. 2014. "种推断未知来源个体汉、藏、维群体来源的方法和系统" [Method and System Used for Inferring Han, Tibetan and Uyghur Population as Source of Individual with Unknown Identity]. People's Republic of China patent application CN103146820B, filed 22 February 2013; issued 2 July 2014. https://web.archive.org/web/20230114163511if_/https://patentimages.storage.googleapis.com/a7/b7/f2/8f1fdb1bd5728a/CN103146820B.pdf.

Li Caixia, Wei Yiliang, and Ye Jian. 2016. "种对未知来源个体进行非、欧、东亚群体遗传主成分分析的方法和系统" [Method and System for Applying

African, European and East Asian Population Genetic Principal Component Analysis to an Individual from an Unknown Source]. People's Republic of China patent application CN104212886B, filed 25 July 2014; issued 22 June 2016. https://web.archive.org/web/20230114163825if_/https://patentimages.storage.googleapis.com/ef/58/b1/224edd169933f3/CN104212886B.pdf.

Li Caixia, Xu Youchun, Liu Jing. 2021. 72个snp位点及相关引物在鉴定或辅助鉴定人类族群中的应用 [Application of 72 SNPs and Associated SNP Primers to Help in Identification of Ethnic Groups]. People's Republic of China patent application CN108411008B, filed 1 June 2018; issued 27 July 2021. https://web.archive.org/web/20230114171950/https://patentimages.storage.googleapis.com/c6/d8/ed/5f1f2cff20fe59/CN108411008B.pdf.

Li Haiyan. 2016a. "新一代测序技术在疑难案件中的应用" [Application of Next-Generation Sequencing Technology in Difficult Cases]. 2016年法庭科学新技术应用高峰论坛 [2016 Forensic Science New Technology Application Summit Forum], *Thermo Fisher* (China), 16–18 November 2016, Foshan, Guangdong China. https://web.archive.org/web/20200806011645/https://www.ThermoFisher.com/content/dam/LifeTech/Documents/PPT/China/%E6%96%B0%E4%B8%80%E4%BB%A3%E6%B5%8B%E5%BA%8F%E6%8A%80%E6%9C%AF%E5%9C%A8%E6%A1%88%E4%BB%B6%E4%B8%AD%E7%9A%84%E5%BA%94%E7%94%A8-%E6%9D%8E%E6%B5%B7%E7%87%95.ppt.

– 2016b. "2016年法庭科学新技术应用高峰论坛" [2016 Forensic Science New Technology Application Summit]. Thermo Fisher (China). 16–18 November 2016, Foshan, Guangdong, China. https://web.archive.org/web/20200806152540/https://webcache.googleusercontent.com/search?q=cache%3AbYd-VPcb7J3YJ%3Ahttps%3A%2F%2Fwww.thermofisher.com%2Fcn%2Fzh%2Fhome%2Fclinical%2Fclinical-translational-research%2FForensic-Science-New-Technology-Applicaiton-Summit-2016.html.

Li Hui, Kelly Cho, Judith R. Kidd, and Kenneth K. Kidd. 2009. "Genetic Landscape of Eurasia and 'Admixture' in Uyghurs." *The American Journal of Human Genetics* 85:929–45.

Li Yi, Zhao Wenting, Li Dan, Tao Xianming, Xiong Ziyi, Liu Jing, Zhang Wei, Liu Haibo, Ji Anquan, Tang Kun, Liu Fan, and Li Caixia. 2018. "The Effect of EDARV370A on Facial and Ear Morphologies in Uyghur Population." *Hereditas* 40 (11): 1024–32. Doi:10.16288/j.yczz.18–268.

Li Yi, Zhao Wenting, Li Dan, Tao Xianming, Xiong Ziyi , Liu Jing, Zhang Wei, Ji Anquan, Tang Kun, Fan Liu, and Li Caixia. 2019. "EDAR , LYPLAL1, PRDM16, PAX3, DKK1, TNFSF12, CACNA2D3, and SUPT3H Gene Variants Influence Facial Morphology in a Eurasian Population." *Human Genetics* 138 (6): 681–9.

– 2021. "Retraction Note: EDAR, LYPLAL1, PRDM16, PAX3, DKK1, TNFSF12, CACNA2D3, and SUPT3H Gene Variants Influence Facial Morphology in a Eurasian Population." *Human Genetics* 140:1499.

Linnemann, Travis. 2017. "Proof of Death: Police Power and the Visual Economies of Seizure, Accumulation and Trophy." *Theoretical Criminology* 21 (1): 57–77.

Linnemann Travis, and Corina Medley. 2017. "Fear the Monster: Racialised Violence, Sovereign Power and the Thin Blue Line." In *Routledge Handbook of Fear of Crime*, edited by Murray Lee and Gabe Mythen, 65–81. London: Routledge.

Littlemore, Jeannette. 2015. *Metonymy*. Cambridge, UK: Cambridge University Press.

Liu Fan. 2017. "Fan Liu's Curriculum Vitae." Erasmus University Medical Center. https://web.archive.org/web/20190307164220/https://www6.erasmusmc.nl/47743/535996/FanLiu_CV_Sept2017.pdf.

Liu Fan, Chen Yan , Zhu Gu , Pirro G. Hysi, Wu Sijie, Kaustubh Adhikari, Krystal Breslin, et al. 2018. "Meta-Analysis of Genome-Wide Association Studies Identifies 8 Novel Loci Involved in Shape Variation of Human Head Hair." *Human Molecular Genetics* 27 (3): 559–75.

Liu Jing, Li Caixia, Zhao Wenting, Jiang Li, and Hao Weiqi. 2020. "对未知来源个体进行五大洲际族群来源推断的方法和系统" [Method and System for Inferring Five-Intercontinental Ethnic Group Sources of Unknown Source Individuals]. People's Republic of China patent application CN107419017B, filed 25 July 2017; issued 8 September 2020. https://web.archive.org/web/20230114171721/https://patentimages.storage.googleapis.com/23/5e/a4/85e6efad519447/CN107419017B.pdf.

Liu Jing, Ma Mi, and Li Caixia. 2020. "一种基于毛细管电泳检测的东亚南北方人群推断体系" [Southeast Asia-North Population Inference System Based on Capillary Electrophoresis Detection]. People's Republic of China patent application CN111073888A, priority date 9 January 2020. https://web.archive.org/web/20230114173046/https://patentimages.storage.googleapis.com/8f/b2/62/76caf426d0426e/CN111073888A.pdf.

Liu Jing, Qiao Lu, Zhao Wenting, Jiang Li, Ji Anquan, Wang Guiqiang, Ye Jian, Tang Kun, and Li Caixia. 2017. "Experimentation of Human Facial Predication by Relevant DNA SNPs." *Forensic Science and Technology* 42 (4): 264–9. https://web.archive.org/web/20170923085818/http://www.xsjs-cifs.com/article/2017/1008-3650-42-4-264.html.

Liu Ming, Li Yi, Yang Yafang, Yan Yuwen, Liu Fan, Li Caixia, Zeng Faming, and Zhao Wenting. 2020. "Human Facial Shape Related SNP Analysis in Han Chinese Populations." *Hereditas* 42 (7): 680–90. https://web.archive.

org/web/20211025222301/http://www.chinagene.cn/CN/article/downloadArticleFile.do?attachType=PDF&id=5519.

Liu Yang, Sun Changchun, Ma Mi, Wang Ling, Zhao Wenting, Ma Quan, Ji Anquan, Liu Jing, and Li Caixia. 2020. "The Ancestry Inference of Chinese Populations Using 74-Plex SNPs System." *Hereditas* 42 (3): 296–308. https://web.archive.org/web/20200822204343/http://www.chinagene.cn/EN/article/downloadArticleFile.do?attachType=PDF&id=5482.

Lorenz, Joseph G., and David G. Smith. 1994. "Distribution of the 9-bp Mitochondrial DNA Region V Deletion among North American Indians." *Human Biology* 66 (5): 777–88.

Lu Ru-Band, Lu Huei-Chen, Ko Fong-Ming, Chang Carmela, M. Castiglione, Gloria Schoolfield, Andrew J. Pakstis, Judith R. Kidd, and Kenneth K. Kidd. 1996. "No Association between Alcoholism and Multiple Polymorphisms at the Dopamine D2 Receptor Gene (DRD2) in Three Distinct Taiwanese Populations." *Biological Psychiatry* 39 (6): 419–29.

Machado, Helena, and Rafaela Granja. 2020. *Forensic Genetics in the Governance of Crime*. Singapore: Palgrave Pivot. https://web.archive.org/web/20200505231656/http://repositorium.sdum.uminho.pt/bitstream/1822/63912/1/2020_Book_ForensicGeneticsInTheGovernanc.pdf.

Macpherson, Mike, Greg Werner, Iram Mirza, Marcela Miyazawa, Chris Gignoux, and Joanna Mountain. 2008. "White Paper 23–04: Global Similarity's Genetic Similarity Map." *23andMe*. https://web.archive.org/web/20160721180549/https://www.23andme.com/res/pdf/ZjMVeYK-4_v2rosyQPoklw_23-04_Genetic_Similarity_Map.pdf.

Marchal, Jules. 2008. *Lord Leverhulme's Ghosts: Colonial Exploitation in the Congo*. London: Verso.

Marks, Jonathan. 2003. "Human Genome Diversity Project: Impact on Indigenous Communities." *Encyclopedia of the Human Genome*, 1–4 . London: Macmillan. https://web.archive.org/web/20070628211759/personal.uncc.edu/jmarks/pubs/EncHumGen.pdf.

Marshall, Jack N.G., Ana Illera Lopez, Abigail L. Pfaff, Sulev Koks, John P. Quinn, and Vivien J. Bubb. 2021. "Variable Number Tandem Repeats – Their Emerging Role in Sickness and Health." *Experimental Biology and Medicine* 246 (12): 1368–76.

Marvin, Carolyn, and David Ingle. 1999. *Blood Sacrifice and the Nation: Totem Rituals and the American Flag*. Cambridge, UK: Cambridge University Press.

Marx, Karl. [1867] 1996. *Capital. Vol. 1*. Vol. 35 of the *Collected Works of Karl Marx and Frederick Engels*. New York: International Publishers.

Max Planck Gesellschaft. 2014. “Science Connects.” Max Planck Gesellschaft (Society; Germany), 12 May 2014. https://web.archive.org/web/20181212164515/https://www.mpg.de/8198951/40-years-cas-mpg.

– 2020. “CAS-MPG Partner Institute for Computational Biology: About the Institute.” Max Planck Gesellschaft (Society; Germany). https://web.archive.org/web/20200928011812/https://www.mpg.de/273222/CAS-MPG_Partner_Institute_for_Computational_Biology.

Mbembe, Achille. 2003. “Necropolitics.” *Public Culture* 15 (1): 11–40.

M’charek, A. 2008. “Silent Witness, Articulate Collectives: DNA Evidence and the Inference of Visible Traits.” *Bioethics* 22:519–28.

M’charek, A., V. Toom, and B. Prainsack. 2012. “Bracketing off Population Does Not Advance Ethical Reflection on EVCs: A Reply to Kayser and Schneider.” *Forensic Science International: Genetics* 6 (1): 16–17.

McLaughlin, Kathleen. 2016. “Bringing Legends to Life.” *Science* 354 (6316): 1094–5.

McQuade, Brendan. 2020. “The Prose of Pacification: Critical Theory, Police Power, and Abolition Socialism.” *Social Justice* 47 (3): 55–76.

Mead, Aroha. 2007. “The Polynesian Excellence Gene and Life Patent Bottom Trawling.” In *Pacific Genes & Life Patents: Pacific Indigenous Experiences & Analysis of the Commodification & Ownership of Life*, edited by Aroha Te Pareake Mead and Steven Ratuva, 34–59. Call of the Earth (Llamado de la Tierra) and the United Nations University Institute of Advanced Studies. https://web.archive.org/web/20211214071732if_/https://calloftheearth.files.wordpress.com/2009/07/coe-publication-final.pdf.

Meriwether, D. Andrew. 1999. “Freezer Anthropology: New Uses for Old Blood.” *Philosophical Transactions of the Royal Society B: Biological Sciences* 354:121–9.

Miller, G. 1982. “Immortalization of Human Lymphocytes by Epstein-Barr Virus.” *The Yale Journal of Biology and Medicine* 55 (3–4): 305–10.

Millward, James A. 2003. *Eurasian Crossroads: A History of Xinjiang*. New York: Columbia University Press.

– 2007. *Eurasian Crossroads: A History of Xinjiang*. New York: Columbia University Press.

Ministry of Foreign Affairs. 2017. “Foreign Ministry Spokesperson Lu Kang’s Regular Press Conference on December 13, 2017.” Ministry of Foreign Affairs of the People’s Republic of China. https://web.archive.org/web/20181208151955/https://www.fmprc.gov.cn/mfa_eng/xwfw_665399/s2510_665401/2511_665403/t1519191.shtml.

– 2020. “Foreign Ministry Spokesperson Zhao Lijian’s Regular Press Conference on May 25, 2020.” Ministry of Foreign Affairs of the People’s Republic of China. https://web.archive.org/web/20200720201739/https://

www.fmprc.gov.cn/mfa_eng/xwfw_665399/s2510_665401/2511_665403/t1782571.shtml.

Ministry of Science and Technology. 2016. "公共安全风险防控与应急技术装备"重点专项2017年度项目申报指南" [Public Safety Risk Prevention and Control and Emergency Technical Equipment 2017 Project Application Guidelines]. Chinese Ministry of Science and Technology, 14 October 2016. https://web.archive.org/web/20190405145253/http://service.most.gov.cn/u/cms/static/201610/14103934ugxf.pdf.

– 2017. "'公共安全风险防控与应急技术装备"重点专项'、'案事件现场勘验与目标关联分析关键技术研究', '项目启动暨实施方案论证会在北京召开'" [The Launch and Implementation Plan Demonstration Meeting of the Key Special Items of "Public Safety Risk Prevention and Control and Emergency Technology and Equipment" and "Research on Key Technologies for Fine Characterization and Accurate Identification of Criminal Suspects" and "Research on Key Technologies for On-site Investigation and Target Related Analysis of Crime Events" Was Held in Beijing]. Chinese Ministry of Science and Technology, 19 October 2017. https://web.archive.org/web/20190402211358/http://www.most.gov.cn/kjbgz/201710/t20171019_135448.htm.

Moffette, David, and William Walters. 2018. "Flickering Presence: Theorizing Race and Racism in the Governmentality of Borders and Migration." *Studies in Social Justice* 12 (1): 92–110.

Moreau, Yves. 2019. "Crack Down on Genomic Surveillance." *Nature* 576:36–8.

– n.d. "Letter to Office of the High Commissioner for Human Rights." United Nations. https://web.archive.org/web/20200805222615/https://www.ohchr.org/Documents/HRBodies/HRCouncil/WGTransCorp/Session4/SubmissionLater/YvesMoreau.pdf.

Mozur, Paul (@paulmozur). 2019. "I took the videos out of the cab and immediately hid them. Later police forced us to delete all videos from Tumxuk. In the footage you can see white poles where surveillance cameras would normally go. After they destroyed the houses, they took the cameras but left the poles." Twitter, 3 December 2019, 5:26 a.m. https://web.archive.org/web/20230112222312/https://twitter.com/paulmozur/status/1201794781631770624?lang=en.

Mueller, Laurence D. 1993. "The Use of DNA Typing in Forensic Science." *Accountability in Research* 3 (1): 55–67.

Müller, Martin. 2015. "Assemblages and Actor-Networks: Rethinking Socio-Material Power, Politics and Space." *Geography Compass* 9 (1): 27–41.

Munsterhjelm, Mark. 2013. "The Political Economy of Hope and Authoritarian Liberalism in Genetics Research." *Borderlands* 12 (1). https://

web.archive.org/web/20140216104922/http://borderlands.net.au/vol12no1_2013/munsterhjelm_political.pdf.

– 2014. *Living Dead in the Pacific: Contested Sovereignty and Racism in Genetic Research on Taiwan Aborigines*. Vancouver: University of British Columbia Press.

– 2015. "Beyond the Line: Violence and the Objectification of the Karitiana Indigenous People as Extreme Other in Forensic Genetics." *International Journal for the Semiotics of Law* 28 (2): 289–316.

– 2018. "Authoritarianism and Indigenous Peoples in the Development of Forensic Genetic Technologies." *Transmissions: A Social Studies of Science Companion Blog*. https://web.archive.org/web/20200201010046/https://sites.library.queensu.ca/transmissions/authoritarianism-and-indigenous-peoples-in-the-development-of-forensic-genetic-technologies/.

– 2019. "Scientists Are Aiding Apartheid in China." *Just Security*. Reiss Center on Law and Security, New York University School of Law. https://web.archive.org/web is/20220402094728/https://www.justsecurity.org/64605/scientists-are-aiding-apartheid-in-china/.

Murphy, Erin E. 2015. *Inside the Cell: The Dark Side of Forensic DNA*. New York: Nation Books.

National Human Genome Research Institute. 2000. "June 2000 White House Event." National Human Genome Research Institute. https://web.archive.org/web/20220803091523/https://www.genome.gov/10001356/june-2000-white-house-event.

National Institute of Justice. n.d.a. "Population Genetics of SNPs for Forensic Purposes. Award #: 2004-DN-BX-K025." National Institute of Justice. https://web.archive.org/web/20200731190525/https://nij.ojp.gov/funding/awards/2004-dn-bx-k025.

– n.d.b. "The Enhancement of the Native American CODIS STR Database for Use in Forensic Casework. Number 2014-DN-BX-K024." National Institute of Justice. https://web.archive.org/web/20210324205929/https://nij.ojp.gov/funding/awards/2014-dn-bx-k024.

– 2006. "National Institute of Justice DNA Grantees Meeting Presenter Biographies." National Institute of Justice. https://ia904707.us.archive.org/27/items/2006-dnagrantees-biographies/2006DNAGranteesBiographies.pdf.

National Research Council. 1996. *The Evaluation of Forensic DNA Evidence*. Washington, DC: National Academy Press. https://www.ncbi.nlm.nih.gov/books/NBK232610/pdf/Bookshelf_NBK232610.pdf.

National Science Foundation. 1989. "Award Abstract #8813234: DNA Markers and Genetic Variation in the Human Species." US National Science Foundation. https://web.archive.org/web/20200824030400/https://www.nsf.gov/awardsearch/showAward?AWD_ID=8813234.

– 1990. "Award Abstract #8718775 Mitochondrial Genes and Human Origins." US National Science Foundation. https://web.archive.org/web/20200824163306/https://www.nsf.gov/awardsearch/showAward?AWD_ID=8718775&HistoricalAwards=false.

Nature Index. 2019. "The Top 10 Global Institutions for 2018." Nature (website). https://web.archive.org/web/20211102212148/https://www.nature.com/articles/d41586-019-01922-z.

– 2020. "Institution Outputs. 1 August 2019–31 July 2020." Nature (website). https://web.archive.org/web/20201118210541/https://www.natureindex.com/institution-outputs/generate/All/global/All/score.

Neel, James V., Heney Gershowitz, Harvey W. Mohrenweiser, Bernard Amos, Donna D. Kostyu, Francisco M. Salzano, Moacyr A. Mestriner, et al. 1980. "Genetic Studies on the Ticuna, an Enigmatic Tribe of Central Amazonas." *Annals of Human Genetics* 44: 37–54.

Neocleous, M. 2000. *The Fabrication of Social Order: A Critical Theory of Police Power*. London: Pluto Press.

– 2008. *Critique of Security*. Edinburgh: Edinburgh University Press.

– 2013. "The Dream of Pacification: Accumulation, Class War, and the Hunt." *Socialist Studies* 9 (2): 7–31.

– 2014. *War Power, Police Power*. Edinburgh: Edinburgh University Press.

– 2015. "The Universal Adversary Will Attack: Pigs, Pirates, Zombies, Satan and the Class War." *Critical Studies on Terrorism* 8 (1): 15–32.

Nerlich, Brigitte, and David C. Clarke. 1999. "Synecdoche as a Cognitive and Communicative Strategy." In *Historical Semantics and Cognition*, edited by Andreas Blank and Peter Koch, 197–214. Berlin: De Gruyter Mouton.

Netherlands Press Council (Raad voor de Journalistiek). 2022. Details of Case "2022/20." https://web.archive.org/web/20220713231747/https://www.rvdj.nl/2022/20.

Nimuendajú, Curt. 1952. *The Tukuna*. Berkeley: University of California Press.

Nussbaum, Martha. 1995. "Objectification." *Philosophy & Public Affairs* (4): 249–91.

Obasogie, Osagie K. 2012. "The Return of Biological Race? Regulating Race and Genetics Through Administrative Agency Race Impact Assessments." *Southern California Interdisciplinary Law Journal* 22 (1). https://web.archive.org/web/20200322184351if_/https://repository.uchastings.edu/cgi/viewcontent.cgi?article=2360&context=faculty_scholarship.

Oefner, Peter J. 2002. "United States Patent 645-3244: Detection of Polymorphisms by Denaturing High-Performance Liquid Chromatography." United States Patent and Trademark Office, filed 10 February 2000; issued 17 September 2002.

Oefner, Peter J., and Peter A. Underhill. 1998. "US 579–5976 Patent: Detection of Nucleic Acid Heteroduplex Molecules by Denaturing High-Performance Liquid Chromatography and Methods for Chromatography and Methods for Comparative Sequencing." United States Patent and Trademark Office, filed 8 August 1995; issued 18 August 1998.

– 2005. "United States Patent 6929911: Method for Determining Genetic Affiliation, Substructure and Gene Flow Within Human Populations." United States Patent and Trademark Office, filed 1 November 2000; issued 16 August 2005.

O'Gorman, Ned. 2005. "Aristotle's Phantasia in the Rhetoric: Lexis, Appearance, and the Epideictic Function of Discourse." *Philosophy and Rhetoric* 38 (1): 16–40.

Ong, Aihwa. 2005. "Ecologies of Expertise: Assembling Flows, Managing Citizenship." In *Global Assemblages: Technology, Politics, and Ethics as Anthropological Problems*, edited by Aihwa Ong and Stephen J. Collier, 337–53. Malden, MA: Blackwell.

– 2006. *Neoliberalism As Exception: Mutations in Citizenship and Sovereignty*. Durham, NC: Duke University Press.

– 2013. "A Milieu of Mutations: The Pluripotency and Fungibility of Life in Asia." *East Asian Science, Technology and Society* 7:69–85.

Opitz, S. 2010. "Government Unlimited: The Security Dispositif of Illiberal Governmentality." In *Governmentality: Current Issues and Future Challenges*, edited by Ulrich Bröckling, Susanne Krasmann, and Thomas Lemke, 93–114. London: Routledge.

Pakstis, Andrew J., Eva Haigh, Lotfi Cherni, Amel Ben Ammar El Gaaied, Alison Barton, Baigalmaa Evsanaa, Ariunaa Togtokh, et al. 2015a. "52 Additional Reference Population Samples for the 55 AISNP Panel." *Forensic Science International: Genetics* 19:269–71.

– 2015b. "Population Samples Studied for 55 Ancestry Informative SNPs." Supplemental Table. *Forensic Science International: Genetics* 19.

Pakstis, Andrew J., Longli Kang, Lijun Liu, Zhiying Zhang, Tianbo Jin, Elena L. Grigorenko, Frank R. Wendt, et al. 2017. "Increasing the Reference Populations for the 55 AISNP Panel: The Need and Benefits." *International Journal of Legal Medicine* 131:913–17.

Pakstis, Andew J., William C. Speed, Usha Soundararajan, Haseena Rajeevan, Judith R. Kidd, Hui Li, and Kenneth K. Kidd. 2019. "Population Relationships Based on 170 Ancestry SNPs from the Combined Kidd and Seldin Panels." *Scientific Reports* (*Nature Research*) 9:18874.

Pan Siyu, Chen Shiting, Tang Kun, Li Caixia, Liu Jing, Ye Jian, and Zhao Wenting. 2018. "Age Estimation and Age-related Facial Reconstruction of Xinjiang Uygur Males by Three-dimensional Human Facial Images."

Journal of Forensic Medicine 34 (4): 363–9. https://web.archive.org/web/20200507212629/http://www.fyxzz.cn/EN/article/downloadArticleFile.do?attachType=PDF&id=22907.

Pan Siyu, Zhao Wenting, Feng Rui, Li Qiong, Jing Xiaoxii, Gao Xing Jian, Chen Yan, et al. 2020. "Predicting Male Pattern Baldness from DNA Variants in a Eurasian Population." *Progress in Biochemistry and Biophysics* 47 (10): 1069–79. https://web.archive.org/web/20210926184144/http://www.pibb.ac.cn/pibbcn/ch/reader/download_pdf.aspx?file_no=20200095&year_id=2020&quarter_id=10&falg=1.

Pan Siyu, Zhao Wenting, Tang Kun, Ma Xin, Ye Jian, Li Caixia. 2017. "Research Progress on Human Facial Age Estimation and Synthesis of Age-Correlated Appearance." *Forensic Science and Technology* 42 (4): 270–6. https://web.archive.org/web/20170923085434/http://www.xsjs-cifs.com:80/article/2017/1008-3650-42-4-270.html.

Pang Jingbo, Rao Min, Chen Qingfeng, Ji Anquan, Zhang Chi, Kang Kelai, Wu Hao, Ye Jian, Nie Sheng-Jie, and Wang Le. 2020 "A 124-plex Microhaplotype Panel Based on Next-Generation Sequencing Developed for Forensic Applications." *Scientific Reports* 10. https://doi.org/10.1038/s41598-020-58980-x.

Panofsky, Aaron, and Catherine Bliss. 2017. "Ambiguity and Scientific Authority: Population Classification in Genomic Science." *American Sociological Review* 82 (1): 59–87.

Peebles, P., T. Trisch, and A. Papageorge. 1978. "727 Isolation of Four Unusual Pediatric Solid Tumor Cell Lines." *Pediatric Research* 12 (485). https://doi.org/10.1203/00006450–197804001–00732.

Peng Qianqian, Li Jinxi, Tan Jingze, Yang Yajun, Zhang Manfei, Wu Sijie, Liu Yu, et al. 2016. "EDARV 370 A Associated Facial Characteristics in Uyghur Population Revealing Further Pleiotropic Effects." *Human Genetics* 135 (1): 99–108.

Perdue, Peter C. 2005. *China Marches West: The Qing Conquest of Central Eurasia*. Cambridge, MA: Belknap Press of Harvard University Press.

Perelman, C., and L. Olbrechts-Tyteca. 1969. *The New Rhetoric: A Treatise on Argumentation*. Notre Dame: University of Notre Dame Press.

Petrone J. 2011. "Illumina to Remain Secure If NIH Budget Shrinks as Projected, CEO Predicts." GenomeWeb, 21 September 2011. https://web.archive.org/web/20220926215427/https://www.genomeweb.com/arrays/illumina-remain-secure-if-nih-budget-shrinks-projected-ceo-predicts.

Phillips, Dom. 2018. "Footage of Sole Survivor of Amazon Tribe Emerges." *The Guardian* (UK), 19 July 2018. https://web.archive.org/web/20180803082701/https://www.theguardian.com/world/2018/jul/19/footage-sole-survivor-amazon-tribe-emerges-brazil.

Pośpiech, Ewelina, Chen Yan, Magdalena Kukla-Bartoszek, Krystal Breslin, Anastasia Aliferi, Jeppe Dyrberg Andersen, David Ballard, et al. 2018. "Towards Broadening Forensic DNA Phenotyping beyond Pigmentation: Improving the Prediction of Head Hair Shape from DNA." *Forensic Science International: Genetics* 37:241–51.

Prainsack, Barbara. 2015. "Unchaining Research Processes of Dis/Empowerment and the Social Study of Criminal Law and Investigation." In *Knowledge, Technology and Law*, edited by Emile Cloatre and Martyn Pickersgill, 71–85. New York: Routledge.

Prison Policy Initiative. 2014. "Incarceration Rates in Arizona." Prison Policy Initiative (US). https://web.archive.org/web/20200414235625/https://www.prisonpolicy.org/graphs/2010rates/AZ.html.

Pugliese, Joseph. 1999. "Identity in Question: A Grammatology of DNA and Forensic Genetics." *International Journal for the Semiotics of Law* 12:419–44.

– 2010. *Biometrics: Bodies, Technologies, Biopolitics*. New York: Routledge.

Qi Zhonghao. 2018. "有新闻！公安部物证鉴定中心援建兵团首个DNA实验室在师市交付启用" [There is News! The First DNA Laboratory of the Corps Aided by the Physical Evidence Identification Center of the Ministry of Public Security Was Delivered and Put Into Use in Division City]. *Xuehua News*, 22 June 2018. https://web.archive.org/web/20200630162253/https://www.xuehua.us/a/5eb65bbc86ec4d63e6aa82e4.

Qian Yu, Xiong Ziyi, Li Yi, Manfred Kayser, Liu Lei, and Liu Fan. 2021. "The Effects of Tbx15 and Pax1 on Facial and Other Physical Morphology in Mice." *FASEB BioAdvances* 3 (12): 1011–9.

Qiao Long. 2017. "Former Chinese Judge Jailed over Alleged 'Ethnic Hatred,' Porn on Phone." *Radio Free Asia*, 15 March 2017. https://web.archive.org/web/20190831005904/https://www.rfa.org/english/news/uyghur/judge-03152017120944.html.

Qiao Lu, Yang Yajun, Fu Pengcheng, Hu Sile, Zhou Hang, Tan Jingze, Lu Yan, Lou Haiyi, Lu Dongsheng, Wu Sijie, Guo Jing, Peng Shouneng, Jin Li, Guan Yaqun, Wang Sijia, Xu Shuhua, and Tang Kun. 2016. "Detecting Genome-Wide Variants of Eurasian Facial Shape Differentiation: DNA Based Face Prediction Tested in Forensic Scenario." bioRxiv, Cold Spring Harbor Laboratory (US). https://web.archive.org/web/20230112231330/https://www.biorxiv.org/content/10.1101/062950v3.full.pdf.

Qiao Lu, Yang Yajun, Fu Pengcheng, Hu Sile, Zhou Hang, Peng Shouneng, Tan Jingze, Lu Yan, Lou Haiyi, Lu Dongsheng, Wu Sijie, Guo Jing, Jin Li, Guan Yaqun, Wang Sijia, Xu Shuhua, and Tang Kun. 2018. "Genome-Wide Variants of Eurasian Facial Shape Differentiation and a Prospective Model of DNA Based Face Prediction." *Journal of Genetics and Genomics* 45 (8): 419–32.

Qing Jingyan. 2015." 法制教育培訓學校正式開課" [Legal Education Training School Officially Opened]. *Bingtuan Television* (owned by the Bingtuan in Xinjiang), 4 August 2015. https://web.archive.org/web/20190615154544/https://kknews.cc/education/9m56n5l.html.

Queensland Medical Research Institute. 2020. "Genetic Epidemiology." Queensland Institute of Medical Research (Australia). https://web.archive.org/web/20200311054720/https://www.qimrberghofer.edu.au/lab/genetic-epidemiology/.

Radden, Günter, and Zoltán Kövecses. 1999. "Towards a Theory of Metonymy." In *Metonymy in Language and Thought*, edited by Klaus-Uwe Panther and Günter Radden, 17–59. Amsterdam: John Benjamins.

Radin, Joanna. 2017. *Life on Ice: A History of New Uses for Cold Blood*. Chicago: University of Chicago Press.

Ramzy, Austin, and Chris Buckley. 2019. "'Absolutely No Mercy': Leaked Files Expose How China Organized Mass Detentions of Muslims." *New York Times*, 16 November 2019. https://www.nytimes.com/interactive/2019/11/16/world/asia/china-xinjiang-documents.html.

Reardon, Jenny. 2005. *Race to the Finish: Identity and Governance in an Age of Genomics*. Princeton, NJ: Princeton University Press.

Reardon, Jenny, and Kim TallBear. 2012. "Your DNA Is Our History: Genomics, Anthropology, and the Construction of Whiteness as Property." *Current Anthropology* 53 (S5): S233–S245. https://doi.org/10.1086/662 6292012.

Reardon, Sara. 2014. "Mugshots Built from DNA Data: Computer Program Crudely Predicts a Facial Structure from Genetic Variations." *Nature News* (website), 20 March 2014. https://web.archive.org/web/20140325052609/https://www.nature.com/news/mugshots-built-from-dna-data-1.14899.

Recruitment Program of Global Experts. n.d. "The Recruitment Program for Innovative Talents (Long Term)." China Uni Jobs. https://web.archive.org/web/20190402194707/http://www.1000plan.org/en/.

Ren Ping, Liu Jing, Zhao Hong, Fan Xiuping, Xu Youchun, and Li Caixia. 2019. "Construction of a Rapid Microfluidic-based SNP Genotyping (MSG) Chip for Ancestry Inference." *Forensic Science International: Genetics* 41:145–51.

Reynolds, Laurence, and Bronislaw Szerszynski. 2012. "Neoliberalism and Technology: Perpetual Innovation or Perpetual Crisis?" In *Neoliberalism and Technoscience. Theory, Technology and Society*,

edited by Luigi Pellizzoni and Marja Ylönen, 27–46. Farnham, UK: Ashgate.
Risch, Neil, and Bernie Devlin. 1992. "Letter to the Editor." *Science* 256:1745–6.
Roberts, Dorothy. 2011a. *Fatal Invention: How Science, Politics, and Big Business Re-Create Race in the Twenty-First Century*. New York: The New Press.
– 2011b. "Collateral Consequences, Genetic Surveillance, and the New Biopolitics of Race." *Howard Law Journal* 54 (3): 567–86.
– 2013. "Law, Race, and Biotechnology: Toward a Biopolitical and Transdisciplinary Paradigm." *Annual Review of Law and Social Science* 9 (1): 149–66
Roberts, Leslie. 1991a. "A Genetic Survey of Vanishing Peoples." *Science* (American Association for the Advancement of Science) 252 (5013): 1614–17.
– 1991b. "Was *Science* Fair to Its Authors?" *Science* 254 (5039): 1722.
– 1992a. "Prosecutor v. Scientist: A Cat and Mouse Relationship." *Science* 257:733–6.
– 1992b. "Science in Court: A Culture Clash." *Science* 257:732; 735.
Roberts, Sean. 2018a. "The Narrative of Uyghur Terrorism and the Self-Fulfilling Prophecy of Uyghur Militancy." In *Terrorism and Counter-Terrorism in China,* edited by Michael Clarke, 99–127. Oxford: Oxford University Press.
– 2018b. "The Biopolitics of China's 'War on Terror' and the Exclusion of the Uyghurs." *Critical Asian Studies* 50 (2): 232–58.
– 2020. *The War on the Uyghurs: China's Internal Campaign against a Muslim Minority*. Princeton, NJ: Princeton University Press.
Rodrigues, Thiago, and Mariana Kalil. 2021. "A Military-Green Biopolitics: The Brazilian Amazon Between Security and Development." *Brazil Political Science Review* 15 (2): e0005. https://web.archive.org/web/20210519181824/https://brazilianpoliticalsciencereview.org/wp-content/uploads/articles_xml/1981-3821-bpsr-15-2-e0005/1981-3821-bpsr-15-2-e0005.x89995.pdf.
Rodríguez-Merino, Pablo A. 2019. "Old 'Counterrevolution,' New 'Terrorism': Historicizing the Framing of Violence in Xinjiang by the Chinese State." *Central Asian Survey* 38 (1): 27–45.
Rohter, L. 2007. "In the Amazon, Giving Blood but Getting Nothing." *New York Times*, 20 June 2007. https://web.archive.org/web/20150605071915/http://www.nytimes.com/2007/06/20/world/americas/20blood.html.
Rollins, Oliver. 2018. "Risky Bodies: Race and the Science of Crime and Violence." In *Living Racism: Through the Barrel of the Book*, edited by

Theresa Rajack-Talley and Derrick R. Brooms, 97–118. Lanham, MD: Lexington Books.

Rose, Nikolas. 1999. *Powers of Freedom: Reframing Political Thought*. Cambridge: Cambridge University Press.

Rossi, Faust F. 1991. *Expert Witnesses*. Chicago: American Bar Association.

Rural Action Foundation International. 1993. "Indigenous People Protest U.S. Secretary of Commerce Patent Claim on Guaymi Indian Cell Line," press release, Rural Action Foundation International, 25 October 1993.

Ryan, Hollie. 2009. "Biometrics and the 'Multiple Use Dilemma': Enabling Post 9/11 National Security by Understanding a New Technology Tool." In *Military Intelligence Professional Bulletin*, 59–66, U.S. Army Intelligence Center, Fort Huachuca, AZ. https://web.archive.org/web/20180918034121/https://www.governmentattic.org/21docs/MIPBjan-Mar2009.pdf.

Sachs, Jessica Snyder. 2004. "DNA and a New Kind of Racial Profiling." *Popular Science*, 11 June 2004. https://web.archive.org/web/20090803100814/https://www.popsci.com/scitech/article/2004-06/dna-and-new-kind-racial-profiling/#.

Santos, Ricardo Ventura. 2002. "Indigenous People, Postcolonial Context, and Genomic Research in the Late 20th Century: A View from Amazonia (1960–2000)." *Critique of Anthropology* 22 (1): 81–104.

Sarat, Austin. 1995. "Violence, Representation, and Responsibility in Capital Trials: The View from the Jury." *Indiana Law Journal* 70 (4): Article 3. https://web.archive.org/web/20200319180955if_/https://www.repository.law.indiana.edu/cgi/viewcontent.cgi?referer=&httpsredir=1&article=1708&context=ilj.

Sassen, Saskia. 2006. *Territory, Authority, Rights: From Medieval to Global Assemblages*. Princeton, NJ: Princeton University Press.

Schmitt, Carl. [1950] (2003) *The Nomos of the Earth in the International Law of the Jus Publicum Europaeum*. New York: Telos Press.

Schwarck, E. 2018. "Intelligence and Informatization: The Rise of the Ministry of Public Security in Intelligence Work in China." *The China Journal* 80 (1): 1–23.

Sciping.com. 2018. "2017年国家重点研发计划重点专项公示立项项目清单（excel 截止2018.5.4，49个专项，1309个项目，超264亿元）" [2017 National Key Research and Development Program Key Special Public Announcement Project List (excel as of 2018.5.4, 49 special projects, 1309 projects, over 26.4 billion yuan)]. *Sciping.com* (China), 17 November 2018. https://web.archive.org/web/20201217105048/https://www.sciping.com/22487.html.

Seigel, Micol. 2018. *Violence Work: State Violence and the Limits of Police*. Durham, NC: Duke University Press,

Sepich, Jayann. 2017. "DNA Databases: Solving Crimes and Saving Lives…. A Mother's Story" (ellipsis in original). PowerPoint presentation. Human Identification Solutions Conference, Vienna, Austria, 3–4 May 2017. https://web.archive.org/web/20200823145241/https://www.thermofisher.com/content/dam/LifeTech/Documents/PDFs/hids/04%20Jayann%20Sepich.pdf.

Seto, Ken-ichi. 1999. "Distinguishing Metonymy from Synecdoche." In *Metonymy in Language and Thought*, edited by Klaus-Uwe Panther and Günter Radden, 91–120. Amsterdam : John Benjamins.

Shanghai Institute of Nutrition and Health. n.d. "People: Wang Sijia." Shanghai Institute of Nutrition and Health. https://web.archive.org/web/20220303093717/http://english.sinh.cas.cn/people/fs/201809/t20180920_197703.html.

Shea, E. 2008. *How the Gene Got Its Groove: Figurative Language, Science, and the Rhetoric of the Real*. New York: New York University Press.

Sheard, Cynthia M. 1996. "The Public Value of Epideictic Rhetoric." *College English* 58 (7): 765–94.

Skidmore, Thomas E. 1988. *The Politics of Military Rule in Brazil, 1964–85*. New York: Oxford University Press.

Smaling, Elmer. 2021. "Erasmus MC Can't Just Throw the Uyghurs under the Bus, Can It?" *Erasmus Magazine* (the Netherlands), 22 September 2021. https://web.archive.org/web/20211015180120/https://www.erasmusmagazine.nl/en/2021/09/22/erasmus-mc-cant-just-throw-the-uyghurs-under-the-bus-can-it/.

Smith, Anthony D. 2000. "The 'Sacred' Dimension of Nationalism." *Millennium: Journal of International Studies* 29 (3): 791–814.

Smith, David Glenn, Joseph Lorenz, Becky K. Rolfs, Robert L. Bettinger, Brian Green, Jason Eshleman, Beth Schultz, and Ripan Malhi. 2000. "Implications of the Distribution of Albumin Naskapi and Albumin Mexico for New World Prehistory." *American Journal of Physical Anthropology* 111:557–72.

Smith Finley, Joanne. 2019. "The Wang Lixiong Prophecy: 'Palestinization' in Xinjiang and the Consequences of Chinese State Securitization of Religion." *Central Asian Survey* 38 (1): 81–101.

Smith, Linda Tuhiwai. 1999. *Decolonizing Methodologies: Research and Indigenous Peoples*. London: Zed Books.

Smith, Lindsay A., and Vivette García-Deister. 2021. "Genetic Syncretism: Latin American Forensics and Global Indigenous Organizing." *BioSocieties* 16 (4): 447–69.

Soares, Marília Facó. 2018. "Ticuna." Povos Indígenas no Brasil. https://web.archive.org/web/20200824014256/https://pib.socioambiental.org/en/Povo:Ticuna.

Soundararajan, Usha, Yun Libing, Shi Meisen, and Kenneth K. Kidd. 2016. "Minimal SNP Overlap among Multiple Panels of Ancestry Informative Markers Argues for More International Collaboration." *Forensic Science International: Genetics* 23:25–32.

Southern Medical University. 2020. "'首届南方法医高峰论坛–暨网络云讲堂' - 法庭科学云峰会成功举办)" ['The First Southern Forensic Medicine Summit Forum' – and 'Network Cloud Lecture Hall' – Forensic Science Cloud Summit was successfully held]. Southern Medical University (China), 27 May 2020. https://web.archive.org/web/20211025190417/https://www.smu-sfjdzx.com/archives/1774.

Star, Susan, and James Griesemer. 1989. "Institutional Ecology, 'Translations' and Boundary Objects: Amateurs and Professionals in Berkeley's Museum of Vertebrate Zoology, 1907–39." *Social Studies of Science* 19 (3): 387–420.

State Council Information Office of the People's Republic of China (2019a). "The Fight Against Terrorism and Extremism and Human Rights Protection in Xinjiang." State Council Information Office of the People's Republic of China, 18 March 2019. https://web.archive.org/web/20190420013936/http://english.gov.cn/r/Pub/GOV/ReceivedContent/Other/2019-03-18/WhitePaper.docx.

– 2019b. "Vocational Education and Training in Xinjiang." White Paper by Chinese Government. http://web.archive.org/web/20191209065936/http://www.xinhuanet.com/english/download/VocationalEducationandTrainingin Xinjiang.docx.

Steen, Gerard. 2005. "Metonymy." In *The Routledge Encyclopedia of Narrative Theory*, edited by D. Herman, M. Jahn, and M. L. Ryan, 307–8. London: Routledge.

St Louis, Brett. 2022. "Race as Technology and the Carceral Methodologies of Molecular Racialization." *The British Journal of* Sociology, 73l(1): 206–19.

Sullivan Patrick J. 1992. "DNA Fingerprint Matches." *Science* 256 (5065): 1743–6.

Statistics Bureau of Xinjiang Uyghur Autonomous Region. 2017. "3–7 各地、州、市、县(市)分民族人口数" [3–7 Population by Nationality by Prefecture, Prefecture, City and County]. Xinjiang provincial government, 15 March 2017. https://web.archive.org/web/20191102062155/http://www.xjtj.gov.cn/sjcx/tjnj_3415/2016xjtjnj/rkjy/201707/t20170714_539450.html.

Stephens, Kathryn M., Cydne Holt, Carey Davis, Anne Jager, Paulina Walichiewicz, Yonmee Han, David Silva, et al. 2019. "Methods and Compositions for DNA Profiling." US patent US10422002B2, filed 13 February 2015; issued 24 September 2019.

Storto, Luciana, and Felipe F. Vander Velden. 2018. "Karitiana." Povos Indígenas no Brasil. https://web.archive.org/web/20181209123815/https://pib.socioambiental.org/en/Povo:Karitiana.

Strumpf, Dan, and Natasha Khan. 2018a. "US Lawmakers Say U.S. Tech Is Being Used in Abusive Chinese Crackdowns." *Wall Street Journal*, 15 May 2018. Factiva Database.

– 2018b. "U.S. Tech Sent to Chinese Police Was Within Rules, Commerce Dept. Says." *Wall Street Journal*, 26 July 2018. Factiva Database.

Sun Jian. 2017. "为了祖国利益 向"三股势力"全面宣战!" [For the Sake of the Motherland, We Declare Total War on the Three Forces!]. *Sina.com.cn*, 20 April 2017. https://web.archive.org/web/20170421021321/http://news.sina.com.cn/c/2017-04-21/doc-ifyepnea4367007.shtml.

Sun Qifan, Zhao Lei, Jiang Li, Quan Yangke, Zhao Xingchun, and Li Caixia. 2015. "Forensic Application of Human Characters Inferred from DNA." *Forensic Science and Technology* (China) 40 (3). https://web.archive.org/web/20200811160009/http://www.xsjs-cifs.com/article/2015/1008-3650-40-3-232.html.

TallBear, Kim. 2013. *Native American DNA: Tribal Belonging and the False Promise of Genetic Science*. Minneapolis: University of Minnesota Press.

Tang Kun. 2014. "Seeing Faces and History through Human Genome Sequences." *Bulletin of the Chinese Academy of Sciences* 28 (2): 161–5. https://web.archive.org/web/20190422014925/http://english.cas.cn/bcas/2014_2/201411/P020141121529624692677.pdf.

Tang Kun, and Guo Jianya. 2016. "三维人脸的自动定位方法、曲面配准方法和系统" [Three-Dimensional Face Automatic Positioning Method, Curved Surface Registration Method and System]. PRC patent application CN105740851A, filed 16 March 2016.

Tavares, Paulo. 2013. "The Geological Imperative: On the Political Ecology of the Amazonia's Deep History." In *Architecture in the Anthropocene: Encounters Among Design, Deep Time, Science and Philosophy*, edited by Etienne Turpin, 209–39. Ann Arbor: Open Humanities Press.

Taylor, Ian (dir.). 1995. *The Gene Hunters*. Films for the Humanities & Sciences.

The People of the State of California vs. Paul Eugene Robinson. 2002. Testimony of Ranajit Chakraborty. Reporters Transcript of Daily Proceedings. National Legal Aid and Defender Association, 30 December 2002. https://ia601408.us.archive.org/9/items/2002.-people.v.-robinson/2002.People.v.Robinson.pdf.

Thermo Fisher. 2016. "HIDS 2016–Our Speakers." Thermo Fisher. https://web.archive.org/web/20211114005449/https://www.thermofisher.com/us/en/

home/industrial/forensics/human-identification/human-identification-solutions-conference-2016/hids-conference-speakers-2016.html.
– 2017a. "法庭科学新技术应用高峰论坛" [Revolutionary Forensics: Answers from CE to NGS]. Bioon Conference, virtual, 1–3 November 2017. https://web.archive.org/web/20200407184312/http://www.bioon.com/z/thermofisher_forensicscience/index.html.
– 2017b. "高峰论坛回顾" [Summit Forum Review]. QQ video platform. https://v.qq.com/x/page/s0508v8r3iz.html.
– 2018a. *2017 Annual Report*. Thermo Fisher. https://web.archive.org/web/20200805230110/https://s1.q4cdn.com/008680097/files/doc_financials/annual/2017/2017-Annual-Report-TMO.pdf.
– 2018b. Human Identification Symposium. Thermo Fisher. https://web.archive.org/web/20180208234242/https://www.thermofisher.com/ca/en/home/industrial/forensics/human-identification/human-identification-solutions-conference-2018.html.
– 2019. "Updated Statement on Xinjiang," contained in a letter by Yves Moreau to the Office of the High Commissioner for Human Rights, United Nations, 20 February. https://web.archive.org/web/20200805222615/https://www.ohchr.org/Documents/HRBodies/HRCouncil/WGTransCorp/Session4/SubmissionLater/YvesMoreau.pdf.
– 2020. "Company Facts." Thermo Fisher. https://web.archive.org/web/20200623090730/https://ThermoFisher.mediaroom.com/fact-sheet.
– 2021. Human Identification Symposium. Virtual conference website. Hosted by labroots.com. https://web.archive.org/web/20220127121944/https://www.labroots.com/ms/virtual-event/hids-2021#/.
Thompson, W.C. 1993. "Evaluating the Admissibility of New Genetic Identification Tests: Lessons from the DNA War." *Journal of Criminal Law and Criminology* 84 (1): 22–104.
Thorup, Mikkel. 2009. "Enemy of Humanity: The Anti-Piracy Discourse in Present-Day Anti-Terrorism." *Terrorism and Political Violence* 21 (3): 401–11.
Tobin, D. 2020. "A 'Struggle of Life or Death': Han and Uyghur Insecurities on China's North-West Frontier." *The China Quarterly* 242:301–23.
Toom, Victor. 2012. "Bodies of Science and Law: Forensic DNA Profiling, Biological Bodies, and Biopower." *Journal of Law and Society* 39 (1): 150–66.
Toom, V., M. Wienroth, A. M'charek, B. Prainsack, R. Williams, T. Duster, T. Heinemann, C. Kruse, H. Machado, and E. Murphy. 2016. "Approaching Ethical, Legal and Social Issues of Emerging Forensic DNA Phenotyping (FDP) Technologies Comprehensively: Reply to 'Forensic DNA Phenotyping: Predicting Human Appearance from Crime Scene Material for Investigative

Purposes' by Manfred Kayser." *Forensic Science International: Genetics* 22:e1–e4.

Torronen, J. 2000. "The Passionate Text. The Pending Narrative as a Macrostructure of Persuasion." *Social Semiotics* 10 (1): 81–98.

Track, Rowena K., Florence C. Ricciuti, and Kenneth K. Kidd. 1989. "Information on DNA Polymorphisms in the Human Gene Mapping Library." In *DNA Technology and Forensic Science*, edited by Jack Ballantyne, George Sensabaugh, and Jan A. Witkowski, 335–45. Cold Spring Harbor: Cold Spring Harbor Laboratory.

Tsosie, Rebecca. 2011. "Indigenous Peoples and Epistemic Injustice: Science, Ethics, and Human Rights." *Washington Law Review* 87:1133–1201.

Tully, James. 1980. *A Discourse on Property: John Locke and His Adversaries*. Cambridge: Cambridge University Press.

Tully, John Andrew. 2011. *The Devil's Milk: A Social History of Rubber*. New York: Monthly Review Press.

Turner, Victor W. 1969. *The Ritual Process: Structure and Anti-Structure*. Chicago: Aldine Publishing.

TwinsUK. 2020. "About Us." King's College London UK. https://web.archive.org/web/20210410093021/https://twinsuk.ac.uk/about-us/what-is-twinsuk/.

uighurbiz.net. 2013. "图木舒克房屋遭强拆的维吾尔兄弟二人遭抓捕. 2013 [Two Uighur Brothers Whose House Was Demolished in Tumxuk Were Arrested]. *Uighurbiz.net* (China), 26 July 2013. https://web.archive.org/web/20130731010449/http://www.uighurbiz.net/archives/17575.

UK Research and Innovation. n.d. "Award Details: The Genetics of Human Physical Appearance (BB/I021213/1)." UK Research and Innovation. https://web.archive.org/web/20190325024727/https://gtr.ukri.org/projects?ref=BB%2FI021213%2F1.

Underhill, Peter A., Jin Li, Alice A. Lin, S. Qasim Mehdi, Trefor Jenkins, Douglas Vollrath, Ronald W. Davis, L. Luca Cavalli-Sforza, and Peter J. Oefner. 1997. "Detection of Numerous Y Chromosome Biallelic Polymorphisms by Denaturing High-Performance Liquid Chromatography." *Genome Research* 7:996–1005. Cold Spring Harbor Laboratory Press. https://web.archive.org/web/20200907232804/https://genome.cshlp.org/content/7/10/996.full.pdf+html.

United Nations Declaration on the Rights of Indigenous Peoples. 2007. United Nations. https://web.archive.org/web/20220402133240/http://www.un.org/esa/socdev/unpfii/documents/DRIPS_en.pdf.

United States v. Bonds. 1993. 12 F.3d 540 (6th Cir.). https://web.archive.org/web/20220320100643/https://casetext.com/case/us-v-bonds-4.

United States v. Yee. 1991. 134 F.R.D. 161. https://web.archive.org/web/20230114143334/https://casetext.com/case/us-v-yee/.

University of California San Diego. 1976. "Year-Long, Seven-Phase Study of Physical and Biological Phenomena along the Amazon River." University of California at San Diego University Archives, 3 March 1976. https://web.archive.org/web/20200823232331/https://library.ucsd.edu/dc/object/bb7032041c/_2.pdf.

US Department of Commerce. 2020. "Commerce Department to Add Nine Chinese Entities Related to Human Rights Abuses in the Xinjiang Uighur Autonomous Region to the Entity List." US Department of Commerce, 22 May 2020. https://web.archive.org/web/20200605144836/https://www.commerce.gov/news/press-releases/2020/05/commerce-department-add-nine-chinese-entities-related-human-rights.

US Department of Commerce Bureau of Industry and Security. 2020. "Supplement No. 4 to Part 744 – Entity List 2020–06–05." https://web.archive.org/web/20210614171236/https://www.bis.doc.gov/index.php/documents/regulations-docs/2347-744-supp-4-6/file.

US Department of Justice. n.d. "Alternative Genetic Markers." US Department of Justice. https://web.archive.org/web/20090812230143/http://dna.gov/research/alternative_markers/.

– 2003. "Advancing Justice through DNA Technology: Executive Summary." US Department of Justice. https://web.archive.org/web/20100120075642/http://www.justice.gov/ag/dnapolicybook_exsum.htm.

US Senate. 1993. "Human Genome Diversity Project." Hearing Before the Committee on Governmental Affairs United States Senate, 103rd Congress, First Session, 26 April 1993. ISBN 0–16–043334–7. https://ia800200.us.archive.org/17/items/humangenomediver00unit/humangenomediver00unit.pdf.

US Treasury Department. 2020. "Xinjiang Supplied Chain Business Advisory: Risks and Considerations for Businesses with Supply Chain Exposure to Entities Engaged in Forced Labor and other Human Rights Abuses in Xinjiang." US State Department, 1 July 2020. https://web.archive.org/web/20200702013916/https://www.treasury.gov/resource-center/sanctions/Programs/Documents/20200701_xinjiang_advisory.pdf.

Uyghur Human Rights Project. 2018. "The Bingtuan: China's Paramilitary Colonizing Force in East Turkestan." Uyghur Human Rights Project. https://web.archive.org/web/20200604200135/https://docs.uhrp.org/pdf/bingtuan.pdf.

Valverde, Mariana. 1996. "'Despotism' and Ethical Liberal Governance." *Economy and Society* 25:357–72.

Vander Velden, Felipe Ferreira. 2004. "Por onde o sangue circula: os Karitiana e a intervenção biomedical." [Where Blood Circulates: The Karitiana and Biomedical Intervention]. Master's thesis, Universidade Estadual de

Campinas. https://web.archive.org/web/20230114152403/http://repositorio.unicamp.br/Busca/Download?codigoArquivo=496836.
– 2005. "Corpos que sofrem: uma interpretação Karitiana dos eventos de coleta de seu sangue" [Bodies That Suffer: A Karitiana Interpretation of the Events of their Blood Collection]. Working document for Centro de Estudos em Saúde do índio de Rondônia, Federal University of Rondônia (Brazil). https://web.archive.org/web/20200824154411/http://www.cesir.unir.br/pdfs/doc12.pdf.
– 2010. "De volta para o passado: territorialização e 'contraterritorialização' na história Karitiana" [Back to the Past: Territorialization and 'Counter Territorialization' in Karitiana History]. *Sociedade e Cultura* 13:55–65. https://web.archive.org/web/20211202003411/https://revistas.ufg.br/fcs/article/download/11173/7336/43551.
Van Noorden, Richard, and Davide Castelvecchi. 2019. "Science Publishers Review Ethics of Research on Chinese Minority Groups." *Nature* 576 (6 December): 192–3.
Van Veeren, Elspeth. 2020. "The Hunt as Security Logic: New Reflections on the Power and Politics of Hunting as Secrecy Practice through Twenty Years of the Global War on Terror." Working Paper, University of Bristol. https://web.archive.org/web/20220808131408/http://www.bristol.ac.uk/media-library/sites/spais/Van%20Veeren%202020%20Hunting%20for%2comments%20(final).pdf.
Velasco-Sacristán, Marisol. 2010. "Metonymic Grounding of Ideological Metaphors: Evidence from Advertising Gender Metaphors." *Journal of Pragmatics* 42 (1): 64–96.
Verogen. 2022. "First Criminal Conviction Secured with Next-Gen Forensic DNA Technology." LinkedIn (US). https://www.linkedin.com/company/verogen.
VISAGE. n.d.. "About the VISAGE Consortium." (European Union website). https://web.archive.org/web/20200721223801/http://www.visage-h2020.eu/.
– 2019a. "D2.2 Report on New Markers for Predicting as Detailed as Possible Appearance, Age and Ancestry from DNA." Report for (Epi)Genetic Markers for Constructing Composite Sketches from DNA Project. European Commission. https://web.archive.org/web/20200904013636/https://ec.europa.eu/research/participants/documents/downloadPublic/SopEMW90QmtkWGhSSmsvUotqRkQyZWQzLzBDWk5OWE1Vb1I3NnFTZThUc3MoRoNBQldPdodBPTo=/attachment/VFEyQTQ4M3ptUWZzdk9qWkVNWG80U2hyUHlxT3ZXUHM=.
– 2019b. "D4.1 Report on New Integrative Statistical Framework for Combined Appearance, Age, and Ancestry rediction from DNA." Report for Integrative Statistical Framework with Prototype Software for

Constructing Composite Sketches from DNA Project. European Commission. https://web.archive.org/web/20200904021547/https://ec.europa.eu/research/participants/documents/downloadPublic/bDNUMldScVc1boN3ekx2WWswMmNwNitSeS9FcHMvS2tnVzZGU2M5OUZtNFNYaWNXY2xYSHFRPT0=/attachment/VFEyQTQ4M3ptUWU5N2tUZ2FQNFVidGVweGJJYkJmVmE=.

VisiGen Consortium. 2010a."The Visigen Consortium." VisiGen Consortium. https://web.archive.org/web/20190506164605/https://sites.google.com/site/visigenconsortium/home.

– 2010b. "Recent Site Activity." VisiGen Consortium. https://web.archive.org/web/20190506164638/https://sites.google.com/site/visigenconsortium/system/app/pages/recentChanges.

Volleth, Marianne, Martin Zenker, Ivana Joksic, and Thomas Liehr. 2020. "Long-Term Culture of EBV-Induced Human Lymphoblastoid Cell Lines Reveals Chromosomal Instability." *Journal of Histochemistry and Cytochemistry* 68 (4): 239–51.

Wachowski, Wojciech. 2019. *Towards A Better Understanding of Metonymy*. Oxford, UK: Peter Lang.

Walker, Robert S., Lisa Sattenspiel, and Kim R. Hill. 2015. "Mortality from Contact Related Epidemics among Indigenous Populations in Greater Amazonia." *Scientific Reports* (Nature Research) 5:14032.

Walsh, John J. 2006. "R v. Allan Joseph Legere and DNA Evidence: Reminiscences." Paper prepared for Digital Law Collection Project, University of New Brunswick Law School. https://web.archive.org/web/20190310060830/https://www.unb.ca/fredericton/law/library/_resources/pdf/legal-materials/allan-legere/comms_bibliography/legere_trial_digital_collection__r__v__allan_joseph_legere_.pdf.

Walsh, Susan. 2019a. "Curriculum Vitae." Indiana University's Walsh Lab. https://web.archive.org/web/20200521205659/https://walshlab.sitehost.iu.edu/WalshCV_0119.pdf.

– 2019b. "2014-DN-BX-K031: Improving the Prediction of Human Quantitative Pigmentation Traits such as Eye, Hair and Skin Color using a Worldwide Representation Panel of US and European Individuals." National Criminal Justice Reference Service (US). https://web.archive.org/web/20190910185537/https://www.ncjrs.gov/pdffiles1/nij/grants/253066.pdf.

Wang Yan. 2012. "耶鲁大学基德教授作客"睿德信全球法学家论坛" [Professor Kidd from Yale University was a Guest at the 'Ridexin Global Jurist Forum']. Southwest University of Political Science and Law (China). https://web.archive.org/web/20180602010051/http://news.swupl.edu.cn/xzxz/18464.htm.

Weber, Max. 1978. *Economy and Society: An Outline of Interpretive Sociology*. Edited by Guenther Roth and Claus Wittich. 2 vols. Berkeley: University of California Press.

Wee Sui-Lee. 2019. "American DNA Expertise Helps Beijing Crack Down." *New York Times*, 21 February 2019. https://web.archive.org/web/20190221234632/https://www.nytimes.com/2019/02/21/business/china-xinjiang-uighur-dna-thermo-fisher.html.

– 2021. "Two Scientific Journals Retract Articles Involving Chinese DNA Research." *New York Times*, 9 September 2021. https://web.archive.org/web/20220819224628/https://www.nytimes.com/2021/09/09/business/china-dna-retraction-uyghurs.html.

Wee Sui-Lee, and Paul Mozur. 2019a. "China Mines DNA To Map Out Faces with West's Help: Foreign Desk." *New York Times*, 3 December 2019. https://web.archive.org/web/20230113103019/https://www.nytimes.com/2019/12/03/business/china-dna-uighurs-xinjiang.html.

– 2019b. "China's Genetic Research on Ethnic Minorities Sets Off Science Backlash" *New York Times*, 5 December 2019. Https://web.archive.org/web/20191205001115/https://www.nytimes.com/2019/12/04/business/china-dna-science-surveillance.html.

Weheliye, Alexander G. 2014. *Habeas Viscus: Racializing Assemblages, Biopolitics, and Black Feminist Theories of the Human*. Durham, NC: Duke University Press.

Wei Xiaogang. 2017. "第三师图木舒克市公安局被确定为公安部物证" [February 22 Report on Coop Agreement Tumxuk DNA Lab Proposed]. Tumushukezx.com (Tumxuk). https://web.archive.org/web/20200812224941/http://www.tumushukezx.com/tmsksls/5.html.

Wei Xiaogang, and Wang Yongqiang. 2018. "三师公安司法鉴定中心DNA实验室交付使用" [Third Division DNA Judicial Identification Lab Delivered]. Xinjiang Bingtuan (website), 7 February 2018. https://web.archive.org/web/20190815132825/http://www.xjbtnss.gov.cn/xinwen/shishi xinwen/20180207/53553.html.

Wei Yi-Liang, Wei Li, Zhao Lei, Sun Qi-Fan, Jiang Li, Zhang Tao, Liu Hai-Bo, et al. 2016a. "A Single Tube 27-Plex SNP Assay for Estimating Individual Ancestry and Admixture from Three Continents." *International Journal of Legal Medicine* 130 (1): 27–37.

Wei Yi-Liang, Sun Qi-Fan, Li Qing, Yi Jun-Ling, Zhao Lei, Ou Yuan, Jiang Li, et al. 2016b. "Genetic Structure and Differentiation Analysis of a Eurasian Uyghur Population by Use of 27 Continental Ancestry Informative SNPs." *International Journal of Legal Medicine* 130 (4): 897–903.

– 2016c. "Supplemental Materials: Table S1 – Data and Sample Information for 'Genetic Structure and Differentiation Analysis of a Eurasian Uyghur

Population by Use of 27 Continental Ancestry Informative SNPs.'" *International Journal of Legal Medicine* 130 (4).

Weiss, Kenneth M., Kenneth K. Kidd, and Judith M. Kidd. 1992. "Human Genome Diversity Project." *Evolutionary Anthropology* 1:80–2.

Weiqi Hao, Liu Jing, Jiang Li, Huang Meisha, Li Jiuling, Ma Quan, Liu Chao, Li Caixia, and Wang Huijun. 2018. "The Study of a SNP-Multiplex for the Ancestry Inference of Five Continental Populations." *Acta Universitatis Medicinalis Nanjing (Natural Science)*, 38 (3): 331–7. http://web.archive.org/web/20201212173302/http://jnmu.njmu.edu.cn/zr/ch/reader/create_pdf.aspx?file_no=aumn180310&flag=1&journal_id=aumn&year_id=2018.

Wendt, Frank R., Jennifer D. Churchill, Nicole M.M. Novroski, Jonathan L. King, Jillian Ng, Robert F. Oldt, Kelly L. McCulloh, et al. 2016. "Genetic Analysis of the Yavapai Native Americans from West Central Arizona Using the Illumina MiSeq FGx™ Forensic Genomics System." *Forensic Science International: Genetics* 24:18–23.

Wikoff, Dave and Mike Holmes. 2009. "White Paper: Biometric and Forensic Support to Irregular Warfare." In *Military Intelligence Professional Bulletin*, 7–11. U.S. Army Intelligence Center: Fort Huachuca, AZ. https://web.archive.org/web/20180918034121/https://www.governmentattic.org/21docs/MIPBjan-Mar2009.pdf.

Will, J.F. 2003. "Comment: DNA As Property: Implications on the Constitutionality of DNA Dragnets." *University of Pittsburg Law Review* 65:129–43.

Wolfe, Patrick. 2006. "Settler Colonialism and the Elimination of the Native." *Journal of Genocide Research* 8 (4): 387–409.

Wood, Andrew R., Tonu Esko, Yang Jian, Sailaja Vedantam, Tune H. Pers, Stefan Gustafsson, Audrey Y. Chu, et al. 2014. "Defining the Role of Common Variation in the Genomic and Biological Architecture of Adult Human Height." *Nature Genetics* 46:1173–86.

World Intellectual Property Organization. 2020. "Statistical Country Profiles: China." World Intellectual Property Organization. http://web.archive.org/web/20200516232251/https://www.wipo.int/ipstats/en/statistics/country_profile/profile.jsp?code=CN.

Wu Sijie, Tan Jingze, Yang Yajun, Peng Qianqian, Zhang Manfei, Li Jinxi, Lu Dongsheng, et al. 2016. "Genome-Wide Scans Reveal Variants at EDAR Predominantly Affecting Hair Straightness in Han Chinese and Uyghur Populations." *Human Genetics* 135 (11): 1279–86.

Wu Sijie, Zhang Manfei, Yang Xinzhou, Peng Fuduan, Juan Zhang, Tan Jingze, Yang Yajun, et al. 2018. "Genome-Wide Association Studies and CRISPR/Cas9-Mediated Gene Editing Identify Regulatory Variants

Influencing Eyebrow Thickness in Humans." *PLoS Genetics* 14 (9): e1007640–e1007640.

Xi Jinping. 2018. "Xi Jinping Working Together to Build a Better World." *News from China* 30 (2): 26–31. Embassy of the People's Republic of China in the Republic of India. https://web.archive.org/web/20211113200500/https://www.fmprc.gov.cn/ce/cein/chn/xwfw/zgxw/P020180302710745450671.pdf.

Xinhua. 2020. "Xi Focus: Xi Stresses Building Xinjiang Featuring Socialism with Chinese Characteristics in New Era." *Xinhua News Agency* (Chinese government), 26 September 2020. https://web.archive.org/web/20210204103028/http://www.xinhuanet.com/english/2020-09/26/c_139399524.htm.

Xiong Ziyi, Zhou Haibo, Gabriela Dankova, Laurence J. Howe, Myoung Keun Lee, Pirro G. Hysi, Markus A. de Jong, et al., on behalf of the International Visible Trait Genetics Visible Trait Genetics (VisiGen) Consortium. 2018. "Novel Genetic Loci Affecting Facial Shape Variation in Humans." SSRN repository. https://papers.ssrn.com/sol3/papers.cfm?abstract_id=3307375.

Xiong Ziyi, Gabriela Dankova, Laurence J. Howe, Myoung Keun Lee, Pirro G. Hysi, Markus A. de Jong, Gu Zhu, et al., on behalf of the International Visible Trait Genetics Visible Trait Genetics (VisiGen) Consortium. 2019. "Novel Genetic Loci Affecting Facial Shape Variation in Humans." *eLife* 8:e49898. https://web.archive.org/web/20220120203226/https://elife-sciences.org/download/aHR0cHM6Ly9jZG4uZWxpZmVzY2llbmNlcy5vcmcvYXJ0aWNsZXMvNDk4OTgvZWxpZmUtNDk4OTgtdjMucGRm-P2Nhbm9uaWNhbFVyaT1odHRwczovL2VsaWZlc2NpZW5jZXMub3JnL-2FydGljbGVzLzQ5ODk4/elife-49898-v3.pdf?_hash=nhS3JA1GoOQt4QZpnP6go8rJQDtg47V1jKCyw658N5M%3D.

Xu Aiting, Cheng Caijuan, Qiu Keyang, Wang Xiaoxu, and Zhu Yuhan. 2021. "Innovation Policy and Firm Patent Value: Evidence from China." *Economic Research-Ekonomska Istraživanja*. DOI: 10.1080/1331677X.2021.1970607.

Xu Shuhua, Huang Wei, Qian Ji, and Jin Li. 2008. "Analysis of Genomic Admixture in Uyghur and Its Implication in Mapping Strategy." *American Journal of Human Genetics* 82:883–94.

Xu Shuhua, and Jin Li. 2008. "A Genome Wide Analysis of Admixture in Uyghurs and a High-Density Admixture Map for Disease Gene Discovery." *American Journal of Human Genetics* 83 (3): 322–36.

Yale University. 2022. "Kenneth K. Kidd Profile." Yale University School of Medicine.

https://web.archive.org/web/20221222130010/https://medicine.yale.edu/profile/kenneth-kidd/.

Ye Jian, Esteban J, Parra, Donna M. Sosnoski, Kevin Hiester, P. A. Underhill, and Mark D. Shriver. 2002. "Melting Curve SNP (McSNP) Genotyping: A Useful Approach for Diallelic Genotyping in Forensic Science." *Journal of Forensic Sciences* 47 (3): 593–600.

Young, Iris Marion. 2003. "The Logic of Masculinist Protection: Reflections on the Current Security State." *Journal of Women in Culture and Society* 29 (1): 1–25.

Zenz, Adrian. 2018. "Table 1: List of Government Bids Related to Re-Education Facilities." Jamestown Foundation (US). https://web.archive.org/web/20210427221922/https://jamestown.org/programs/cb/79853-2/.

Zhang Tao, Feng Baoqiang, Zhou Hao, Ma Mi, Sun Qifan, and Liu Haibo. 2019. "Application of 27-plex SNP Race Inference Assay into Tracing a Suspect's Ethnogenesis: Case Report." *Forensic Science and Technology* 44 (3): 276–9. https://web.archive.org/web/20200702150420/http://www.xsjs-cifs.com/article/2019/1008-3650-44-3-276.html.

Zhao Lei, Li Caixia, Feng Lei, Wang Wei. 2019. "一种对未知检材进行个体识别的方法和系统" [Method and system for individual identification of unknown samples]. People's Republic of China patent CN106480198B filed 1 November 2016, issued 17 September 2019. https://web.archive.org/web/20230114164234/https://patentimages.storage.googleapis.com/24/bd/b2/dc6ae9a88657b4/CN106480198B.pdf.

Zhao Shilei, Shi Cheng-Min, Ma Liang, Liu Qi, Liu Yongming, Wu Fuquan, Chi Lianjiang, and Chen Hua. 2019. "AIM SNPtag: A Computationally Efficient Approach for Developing Ancestry Informative SNP Panels." *Forensic Science International: Genetics* 38:245–53.

Zhao Wenting, Jiang Li, Liu Jing, Zhao Lei, Ma Xin, Ji Anquan, and Li Caixia. 2017. "DNA-Based Facial Portrait for Forensic Investigative Purposes." *Forensic Science and Technology* 42 (4): 259–63. https://web.archive.org/web/20170923084008/http://www.xsjs-cifs.com/article/2017/1008-3650-42-4-259.html.

Zhao Xingchun, Meng Qingzhen, Niu Yong, Ji Anquan, Liu Huinian, Ma Xin, and Ye Jian. 2018. "Forensic Scientific Research and Development in the 13th Five-Year Plan." *Forensic Science and Technology* 43 (2): 87–91. https://web.archive.org/web/20200811214816/http://www.xsjs-cifs.com/article/2018/1008-3650-43-2-87.html.

Zhao Yusha. 2017. "Xinjiang Installs Body Scanners at Checkpoints." *Global Times News* (PRC news website), 26 September 2017. https://web.archive.org/web/20171213225415/http://www.globaltimes.cn/content/1068368.shtml.

Zhou Wenting. 2017. "Developing a Collaborative Approach to Groundbreaking Scientific Research." *China Daily* (PRC government news website), 19 October 2017. https://web.archive.org/web/20191013005842/http://www.chinadaily.com.cn/opinion/5yearscorecard/2017-10/19/content_33437996.htm.

Index